Opportunities for increasing crop yields

Association of Applied Biology

Opportunities for increasing crop yields

Edited by R G Hurd, P V Biscoe and C Dennis

Taken from proceedings of a meeting held at the University of Reading, 17–21 September 1979 to celebrate the 75th Anniversary of the Association of Applied Biology.

Pitman Advanced Publishing Program
BOSTON · LONDON · MELBOURNE

PITMAN PUBLISHING LIMITED
39 Parker Street, London WC2B 5PB

PITMAN PUBLISHING INC.
1020 Plain Street, Marshfield, Massachusetts

Associated Companies
Pitman Publishing Pty Ltd., Melbourne
Pitman Publishing New Zealand Ltd., Wellington
Copp Clark Pitman, Toronto

Library of Congress Cataloging in Publication Data

Association of Applied Biologists.
Advances in crop production and protection.
(The Pitman series in applied biology)
Selected papers from the 75th anniversary meeting of the Association of Applied Biologists held at Reading University, Sept. 17-21, 1979.
Includes index.
1. Agriculture–Congresses. 2. Plants, Cultivated–Congresses. 3. Plant physiology–Congresses. 4. Plants, Protection of–Congresses. I. Hurd, Ronald Graham, 1932- II. Biscoe, P. III. Dennis, C. IV. Title V. Title: Crop production and protection. VI. Series: Pitman series in applied biology.
SB16.G7A87 1980 631 79-25924
ISBN 0 273 08481 X

Set in 11/13 pt IBM Press Roman
by LMA Ltd, Southampton, Hants
Printed and bound in Great Britain at
The Pitman Press, Bath

Contents

Preface

On the occasion of the 50th Anniversary of the Association of Applied Biologists in 1954, the words of Swift 'how deserving are those who make two blades of grass grow where only one grew before' were suggested as a suitable motto for the Association. In his inaugural address to that meeting, Sir E. John Russell drew attention to the danger that specialists, working within the general field of Applied Biology, might lose touch with each other and revert to 'a mass of facts of greater or less moment out of which a clear and lucid system is yet to be framed'.

The Association, which celebrated its 75th Anniversary at Reading in September 1979, has since sought to avoid this undesirable situation by stimulating discussion between specialists working in the different disciplines contributing to crop production. It seeks to achieve a balance between general and specialized discussion through its discipline-orientated groups comprising entomologists, pathologists, nematologists, plant breeders and physiologists, as well as those concerned with weeds and post-harvest storage. These groups, while conducting their own meetings, contribute to programmes designed to discuss different aspects of crop culture. We are guided by the knowledge that the continued effectiveness of the applied biologist is governed by the degree to which the attention given to a special line of investigation remains subsidiary to a wider appreciation of commercial crop production. Some would say we are merely resurrecting the philosophy of the old school of natural historians but, be that as it may, the headlong and widespread adoption of new cultural technologies accelerates ecological change so that old problems are posed in a new guise. It is to meet this challenge that the AAB steps out towards its centenary. The motto of the Ohio Agricultural Extension Service, 'discover, develop and disseminate', might equally apply to the aspirations of the Asscoiation.

Our 75th Anniversary meeting was entitled 'Advances in Crop Production and Crop Protection' and the contents of this volume have been drawn from contributions to this theme, particularly from the programme arranged by our plant physiology group. Opportunities were discussed for increasing crop yields by considering the physiology of different crops, the impact of nutritional and environmental constraints, cultural practice, the benefits and potential of mixed cropping systems and a consideration of post-harvest losses. It is hoped that these contributions will both identify some of the opportunities for the 1980s and stimulate others to grasp them.

It is anticipated that future meetings on topical subjects attracting multi-disciplinary interest will provide further material attractive to a wider international audience. This volume is, therefore, intended by both the Association and the publisher to be the first of a series of significant contributions to applied biology.

N W Hussey
Glasshouse Crops Research Institute
President, The Association of Applied Biology

Introduction

The meeting of the Association of applied Biologists in September 1979 at the University of Reading was arranged to celebrate the Association's 75th anniversary. Since its beginning, as the Association of Economic Biologists in 1904, the Association has grown to a current membership of over 1300. Members' interests cover all aspects of research on crop plants and these are catered for by eight specialist working groups. For the Anniversary meeting, each group organized a series of papers under the general title 'Advances in crop production and crop protection'. The plant physiology group's contribution was entitled 'Opportunities for increasing crop yields' and included 15 papers given at the group session with six more delivered at the plenary sessions to open and close the meeting. The present volume contains these contributions together with some of the papers on the same theme from the sessions of the weeds group and the post-harvest biology group.

The first three sections of the book move from the more fundamental to the more applied approach and include some economic aspects of the title. The opening papers are by physiologists who have direct contact with plant breeders. It is repeatedly noted in the book that the physiologist's knowledge of the processes governing crop production will be of increasing relevance to future plant-breeding programmes. The benefits of a close collaboration between physiologists and breeders are clearly demonstrated in the papers by Wilson *et al.* and Robson, where an understanding of the processes of respiration and growth are being exploited in the breeding of higher-yielding grass crops. Research on the physiology of crop production is not only exploited through breeding. Papers on wheat and vegetable crops in Section 2 describe how a knowledge of the nutritional requirements of crops can increase productivity by changing the timing and placement of fertilizers, whilst other cultural and environmental improvements are suggested elsewhere.

Although the aim of the papers in these sections is primarily to indicate how yields can be further improved, this is not necessarily compatible with achieving the optimum economic yield. Two papers in

Section 3 by a leading cereal grower and the Chief Agricultural Officer in the UK Agricultural Advisory Service, are included to help define the industry's needs and to enable scientists to examine critically their research and its relevance to the industry. They mention, for example, that modern marketing methods and highly competitive conditions make it invaluable to have predictable and consistent yields. This is particularly important for top fruit where yields fluctuate violently from year to year. The contributions from the UK fruit research stations were chosen because they are very relevant in this context. They describe, for the first time, analyses of the variations in fruit yield and their possible regulation which could lead to a major breakthrough in this area.

Another economic consideration in improving productivity, which is becoming increasingly important, is the judicious use of inputs which are directly related to energy costs. Two papers in Section 3 deal with this topic. First, Moorby shows how a knowledge of the temperature requirements of the tomato crop can be used to cut energy costs without sacrificing productivity. Second, Pascal considers to what extent the finite resources in energy and minerals available to the world will constitute limitations to crop production – and crop protection – in the future.

It can be difficult, time consuming and expensive for the physiologist to determine the effect of a change in the level of one of the inputs, e.g. fertilizer, on the optimum economic yield, as defined by the growers. One approach which is being increasingly used to cope with the complex of factors which operate during the growth of a crop is the computer-based mathematical model. Several papers in Section 2 describe the use of such models, which are a natural progression of the quantitative approach to physiology that has increasingly characterized research during this century. The need for and validity of the models is established by the authors, often on the basis that it may be the only practical way to obtain an increase of another few per cent in yield, which would be impossible to measure significantly using conventional methods. The considerable relevance to profitability of such a small increase is clearly explained by the cereal grower. The contribution by Spitters on mixed cropping comes from an institute in the Netherlands devoted entirely to this mathematical approach. This and the remaining papers on mixed cropping were selected because this approach to crop production, which is widely adopted in the tropics, merits more attention in this country.

The remaining two sections of the book deal with the physiological aspects of weed control and post-harvest losses. Great changes in agriculture have taken place in recent years as a result of the extensive use of herbicides. As with any such drastic change it is particularly important to review the practices used to make sure that, on balance, the present consequences of the treatments are desirable and that their direction is right for the future. For this reason most of the papers in this section emphasize the longer view.

Post-harvest biology – as we are informed by its practitioners – always seems to be relegated to the end of any set of proceedings such as this. We have no hesitation in repeating this 'rule' for obvious logical reasons. Nevertheless one has only to recognize the enormous losses which take place at this stage in many crops to accept, using another logical pathway, that this section could command a much earlier insertion. Unless the reasons for those losses are fully understood and can be controlled then the advances in productivity that the earlier papers describe or promise will have little relevance to the population at large.

Any publication of this sort is the result of collaborative effort. We are indebted to the AAB programme secretary, Dr M. Thresh, who played a key role in organizing the symposium as a whole, and to the Plant Physiology Group Committee of the AAB who planned the contributions which form the basis of this book, in collaboration with all the prospective authors. The burden for much of this collaboration fell upon Dr M. Robson, and the efficiency with which he completed this task has made our own contribution that much easier. We must also thank the ex-Chairman of the Committee, Dr S. Thomas who, with Dr Robson, discussed with the AAB the concept of the book and obtained their approval for its publication as part of the 75th anniversary celebrations of the Association. Finally we have to thank all the authors for their contributions and for their fortitude in tolerating our suggestions for 'improvement'.

R. G. Hurd
P. V. Biscoe
C. Dennis

Section 1 **Physiological aspects in relation to breeding**

Physiological limitations to cereal yields and ways of reducing them by breeding

R.B. Austin
Plant Breeding Institute, Cambridge, England

Introduction

During the past thirty years the yields of the world's major grain crops, wheat, rice, maize and barley, have increased dramatically in many countries. The period of rapid increase in yield has begun at times varying from country to country, but in most cases it has followed the introduction of new varieties, the increased use of fertilizers and crop-protection chemicals and often an increase in the intensity of mechanization. In Britain, the annual increases in the estimated yields of wheat and barley, 0·069 and 0·054 t ha^{-1} show no sign of levelling off (Austin, 1978) and similar trends are evident in other areas where cereals are grown intensively (e.g. Duvick, 1977). Yet there must be a biological limit to the yield attainable in a given environment, which will be determined by the supplies of solar energy, carbon dioxide, water and nutrients. When this limit is reached no further increase in yield can be expected from breeding improved varieties, though it will be necessary to sustain breeding programmes to maintain an adequate level of resistance to highly adaptable pathogens and pests and to meet the changing demands of quality.

The purpose of this article is to examine the physiological limitations to cereal yields in circumstances where water and nutrient supplies are not limiting and to assess the scope for breeding higher-yielding varieties than those at present in cultivation. Emphasis will be given to wheat, but reference will be made to rice, maize and barley for comparison.

Aspects of development and growth

The general features of the development of the major cereal crops and their dependence on environmental factors are well known (Evans, Wardlaw and Fischer, 1975; Austin and Jones, 1975; Evans and Wardlaw, 1976). Appropriate responses to photoperiod and temperature are a pre-requisite for fitness in a given environment. Fortunately the 'coarse tuning' by selection for appropriate life-cycle timing does not present breeders with serious problems. The importance of 'fine tuning', that is, recognizing and choosing between alternative combinations of response which give apparently similar developmental patterns and maturity, is very imperfectly understood. Thus, the extent to which yield improvement in so-called 'adapted' genotypes is a result of selection for minor differences in environmentally dependent developmental responses is not known and deserves more attention from physiologists.

Developmental processes

In the small-grain cereals tiller bud formation and ear initiation are complete when the plant is a small fraction of its ultimate size and genetic differences in primordium development, which affect grain number per ear, are evident at this stage. Similarly, genetic differences in the rate of emergence of tiller buds are evident and these affect tiller numbers and synchrony. Following its initiation, the further differentiation of the inflorescence takes place as leaves are expanding and emerging and stem growth is proceeding. It has often been claimed that these processes compete with each other for a limited supply of assimilates (Friend, 1965; Bingham, 1969; Patrick, 1972) and that, in selecting for higher yield, breeders have, *inter alia,* selected forms in which at this stage of growth the competitive advantage of the ears has been increased. Although there is much circumstantial evidence for this hypothesis, it remains unproven by rigorous experiment. If it can be shown that floral and vegetative growth are in competition at this stage, then growth in stem weight necessary to support leaves in the best positions for maximum and uniform light interception, and to support the inflorescence, will occur at the expense of yield. The resultant of these processes alone would lead to the expectation of an optimum stem weight (and length) for maximum yield. This optimum weight would be increased if more assimilate were available. In the context of variety improvement this would require the exploitation of genetic differences in crop growth rates.

Late in the process of inflorescence development ovary size is determined, and may act as a limit to grain size. This limitation may result from competition for assimilates within the inflorescence, and be additional to the competition between the inflorescence and the vegetative organs. The existence of such competition may partly explain the negative correlations between grain number and size, and between stem length and grain yield that are often found in wheat. It may also be the reason why cereals are very sensitive to drought and reduced irradiance in the two to three weeks prior to anthesis (Fischer, 1973).

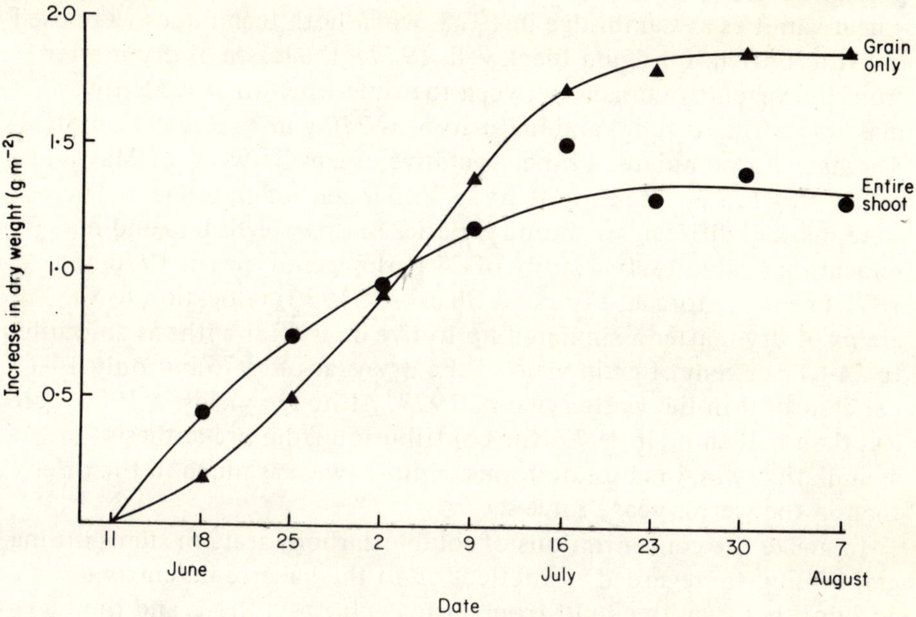

Fig. 1 Increases in shoot (●) and grain (▲) dry weight after anthesis. Data are means of six varieties of winter wheat grown at Cambridge in 1973.

Dry matter production and grain growth

For 15-20 days after fertilization (which normally occurs within a few hours of pollination), the demand for assimilates by the developing grain is generally less than the capacity of the leaves to supply them (see, for example, Fig. 1). In the small-grain cereals growth in structural dry matter is virtually complete by this time. As a consequence there is usually a considerable accumulation of soluble carbohydrate in the stems and leaf sheaths, the concentrations of which may exceed 30 per cent of the dry weight of these organs. In wheat and barley in England the concentrations reach a maximum about 15 days after anthesis, when the crop may contain 2 t ha^{-1} of soluble carbohydrates (Ford, Blackwell, Parker and Austin, 1979). Usually, there is little soluble carbohydrate present in the stems at maturity.

From 15-20 days after anthesis in the small grain cereals, the net dry weight increase of the entire shoot is generally less than the dry weight increase of the grains alone. There has long been controversy as to whether the soluble carbohydrate accumulated in the stems and leaf sheaths is remobilized to the grains, or is lost in respiration. Only by a combination of growth analysis and carbon-14 labelling has it been possible to settle this matter unequivocally. In a study of six winter wheat varieties at Cambridge in 1975, when both techniques were used (Austin, Edrich, Ford and Blackwell, 1977), the losses of dry matter from the vegetative organs between the time when their weight was maximum (on 30 June) and maturity was 270 g m^{-2}. Relocation of dry matter accumulated in the vegetative organs between 21 May and 6 July was 197 g m^{-2}, equivalent to 30 per cent of grain yield. There were marked differences among varieties in dry-weight loss and in relocation (Table 1). In a study of 24 barley genotypes in 1976 and 1977 (Austin, Morgan, Ford and Blackwell, 1980) relocation to the grains of dry matter assimilated up to five days after anthesis amounted to 54-61 per cent of grain yield in the dry year of 1976 but only 14-15 per cent of it in the wetter year of 1977. Although yields in 1976 were less than half those in 1977, the contribution from pre-anthesis assimilation was, in absolute terms, almost twice as much in the drier than in the wetter year (Table 2).

In maize the concentrations of soluble carbohydrate in stems during grain filling are regarded as a reflection of the balance in 'current account' between the input from current photosynthesis and the demand to support grain filling (Daynard, Baron and Tollenaar, 1980).

Table 1. Losses of dry matter from the vegetative organs of wheat and the calculated relocation of dry matter from the vegetative organs to the grain. Data from an experiment at Cambridge in 1975 (from Austin, Edrich, Ford and Blackwell, 1977)

	Mean of six varieties	Templar	Hobbit
Losses of dry wt from vegetative organs 30 June-maturity (g m^{-2})	270	120	357
Above losses as a percentage of the weight of the vegetative organs on 30 June	27	13	36
Relocation to grain from vegetative organs 21 May-6 July (g m^{-2})	197	153	201
Relocation to grain as a percentage of grain yield	30	23	24
Grain yield (g m^{-2})	657	651	841

With potentially high yielding forms, when drought or low temperature reduce photosynthesis to below the requirement for grain filling, soluble carbohydrates in the stems can be reduced to a very low level, a condition associated with lodging. On the other hand, in tropical lowland forms of maize having low yield potential, high concentrations of soluble carbohydrate are present in the stems at maturity (CIMMYT, 1977), even though losses in vegetative dry matter during late grain filling can amount to the equivalent of 40 per cent of grain yield (Goldsworthy and Colegrove, 1974; Goldsworthy, Palmer and Sperling, 1974).

Table 2. Grain yields of spring barley genotypes of different height in two years together with the calculated contributions to grain yields from assimilation up to 5 days after anthesis (from Austin, Morgan, Ford and Blackwell, 1980)

Height class	Grain yield ($g\ m^{-2}$)	Contribution to grain yield from assimilation up to 5 days from anthesis ($g\ m^{-2}$)	(% of yield)
1976 - hot, dry season			
Tall	318	173	54
Single dwarf	296	180	61
Double dwarf	291	178	61
1977 - cool, wet season			
Tall	715	107	15
Single dwarf	654	89	14
Double dwarf	659	91	14

It may be argued that it is immaterial whether the *particular* carbon atoms in soluble carbohydrates in stems are respired or translocated to the grains as long as the overall respiratory losses are low. Carbon assimilated by the flag leaves of wheat during mid-grain filling is used at higher efficiency than that assimilated by lower leaves (Austin and Morgan, unpublished). Similar results have been obtained by Ryle and Powell (1974) for barley and by Palmer, Heichel and Musgrave (1973) for tropical maize. In maize, lower leaves provide most of the assimilate for soluble carbohydrate accumulation in the stems. Therefore in maize at least, stored assimilate is used for grain filling at lower efficiency than current assimilate.

In considering the carbon economy of cereals during grain filling, consumption by the roots is frequently ignored. While net increase in root weight has usually ceased by anthesis (Welbank, Gibb, Taylor and Williams, 1974), a significant amount of carbon is translocated to the roots during grain filling. In unpublished work R. B. Austin and C. L.

Morgan found this translocation amounted to 6 per cent of current assimilation during the mid-grain-filling phase in wheat.

Nitrogen and grain growth

In wheat and barley the uptake of nitrogen is largely complete by anthesis (Spratt and Gasser, 1970; Austin, Ford, Edrich and Blackwell, 1977). Much of the nitrogen present as protein in the grain at maturity is relocated from the vegetative organs during grain filling. Thus, the nitrogen content of the crop at anthesis is an important determinant of grain protein yield. As genetic variation in nitrogen concentration at anthesis seems to be small (Austin, Ford *et al.*, 1977) genetic increases in grain protein yield are likely to be associated with increased biomass at anthesis. However, there is some genetic variation in nitrogen uptake during grain filling in wheat and, in principle, it could be exploited to increase grain protein yield. For this to be possible, there would need to be sufficient available nitrogen and water in the soil after anthesis, conditions which are unusual on a world scale, except where irrigation is practised.

The relocation of nitrogen from the vegetative organs to the grain sets a minimum value to the relocation of dry matter. Taking values for wheat given by Austin, Ford *et al.* (1977), the nitrogen concentration in the vegetative organs at anthesis was 1·4 per cent, and at maturity 0·6 per cent, a decrease of 0·8 per cent. Since the nitrogen concentration of plant protein is about 6·25 per cent, the minimum dry matter corresponding to this decrease is 6·25 x 0·8 per cent or 5 per cent, considerably less than the dry weight losses commonly observed.

From the foregoing outline, it can be seen that there are many physiological limitations to yield even in the uncommon condition where water and nutrients are apparently non-limiting. Broadly the limitations can be considered to be of two kinds, developmental and photosynthetic, the former acting mainly before anthesis and the latter after it. Many limitations, however, may affect development through a restriction on photosynthesis, or photosynthesis through a restriction on development.

The integration of the effects of all these limitations and their interactions determines grain and straw yields at maturity.

Potential yield

There have been several attempts to estimate the potential yields of cereal crops. One way is to calculate the potential photosynthesis during the grain-filling period, taking account of respiratory losses and relocation of materials assimilated prior to grain filling (Murata and Matsushima, 1975; Austin, 1978). For rice, growing in an environment where the daily receipts of solar radiation are 16·7 MJ m^{-2}, assuming an efficiency of 26 per cent in photosynthesis, the net carbohydrate production in a 40-day grain-filling period was calculated to be 16·4 t ha^{-1}, equivalent to a grain yield of 19·3 t ha^{-1} at a moisture content of 15 per cent. Maximum yields in the same environment are little more than 10 t ha^{-1}.

For wheat in England, Austin (1978) used observed values of LAI and photosynthetic rates in high-yielding crops, and allowed for the observed decline in LAI during grain filling. During a 40-day grain-filling period, in which daily solar radiation receipt was 16·7 MJ m^{-2}, the dry matter available for grain filling was calculated to be 9·2 t ha^{-1}. To this was added 7 per cent of the expected net assimilation over the 15-day period prior to anthesis (0·5 t ha^{-1}), giving a potential grain dry-matter yield of 9·7 t ha^{-1}, equivalent to 11·4 t ha^{-1} at 15 per cent moisture. The highest wheat yields recorded in England are close to this calculated potential.

The discrepancy between the results of the calculations arises mainly because the gross photosynthetic product in rice was calculated to be almost 34 t ha^{-1}, whereas in wheat it was taken as 15·3 t ha^{-1}. It seems extremely unlikely that light interception by green tissue and the efficiency of photosynthesis could ever be great enough to enable the higher value to be reached, while the lower value is based on assumptions that are known to be achievable with present cultivars under good conditions and is probably too conservative.

Alternative methods are based on calculating the potential photosynthesis during the whole growth cycle of the crop, making an allowance for respiratory losses and multiplying the results by 0·5, the harvest index of modern high-yielding varieties. These are cruder methods which do not take account of much that is known of the physiology of the crop, though, like the calculations based on the grain-filling period only, the results depend entirely on the assumed efficiency of photosynthesis and respiration.

Yield improvements resulting from the breeding of new varieties, and their physiological basis

From national variety trials carried out in England, Silvey (1979) has calculated that new varieties have been responsible for half of the increase in yield of wheat between 1947 and 1977. As yields have increased by 80 per cent, the varietal component of this increase is estimated to be 40 per cent. In 1978, the yields of old and new winter wheat varieties were compared in two trials at Cambridge (Austin, Bingham *et al.*, in press). In both trials, lodging was prevented and diseases were controlled by the application of fungicides. One trial was carried out on soil of low fertility and was given 38 kg ha^{-1} of fertilizer nitrogen. The other was on a soil of higher fertility and was given an application of 104 kg ha^{-1} of nitrogen. The results (Table 3) show that the new varieties outyielded Little Joss, the oldest variety tested, by close to 40 per cent, both at low and at high fertility. The newer, high-yielding varieties were shorter and reached anthesis earlier than the older ones. The maximum leaf area indices and leaf weights m^{-2} were similar, but the stems were shorter and the stem weight m^{-2} less in the newer varieties. Within each experiment the total dry-matter

Table 3. Yields of grain (15 per cent moisture, fresh weight basis) from old and new varieties of winter wheat in trials on soil of low and high fertility

Variety	Year of introduction	Grain yield (t ha^{-1} at 15% moisture)	
		Low fertility trial (38 kg N ha^{-1})	High fertility trial (104 kg N ha^{-1})
Little Joss	1908	3·88	6·14
Holdfast	1935	3·42	5·83
Cappelle Desprez	1953	4·40	6·89
Maris Widgeon	1964	4·20	6·68
Maris Huntsman	1972	4·75	7·69
Hobbit	1977	5·44	8·58
Mardler	1978	4·79	7·30
370/500*		5·00	7·45
730/3937*		5·12	8·31
989/10*		5·40	8·90
Armada	1978	4·99	8·07
Benoist 10483†		4·47	8·50
s.e.		0·13	0·15
Mean		4·66	7·52

* Advanced selections from the breeding programme of the Plant Breeding Institute.

† Closely related to Talent, introduced in France in 1973.

production varied little, though varieties which gave the highest dry-matter yield on the high-fertility site also did so on the low-fertility site, indicating that there were consistent, though small differences in dry-matter production. Increased grain yield was, on average, associated with a matching decrease in straw yield (Fig. 2). However, the correlation between grain yield and straw yield accounted for only 46 per cent of the variation in grain and straw yield in the trial at low fertility and 21 per cent at high fertility. Thus these results suggest that for this set of varieties much of the increase in grain yield was associated with a decrease in straw yield, though varieties with a higher than average total dry-matter production (grain plus straw) yielded more than would be expected from their grain:straw ratio.

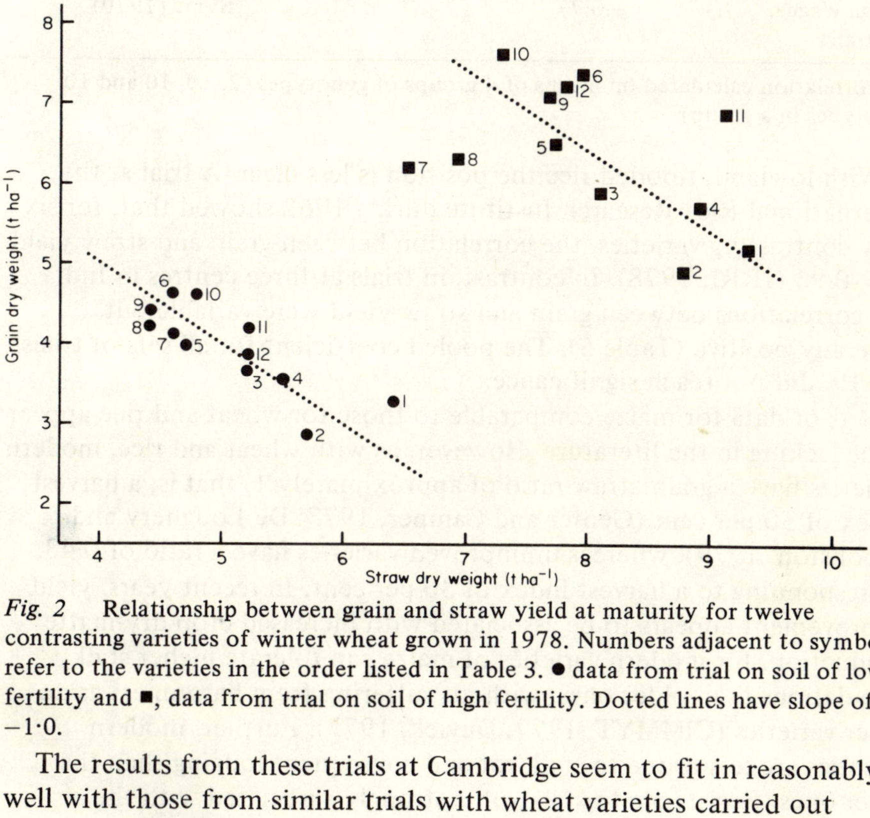

Fig. 2 Relationship between grain and straw yield at maturity for twelve contrasting varieties of winter wheat grown in 1978. Numbers adjacent to symbols refer to the varieties in the order listed in Table 3. ● data from trial on soil of low fertility and ■, data from trial on soil of high fertility. Dotted lines have slope of −1·0.

The results from these trials at Cambridge seem to fit in reasonably well with those from similar trials with wheat varieties carried out elsewhere (Table 4) except those at Guelph. The pooled correlation coefficient between grain and straw yields for five sets of data is -0·48 with 42 degrees of freedom. In the data from Guelph, the range in grain and straw yields was less than in the other cases, despite a similar variation in plant height. Taking all the results together, variation in straw yield accounts for only 23 per cent of the variation in grain yield.

Table 4. Wheat. Correlation coefficients between grain and straw yields in field experiments with well-guarded plots and no lodging or serious disease

Place	Correlation coefficient	No. of varieties in trial	Author(s)
Cambridge, U.K. (low soil fertility)	−0·76	12	Austin, Bingham *et al.* (in press)
Cambridge, U.K. (high soil fertility)	−0·46	12	Austin, Bingham *et al.* (in press)
New Delhi, India	−0·24	20	Jain and Kulshresthra (1976)
Guelph, Canada	+0·89	4*	Singh and Stoskopf (1971)
Wagga Wagga, Australia	−0·77	9	Syme (1970)

* Correlation calculated on means of 4 groups of genotypes (2, 10, 10 and 10 genotypes in a group)

With lowland, flooded rice the position is less clear. A trial at the International Rice Research Institute during 1962 showed that, for six very contrasting varieties, the correlation between grain and straw yield was -0·96 (IRRI, 1978). In contrast, in trials at three centres in India the correlations between grain and straw yield were variable but generally positive (Table 5). The pooled coefficient for all sets of trials, + 0·19, did not reach significance.

Sets of data for maize comparable to those for wheat and rice appear to be lacking in the literature. However, as with wheat and rice, modern varieties have a grain:straw ratio of approximately 1, that is, a harvest index of 50 per cent (Genter and Camper, 1973; De Loughery and Crookston, 1979), whereas unimproved varieties have a ratio of 0·43, corresponding to a harvest index of 30 per cent. In recent years, yield improvement appears to be associated with increased crop dry-matter production, for modern varieties of maize can tolerate higher stalk populations than older ones, without suffering from barrenness as do older varieties (CIMMYT, 1977; Duvick, 1977). Further, modern varieties are less prone to stalk rot and consequent lodging than the older ones. Resistance to stalk rot is strongly associated with the maintenance of soluble carbohydrates in the basal internodes (Mortimore and Ward, 1964). All these differences are consistent with greater biomass production before the onset of grain filling and with the maintenance of photosynthetic activity during grain filling sufficient to provide for the needs of the grain and avoid the undue depletion of stem carbohydrates.

Table 5. Rice. Correlation coefficients between grain and straw yields in field experiments with well-guarded plots and no lodging or serious disease

Place	Correlation coefficient	Number varieties in trial	Author(s)
Hyderabad, India			
rabi season			
early maturing varieties	−0·12	7	Venkateswarlu,
medium maturing varieties	+0·51	7	Suryaprakasha-Rao
late maturing varieties	+0·27	7	and Rao (1977)
Kanpur, India			
early maturing varieties	−0·20	9	Mehrotra, Saxena,
medium maturing varieties	+0·21	10	Mathur and Sinba
late maturing varieties	−0·66	5	(1974)
Cuttack, India			
rabi season	+0·76	11	Murty, Sàhu and
kharif season			Nayak (1974)
low nitrogen	+0·65	6	Sàhu and Murty
high nitrogen	+0·73	6	(1975)
Los Baños, Philippines	−0·96	6*	International Rice Research Institute (1978)

* Data for each variety averaged over 12 sowing dates. Tall varieties may have lodged.

In barley, yield improvement has also been associated with an increase in the grain:straw ratio, there having been little change in total dry-matter production. Old varieties have a harvest index of about 40 per cent, modern ones about 55 per cent (Watson, Thorne and French, 1958; Sandfaer, Jorgensen and Haar, 1965; Sandfaer and Haar, 1975).

Prospects for the breeding of higher-yielding varieties

As the best modern varieties of the four major cereal crops now have a harvest index of about 50 per cent, and there has been, up to the present, a consistent upward trend in harvest index, it is relevant to consider what further improvements might be achieved by a continuation of this trend. For four high-yielding dwarf varieties of wheat (Austin, Bingham *et al.*, in press), the dry weights, in g m^{-2}, of the component organs at maturity were: grain, 707; chaff (rachis, glumes, etc.), 143; leaf lamina, 139 and stem and sheath, 453. The harvest index was therefore 49 per cent. Mechanical considerations make it unlikely that chaff weight could be reduced without a concomitant reduction in grain weight. Physiological considerations make it unlikely that leaf lamina weights could be substantially reduced (but see below). Therefore the only opportunity is to reduce stem and sheath dry weights. Because of the need for leaves to be arranged in the canopy to maximize the interception of light and its uniformity of distribution over their area, it is improbable that stem and sheath weights could be reduced to much less than 50 per cent of their present weight relative to other organs. If this value is assumed, the harvest index would become 58 per cent (if they were totally absent it would be 71 per cent) and would reduce the above ground biomass by 226 g m^{-2}. If constant biomass yield were assumed, as seems likely, and the chaff weight were increased *pro rata* to 'accommodate' the extra grain, the organ weights would become: grain, 895; chaff, 181; leaf lamina, 139; stem and sheaf, 226; giving a harvest index of 62 per cent for an increase in grain yield of 26 per cent. This calculation provides an indication of the scope for further increases in yield in wheat from changing the grain:straw ratio, biomass production being assumed to be invariant.

As discussed earlier, a change in harvest index is likely to have implications for the nitrogen economy of the crop. Taking data for four high-yielding dwarf genotypes (Austin, Bingham *et al.*, in press) the distribution of dry weight and nitrogen in the plants at anthesis and at

Table 6. Distribution of dry matter and nitrogen between plant parts at anthesis and at maturity, g m^{-2}. Observed* = average of four dwarf wheat genotypes; assumed = 50% reduction in stem and sheath dry weight and an equivalent increase in ear and grain weight

		Dry weight		Nitrogen	
		Observed	Assumed	Observed	Assumed
At anthesis					
Leaf laminae,	3·57 % N	167	167	5·96	5·96
Ears,	2·17 % N	143	181	3·10	3·93
Stems & sheaths,	1·27 % N	589	294	7·48	3·73
Total				16·54	13·62
Uptake of N during grain filling. (14·9% of that present at anthesis)				2·46	+
At maturity					
Grain	1·94 % N	707	895	13·71	17·36
Straw	0·72 % N	735	546	5·29	3·93
Total				19·00	21·29

* observed data taken from Austin, Bingham *et al.* (in press)

\+ not calculated.

maturity is shown in Table 6. The table also shows the calculated nitrogen contents assuming constant biomass yield and the dry weights given above for the hypothetical variety having a harvest index of 62 per cent. Even though there would be less nitrogen in the straw at maturity, the nitrogen requirement for grain filling would be greater. This requirement could be provided either by an increase in the concentration of nitrogen in the vegetative organs or by a greater uptake during grain filling. Alternatively, if nitrogen uptake were not increased, grain nitrogen would decrease from 1·94 per cent to 1·68 per cent. It must be emphasized that these calculations are hypothetical and only two extreme alternatives are presented.

An increase in harvest index, implying an increase in carbon assimilation after anthesis and a decrease before it, may have implications for life-cycle timing. Again, only speculation is possible. A lower biomass at anthesis could be achieved by selecting for reduced growth rate, keeping the date of anthesis the same, or by selecting for earlier anthesis (assuming that it is not possible for the plants to 'mark time'). Modern varieties of wheat are up to 8 days earlier in reaching anthesis than the old varieties Little Joss and Holdfast (Austin, Bingham *et al.*, in press), and some continuation of this trend may be possible.

As increase in yield resulting from improving the grain:straw ratio are realized, the only option for achieving further increases will come from increasing biomass production. This might be achievable either by lengthening the life cycle or by increasing the rate of net photosynthesis of the crop. For seasonal and rotational reasons, it is usually impracticable to increase the length of the life cycle. Genetic increases in the rate of net photosynthesis may, in principle, be achieved in various ways. The most obvious would be to exploit differences in the light-saturated rates of photosynthesis of individual leaves. In wheat, it appears that this would be very difficult because genotypes with high rates have small leaves and the negative correlation between leaf size and photosynthetic rate is strong (Evans and Dunstone, 1970), although it is not known whether it is fundamental. The consequence of this correlated variation is that the daily photosynthetic product per leaf is greater for the modern, larger-leaved forms than for the primitive, small-leaved ones. There are few reports of such variation in maize and barley, though in rice, as in wheat, varieties with high rates of photosynthesis have small, thick leaves (Tsunoda, 1962). Increases in crop photosynthesis may also be achievable by selecting types with a morphology which permits the intercepted light to be more uniformly distributed over the photosynthetically active surfaces, and by selecting for increased leaf longevity. Both objectives may be achieved in phenotypes with erect upper, but lax lower leaves (Tanaka, Yamaguchi, Shimazaki and Shibata, 1968; Austin, Ford, Edrich and Hooper, 1976).

Conclusions

This chapter has been concerned with the opportunities for increasing yields of cereal crops when water and nutrients are not limiting. How relevant are the conclusions to the more normal situation where water and nitrogen supply are far less than are required to achieve high yield? In locations where the effects of drought are avoided by the use of short-season, early maturing varieties, these are likely to give lower yields per crop than late maturing ones when compared under conditions where water supply does not limit the growing season. Varieties having high grain:straw ratios will have greater water-use efficiency for grain production than those with lower ratios because water use per unit of dry matter accumulated is similar for both grain and straw dry matter (Day *et al.*, 1978). Thus, as is the general experience, varieties with high harvest index outyield those with low harvest index both in partial drought and when water is plentiful. In the absence of lodging, high harvest index is equally advantageous at low and high levels of nitrogen fertilization. In practice, however, both high nitrogen supply and plentiful water greatly increase stem length and promote lodging so that old, tall varieties with low harvest index are further disadvantaged.

In environments with partial drought, it may prove impossible to breed varieties with greater photosynthesis unless they have the ability to extract more water from the soil than existing ones, or are more efficient in their water use. It is possible that water-use efficiency could be somewhat increased by appropriate modifications to crop morphology and stomatal behaviour (Jones, 1976; Austin, 1977). But, if higher light-saturated rate of photosynthesis associated with decreased mesophyll resistances to CO_2 assimilation could be detected and exploited, grain yields would benefit at all levels of water supply.

In conclusion, it appears likely that limited increases in cereal yields will be possible by further reducing the production of non-grain biomass. The alternative route to increased yield is to increase total biomass production, as seems to have been achieved in recent years in the breeding of rice and maize. Some increase in biomass production seems likely to be achievable by selecting for morphological characters which will improve light interception and the uniformity of its distribution over the leaves but such increases may only be realizable when water is not limiting. Undoubtedly the greatest challenge facing breeders and physiologists is to find – or generate – variation in the light-saturated rate of photosynthesis of individual leaves. The exploitation of this variation in cereal breeding could make possible substantial increases in biomass production, both when water is freely available and when it is in limited supply.

References

Austin, R. B. (1977). Improvement of barley for dry land farming. In *CENTO Symposium on Plant Improvement in Arid Zones*, pp. 12-27. Ed. J. Bahrani. Ankara: CENTO.

Austin, R. B. (1978). Actual and potential yields of wheat and barley in the United Kingdom. *ADAS Quarterly Review* **29**, 76-87.

Austin, R. B. & Jones, H. G. (1975). The physiology of wheat. *Report of the Plant Breeding Institute, Cambridge, for 1974,* pp. 20-73.

Austin, R. B., Edrich, J. A., Ford, M. A. & Blackwell, R. D. (1977). The fate of the dry matter, carbohydrates and ^{14}C lost from the leaves and stems of wheat during grain filling. *Annals of Botany, London* **41**, 1309-1321.

Austin, R. B., Ford, M.A., Edrich, J. A. & Blackwell, R. D. (1977). The nitrogen economy of winter wheat. *Journal of Agricultural Science, Cambridge* **88**, 159-167.

Austin, R. B., Ford, M. A. Edrich, J. A. & Hooper, B. E. (1976). Some effects of leaf posture on photosynthesis and yield in wheat. *Annals of Applied Biology* **83**, 425-446.

Austin, R. B., Morgan, C. L., Ford, M. A. & Blackwell, R. D. (1980). Contributions to grain yield from pre-anthesis assimilation in tall and dwarf barley phenotypes in two contrasting seasons. *Annals of Botany, London* **45**, 309-319.

Austin, R. B., Bingham, J., Blackwell, R. D., Evans, L. T., Ford, M.A., Morgan, C. L. & Taylor, M. (1980). A comparison of the yields of old and new varieties of winter wheat at two levels of soil fertility. *Journal of Agricultural Science, Cambridge* **94**, 675-690.

Bingham, J. (1969). Physiological determinants of grain yield in cereals. *Agricultural Progress* **44**, 30-42.

CIMMYT (1977). *Report on maize improvement: 1975.* Mexico: Centro Internacional de Mejoramiento de Maiz y Trigo.

Day, W., Legg, B. J., French, B. K., Johnson, A. E., Lawlor, D. W. & Jeffers, W. de C. A. (1978). A drought experiment using mobile shelters: the effect of drought on barley yield, water use and nutrient uptake. *Journal of Agricultural Science, Cambridge* **91**, 599-624.

Daynard, T. B., Baron, V. S., Tollenaar, M. (1979). Points of current interest in studies of the physiology of the maize plant. In *Maize Production and Utilisation,* pp. 205-215. Ed. E. S. Bunting. The Hague: Nijhoff.

De Loughery, R. L. & Crookston, R. K. (1979). Harvest index of corn affected by population density, maturity rating and environment. *Agronomy Journal* **71**, 577-580.

Duvick, D. N. (1977). Genetic rates of gain in hybrid maize yields during the past 40 years. *Maydica* **22**, 187-196.

Evans, L. T. & Dunstone, R. L. (1970). Some physiological aspects of evolution in wheat. *Australian Journal of Biological Science* **23**, 725-741.

Evans, L. T. & Wardlaw, I. F. (1976). Aspects of the comparative physiology of grain yield in cereals. *Advances in Agronomy* **28**, 301-359.

Evans, L. T., Wardlaw, I. F. & Fischer, R. A. (1975). Wheat. In *Crop physiology – some case histories*, pp. 101-150. Ed. L. T. Evans. Cambridge: University Press.

Fischer, R. A. (1973). The effect of water stress at various stages of development on yield processes in wheat. In *Plant Response to Climatic Factors,* pp. 271-276. Ed. R. O. Slatyer. Paris: UNESCO.

Ford, M. A., Blackwell, R. D., Parker, M. L. & Austin, R. B. (1979). Associations between stem solidity, soluble carbohydrate accumulation and other characters in wheat. *Annals of Botany, London* **44**, 731-738.

Friend, D. J. C. (1965). Ear length and spikelet number of wheat grown at different temperatures and light intensities. *Canadian Journal of Botany* **43**, 345-353.

Genter, C. F. & Camper, H. M. (1973). Component plant part development in maize affected by hybrids and population density. *Agronomy Journal* **65**, 669-671.

Goldsworthy, P. R. & Colegrove, M. (1974). Growth and yield of highland maize in Mexico. *Journal of Agricultural Science, Cambridge* **83**, 213-221.

Goldsworthy, P. R., Palmer, A. F. E. & Sperling, D. W. (1974). Growth and yield of lowland tropical maize in Mexico. *Journal of Agricultural Science, Cambridge* **83**, 223-230.

IRRI (1978). Harvest index : criterion for selecting for high yield ability. *International Rice Research Institute Annual Report for 1977,* pp. 20-26.

Jain, H. K. & Kulshrestha, V. P. (1976). Dwarfing genes and breeding for yield in bread wheat. *Zeitschrift für Pflanzenzuchtung* **76**, 102-112.

Jones, H. G. (1976). Crop characteristics and the ratio between assimilation and transpiration. *Journal of Applied Ecology* **13**, 605-622.

Mehrotra, O. N., Saxena, H. K., Mathur, R. K., & Sinba, N. C. (1974). Physiological variability in paddy. *Indian Journal of Genetics and Plant Breeding* **34A**, 979-987.

Mortimore, C. G. & Ward, G. M. (1964). Root and stalk rot of corn in south western Ontario. III. Sugar levels as a measure of plant vigour and resistance. *Canadian Journal of Plant Science* **44**, 451-457.

Murata, Y & Matsushima, S. (1975). Rice. In *Crop physiology – some case histories,* pp. 73-99. Ed. L. T. Evans. Cambridge: University Press.

Murty, K. S., Sàhu, G. & Nayak, S. K. (1974). Breeding rice varieties for higher physiological efficiency. *Indian Journal of Genetics and Plant Breeding* **34A**, 995-1000.

Palmer, A. F. E., Heichel, G. H. & Musgrave, R. B. (1973). Patterns of translocation, respiratory loss and redistribution of ^{14}C in maize labelled after flowering. *Crop Science* **13**, 371-376.

Patrick, J. W. (1972). Distribution of assimilate during stem elongation in wheat. *Australian Journal of Biological Sciences* **25**, 455-467.

Ryle, G. J. A. & Powell, C. E. (1974). The utilisation of recently assimilated carbon in graminaceous plants. *Annals of Applied Biology* **77**, 145-158.

Sáhu, G. & Murty, K. S. (1975). Influence of nitrogen on dry matter production, nitrogen uptake and yield in high yielding cultures of rice. *Indian Journal of Plant Physiology* **18**, 115-120.

Sandfaer, J. & Haar, V. (1975). Barley stripe mosaic virus and the yield of old and new barley varieties. *Zeitschrift für Pflanzenzuchtung* **74**, 211-222.

Sandfaer, J., Jorgensen, J. H. & Haar, V. (1965). The effect of nitrogen fertilisation on old and new barley varieties. *Yearbook of the Royal Veterinary and Agricultural College, Copenhagen for 1965*, pp. 153-180.

Silvey, V. (1979). The contribution of new varieties to increasing cereal yield in England and Wales. *Journal of the National Institute of Agricultural Botany* **14**, 367-384.

Singh, I. D. & Stoskopf, N. C. (1971). Harvest index in cereals. *Agronomy Journal* **63**, 224-226.

Spratt, E. D. & Gasser, J. K. R. (1970). Effects of fertiliser nitrogen and water supply on distribution of dry matter and nitrogen between the different parts of wheat. *Canadian Journal of Plant Science* **50**, 613-625.

Syme, J. R. (1970). A high yielding Mexican semi-dwarf wheat and the relationship of yield to harvest index and other varietal characteristics. *Australian Journal of Experimental Agriculture and Animal Husbandry* **10**, 350-353.

Tanaka, A., Yamaguchi, J., Shimazaki, Y. & Shibata, K. (1968). Historical changes of rice varieties in Hokkaido viewed from the point of plant type. *Journal of the Science of Soil and Manure, Japan* **39**, 526-534.

Tsunoda, S. (1962). A developmental analysis of yielding ability in varieties of field crops. IV. *Japanese Journal of Breeding* **12**, 49-56.

Venkateswarlu, B., Suryaprakasha-Rao, J. & Rao, A.V. (1977). Relationship between growth duration and yield parameters in irrigated rice (*Oryza sativa* L). *Indian Journal of Plant Physiology* **20**, 69-76.

Watson, D. J., Thorne, G. N. & French, S. A. W. (1958). Physiological causes of differences in grain yield between varieties of barley. *Annals of Botany, London* **22**, 321-352.

Welbank, P. J., Gibb, M. J., Taylor, P. J. & Williams, E. D. (1974). Root growth of cereal crops. *Rothamsted Experimental Station Report for 1973, Part 2*, pp. 26-66.

The herbage crop and its environment — exploiting physiological and morphological variations to improve yields

D. Wilson, C. F. Eagles and I. Rhodes
Welsh Plant Breeding Station, Aberystwyth, Dyfed

Introduction

The leaves and stems of herbage crops not only have to provide an efficient apparatus for CO_2 assimilation and utilization, able to withstand the stresses imposed by climate, soil and animals, but they also have to exist in a form that is nutritionally acceptable for the grazing animal. Therefore, to improve crops by making effective use of available genetic variation it is becoming increasingly important to complement standard selection procedures with others which are based on knowledge of the morphological, physiological, and indeed biochemical, bases of production and adaptation. For example, recent research in outbreeding herbage grasses and legumes has demonstrated that it is possible to improve considerably potential yield and adaptability of existing productive varieties by selecting for specific predetermined characteristics, which can often be measured in young plants growing in standard conditions (Wilson, 1980). Varieties and ecotypes of these species usually display continuous variation in many morphological, anatomical and physiological traits and the problem facing breeders is often as much concerned with defining and detecting suitable plant types as with difficulties in finding sufficient variation. The use of both intra- and inter-population variation to select new populations with contrasting expression of specific traits is a powerful physiological and agronomic aid in defining plant type. This will be illustrated by describing hybridization and selection studies with forage crops, aimed at increasing potential biomass production and tolerance of climatic extremes.

Potential biomass production

Although by no means the only determinant of economic yield in forage crops, a high potential for dry-matter production lays the foundations upon which other improvements may be based. This is dependent on sustained high levels of crop net photosynthesis. The size and arrangement of the photosynthetically active system, and the extent to which photosynthesis is used for active growth are both important determinants of growth (Cooper, 1976). It should be possible, therefore, to increase biomass production, by (a) improving crop light interception, or (b) reducing any wasteful respiration. Both these possibilities have been examined in forage grasses.

Canopy architecture

The grass and grass-clover sward is a complex canopy of leaves, and other plant parts, whose dimensions, distribution and presentation control the interception, transmission and reflection of light within the crop. Genetic variations in leaf and plant morphology, therefore, affect not only the amount of light available for photosynthesis but also the competitive abilities of individual plants growing in mixed communities. Knowledge of these relationships has been used in developing selection criteria for improving dry-matter yield of monoculture grass crops, and is currently being applied to grass-clover associations.

Grass monocultures

Theoretical predictions that communities with erect forage should be more productive than more prostrate types at a high leaf-area index (LAI) have been borne out in practice in grasses (Rhodes, 1973a). Selection studies have demonstrated that individual characters of the grass canopy – long leaves, erect tillers, rigid leaves – were not only associated with yield increases under infrequent cutting but also showed considerable additive genetic variation within and between populations (Rhodes, 1971, 1973a, b, 1975). Although selecting for long leaves invariably resulted in increased yield in infrequent cutting treatments, selection for erect tillers or rigid leaves alone produced smaller increases, and then only in about 60 per cent of the populations studied. However, selection for all these characters together resulted in lines capable of producing up to 30 per cent more dry-matter than the original populations (Rhodes and Mee, 1979). Furthermore, several of these lines have proved to be more productive than the best existing varieties of their type (Rhodes and Mee, 1979), thus demonstrating that it is possible to clearly define simple morphological ideotypes for increased grass production under infrequent defoliation. These types contrast with those which would exhibit high yield in the spaced plant conditions of the breeding nursery and which would probably have long, lax, leaves and strong competitive ability.

Ryegrass-white clover mixtures

White clover has attracted increasing attention recently because of its value in nitrogen fixation and forage quality. However, a major problem with its use in UK conditions is the unpredictability of its contribution to total sward yield. Many studies have described the contribution of white clover to total yield but little attempt has been made to determine the physiological basis of the variation in its contribution or to formulate selection criteria for improved compatibility. However, on the basis of work in Australia on the annual subterranean clover, the emphasis has been on its ability to compete for light in controlling the contribution of clover under, for example, different cutting frequencies (Black, 1960; Stern and Donald, 1962a, b). These studies have led to the belief that breeding for increased stature should increase the competitive ability of the clover and hence its contribution to the total yield. However, one major difference between subterranean and white clover is the need for the latter to survive the winter in a vegetative state.

The contribution of white clover depends on (a) its yielding ability in monoculture, and (b) the extent to which this yielding ability is modified as a result of its competitive ability relative to the grass. Consequently, it is important to quantify the roles of yielding ability and competitive ability, and to elucidate the characters which are associated with these.

Recent experiments by Rhodes and Mee (1978) and by Rhodes and Harris (1979) designed to examine these relationships have involved mixtures and monocultures of ryegrass and white clovers of contrasting morphology grown under several management conditions. In addition to harvested yield, measurements of total above-ground dry matter, partitioned into plant parts, were made at regular intervals. In general, during the establishment year and under infrequent cutting, both yielding ability and competitive ability of the clovers were positively related to stature, as measured by petiole length. However, with frequent cutting no differences existed in the competitive ability of long and short-petioled types (Rhodes and Harris, 1979). During the first full harvest year (i.e. after the first winter) whilst the contribution of all varieties was reduced under infrequent cutting, this was, suprisingly, due to lower competitive ability in the long-petiole types but lower intrinsic yielding ability in the short-petiole type. The competitive ability of the latter was similar under both cutting

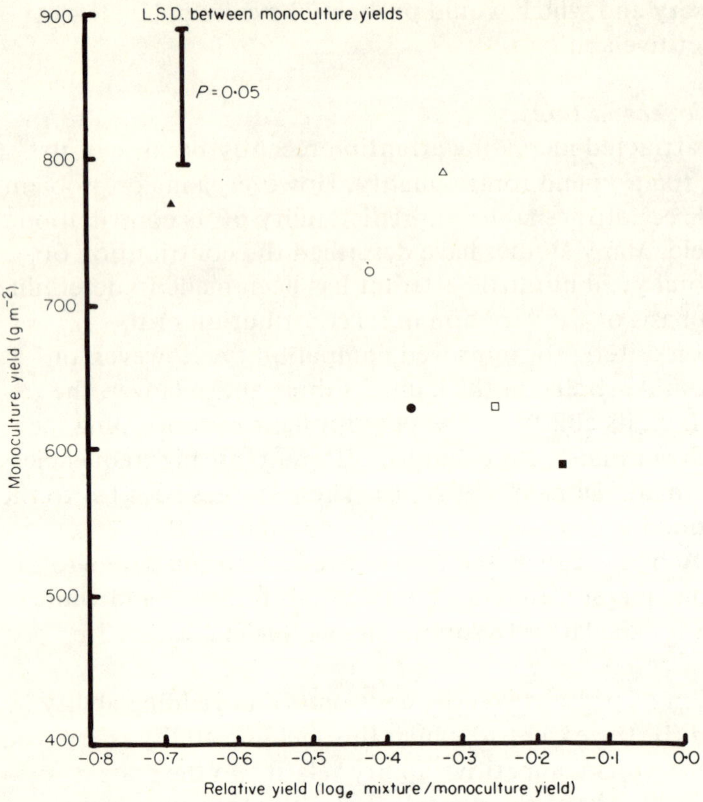

Fig. 1. Yielding ability in monoculture, and competitive ability of three white clover varieties in first full harvest year with frequent (open symbols) and infrequent (closed symbols) cutting. S 184 – short petioles (○,●); S 100 – intermediate petioles (□,■) and Sabeda – long petioles (△,▲). (From Rhodes and Harris, 1979.)

frequencies and greater than the tall type under infrequent cutting (Fig. 1). The intrinsic yielding ability of the taller type was similar under both cutting frequencies and greater than the short-stature clover. Furthermore, after the first winter, yielding ability and competitive ability appeared to be negatively correlated (Fig. 1).

These changes in relative competitive ability after the first winter appeared to be related to differences between varieties, cut at the same height, in the proportion of the above-ground yield which was harvested (Rhodes and Harris, 1979). As a result of this, and because the varieties did not differ in total above ground biomass, residual stolon after harvest was much less in the long- than short-petiole types. This would provide fewer growing points and possibly less carbohydrate reserves which would put them at a competitive disadvantage during the onset of spring growth. It is also likely that their more open habit rendered them more susceptible to winter damage. Recent studies have shown a positive relationship between the amount of spring/early summer growth and the amount of residual stolon (Harris, Rhodes and Mee, unpublished). Consequently, the most desirable white-clover ideotype would probably combine large stature with increased stolon development. Preliminary selection experiments indicate that this combination is possible.

In grass-clover swards, the morphology of the grass component may not only affect its own yield capacity but also that of the clover and even subtle genetic differences in grass-foliage angle for example can result in increased clover yield (Rhodes, Harris and Mee, 1979).

The ultimate aim of this research is to develop mutually compatible grasses and clovers specifically designed to be grown together. This would contrast with the often haphazard mixtures of separately bred components used in mixtures presently available to farmers.

Dark Respiration

Respiratory losses not associated with synthesis of new tissue can account for a substantial proportion of the carbon fixed by photosynthesis of mature crops (Robson, 1973; Biscoe, Scott and Monteith, 1975). This component has been termed 'maintenance' respiration (McCree, 1970; Thornley, 1970; Penning de Vries, 1972) and its reduction by breeding or selection has been suggested as a means of increasing biomass production (McCree, 1974; Penning de Vries, 1974; Wilson, 1975a). Genetic variation in components of respiration are known. For example, much of the variation in yield amongst five grain sorghum varieties growing at high LAI has been attributed to variation in the maintenance coefficient (Fernandez, 1977), calculated according to McCree (1974). Other studies demonstrate negative relationships between whole plant or crop growth and the rate of dark respiration of fully grown tissues (Scheibe and Meyer zu Drewer, 1959; Heichel, 1971; Wilson, 1975a; Jones and Nelson, 1979).

In forage ryegrass, screening and selection from within existing bred varieties, both for slow and for rapid rates of dark respiration of mature leaves, has revealed considerable variation which has a heritability of about 50 per cent in the first generation of selection (Wilson, 1976). Simulated swards of clonal replicates of slow-respiration parental genotypes derived from this programme have consistently shown not only greater crop growth than fast-respiration selection but also improvement over the original variety when grown in warm environments (Wilson, 1975a). In the following generation this improvement occurs mainly during regrowth in sequential cutting systems. During seedling growth of young primary swards, up to LAI of about 7, no distinct advantage for slow respiration has been observed, but in a warm controlled environment (25°/20° C) an improvement of about 20 per cent in crop growth rate during repeated regrowth periods has been achieved by this selection (Wilson, 1976). This agrees with independent carbon exchange studies reported by Robson (p. 45). In field plot trials at Aberystwyth the improvement in annual dry-matter production compared with the original variety has been about 8 per cent (Table 1).

Table 1. Annual dry matter yields of experimental ryegrass populations

Population	Dry matter yield (kg ha^{-1}) Year 1	Year 2
Slow respiration	13 568	15 013
Original variety	12 894	14 190
Fast respiration	12 516	14 101
LSD (5%)	720	641

Although the biochemical basis of these results is not yet understood, they do suggest that the selection produces plants with a reduced energetic requirement for maintenance of the integrity of existing plant tissue. It seems to be a particularly useful selection technique for increasing potential forage biomass production, since the characteristic does not appear to be related to plant anatomy or morphology and it should therefore be possible to combine it, for example, with morphological designs suitable for optimizing light interception by the crop. What is less clear is the possible adaptive significance of a relatively fast respiration rate, since its continued existence in genetically variable populations suggests some advantage in some conditions, although none has been observed.

The need for careful and comprehensive sampling procedures (Wilson, 1975a) means that dark-respiration rate is not suited to screening very large populations, so that its application is most likely in the later stages of a breeding programme.

Adaptability

Breeding for consistency and uniformity of yield, or even for survival, is often more important than a high growth potential. The seasonally variable extremes of water shortage and low temperature are two of the most important limitations to growth of perennial forages. Breeding for avoidance or tolerance of these can often be aided by the use of specific adaptive traits or of environmental simulation in selection.

Efficient use of limited water

Apart from the possibility of adjusting the timing of crop growth and development to coincide with high soil moisture conditions, breeding for avoidance or for tolerance of drought is difficult without an understanding of the adaptive mechanisms involved. These can be concerned with exploration or utilization of water and with the ability to endure physiological stress. Their relative importance depends on the nature and severity of the drought, the existence of water at depth and competitive interactions within the crop (Wilson, 1980). In this paper we are only considering adaptations of the shoot system. These are particularly useful to breeders since they are often easily measured and can be readily incorporated into standard breeding programmes. They may take the form of some inherent feature which exists in the plant even in the absence of stress or of the ability to advantageously modify the phenotype in response to water shortage.

With many crops growing in British conditions, where ability to survive drought at all costs is often of less importance than efficient use of existing water supplies, there seems to be scope for increasing production by breeding for improved water conservation. To benefit from this strategy plants would, of course, have to be grown with others with similar or complementary water-use characteristics. In some crops, this conservation can be achieved by adjusting the harvest index. In small-grain crops for example, breeding for restricted tillering can reduce wasteful use of water (Jones and Kirby, 1977). However, in pasture grasses, where tillers and leaves are the economic end product, stomatal or cuticular control of transpiration is a more useful approach. Fortunately, in C_3 species changes in stomatal conductance can effect changes in transpiration without influencing maximum photosynthesis (Wilson, 1980), so that there is a potential for real increases in water-use efficiency. In ryegrass, improved soil-moisture conservation and dry-matter production during establishment in dry conditions has been achieved by selection for fewer, shorter stomata and for less pronounced leaf ridges (Wilson, 1980). The adaptive value of leaf ridging in this species seems to be to allow transpiration to continue even in adverse conditions (Wilson, 1975b) and may well be useful in warm climates with adequate soil moisture. Specific adaptations such as narrow, erect, waxy leaves and sunken stomata may be useful where water is certain to be frequently limited but they may be associated with other undesirable agronomic traits. For example, stomata in *Dactylis marina,* a sub-species of *D. glomerata* (cocksfoot), are effectively sunken between large papillae cells which reduce transpiration, thus increasing water-use efficiency (Wilson, Abdullah and Trickey, 1977). However, *D. marina* lacks cold tolerance and has a more typically Mediterranean pattern of growth so that it is only by repeated backcrossing and reselection that the papillae cell character can be obtained in a *D. glomerata* background (Borrill, Tyler and Kirby, 1975).

The most recent developments in porometer design now make it practicable to use these instruments to detect genetic variations in physiologically based stomatal responses and stomatal adaptability (Gay, pers. comm.). With persistent water shortage, the inherent ability of the phenotype to adapt to stress during continued development of the photosynthetic apparatus is important for water-use efficiency and herbage productivity. Grass leaves developed during physiological stress tend to have a smaller specific leaf area and this is responsible for much reduction in crop photosynthesis (Leafe, Jones and Stiles, 1977). However, some genotypes with relatively low leaf tensile strength seem capable of continued production of leaves of high specific leaf area under stress (Silcock, 1978). This may at least partly explain the improved field drought tolerance of ryegrass populations selected for this characteristic (Lancashire, Wilson, Bailey, Ulyatt and Singh, 1977).

Growth at low temperature and cold hardiness

One means of broadening the genetic basis for selection in grass-breeding programmes is to introduce populations from contrasting climatic regions. Populations from the Mediterranean region, for example, offer a shift in seasonal distribution of dry-matter production compared with indigenous British genotypes. However, their winter growth potential is normally associated with poor winter hardiness and some plant kill occurs in all but the mildest conditions. In contrast, populations from northern Europe are summer growing with a long period of winter quiescence, often referred to as dormancy, which is associated with a high level of winter hardiness.

The ultimate objective in exploiting genetic variation for physiological characters which control growth at low temperatures and ability to survive sub-zero temperatures is the production of winter-growing winter hardy herbage associated with good summer productivity.

This objective may be achieved by combining the desirable characters in appropriate crosses between contrasting genotypes followed by selection of superior genotypes from within the hybrid. This approach has been used successfully to derive the cocksfoot cultivars Saborto and Calder from the cross between S.37 and a Portuguese population from Sintra near Lisbon. Similar crosses between this Portuguese parent and more extreme north European parents have been less successful in immediately producing a desirable combination of agronomic characters, since winter quiescence is often dominant. Even so, in one such cocksfoot hybrid involving Latar (origin Leningrad) and the Portuguese parent, winter growth was intermediate between the two parents but there was a high degree of winter hardiness (Table 2). However, two

Table 2. Leaf growth in an unheated glasshouse (November-February) and cold tolerance of a hybrid between contrasting cocksfoot populations

Population	Rate of leaf expansion ($mm^2\ d^{-1}$)	Percentage kill at $-8^{\circ}C$ for 24 h
Latar (Leningrad)	14	0
Portuguese	51	75
Hybrid	32	5

generations of selection for rate of leaf area expansion at low temperature, together with parallel selection for cold tolerance, subsequently resulted in a marked increase from 32 $mm^2\ d^{-1}$ to above 60 $mm^2\ d^{-1}$, while survival at $-8^{\circ}C$ for 24 h was maintained above 80 per cent; although this was less in genotypes with leaf area expansion rates greater than 80 $mm^2\ d^{-1}$. Assessment of these lines under field conditions is necessary to determine whether these improved rates of leaf expansion are exhibited during the winter and whether there has been a shift in seasonal productivity.

An alternative approach is to start with agronomically desirable material already adapted to UK conditions and improve certain aspects such as early spring growth by mass selection. The factors controlling leaf expansion in the spring may determine whether this strategy is possible. Low temperatures and short days often appear to interact in determining cessation of active growth of northern populations in the autumn (Eagles, 1971) and the existance of similar control mechanisms for the onset of growth in the spring would complicate selection procedures. However, in experiments where northern populations were allowed to express their normal growth rhythm in the autumn prior to transfer during the winter and spring to controlled environment conditions with greater temperature and/or daylength, it appeared that temperature was the major factor controlling the onset of leaf expansion. Increased daylength had little effect on the initial response. Based on these results a programme is being developed to select genotypes which show an ability to initiate the onset of leaf expansion at lower temperatures in the spring using a temperature-gradient tunnel. Parallel studies of cold tolerance in the selections will be necessary to ensure that there is no decline in survival of these plants as a result of exposure to late spring frosts during active growth. A higher temperature threshold for dehardening, shown by hardy populations (Fuller, 1979), may avoid this problem where the selections are made on inherently hardy types.

In a similar way the distortion of seasonal growth rhythms in favour of autumn productivity resulting from management practices such as late intensive defoliation and heavy fertilizer applications can reduce winter hardiness appreciably. A technique to assess freezing tolerance of mature plant tillers (Fuller, 1979) offers the opportunity to study the effect of intensive autumn management on natural hardening.

References

Biscoe, P. V., Scott, R. K. & Monteith, J. L. (1975). Barley and its environment. III. Carbon budget of the stand. *Journal of Applied Ecology* **12**, 269-293.

Black, J. N. (1960). The significance of petiole length, leaf area and light interception in competition between strains of subterranean clover (*Trifolium subterranean* L.) grown in swards. *Australian Journal of Agricultural Research* **11**, 277-291.

Black, J. N. (1961). Competition between two varieties of subterranean clover (*Trifolium subterranean* L.) as related to the proportions of seed sown. *Australian Journal of Agricultural Research* **12**, 810-820.

Borrill, M., Tyler, B. F. & Kirby, M. (1975). Progress in breeding digestible hybrid cocksfoot. In *Report of the Welsh Plant Breeding Station for 1974,* pp. 13-15.

Cooper, J. P. (1976). Photosynthetic efficiency of the whole plant. In *Food Production and Consumption: the Efficiency of Human Food Chains and Nutrient Cycles* pp. 107-126. Eds. A. N. Duckham, J. G. W. Jones & E. H. Roberts. Amsterdam: North Holland.

Eagles, C. F. (1971). Effect of photoperiod on vegetative growth in two natural populations of *Dactylis glomerata* L. *Annals of Botany* **35**, 75-86.

Fernandez, C. J. (1977). Differences in carbon economy among five grain sorghum cultivars. *M.Sc. Thesis,* Texas A & M University.

Fuller, M. P. (1979). Physiological aspects of cold hardiness in forage grasses. *Ph.D. Thesis,* University of Wales.

Heichel, G. H. (1971). Confirming measurements of respiration and photosynthesis with dry matter accumulation. *Photosynthetica* **5**, 93-98.

Jones, H. G. & Kirby, E. J. M. (1977). Effects of manipulation of number of tillers and water supply on grain yield in barley. *Journal of Agricultural Science* **88**, 391-397.

Jones, R. J. & Nelson, C. J. (1979). Respiration and concentration of water soluble carbohydrate in plant parts of contrasting tall fescue genotypes. *Crop Science* **19**, 367-372.

Lancashire, J. A., Wilson, D., Bailey, R. W., Ulyatt, M. J. & Singh, P. (1977). Improved summer performance of a 'low cellulose' selection from 'Grasslands Ariki'. *New Zealand Journal of Agricultural Research* **20**, 63-67.

Leafe, E. L., Jones, M. B. & Stiles, W. (1977). The physiological effects of water stress on perennial ryegrass in the field. In *Proceedings of the Thirteenth International Grassland Congress,* 18-27 May 1977, pp. 165-184, Leipzig, GDR.

McCree, K. J. (1970). An equation for the rate of respiration of white clover plants grown under controlled conditions. In *Production and Measurement of Photosynthetic Productivity* pp. 221-230. Ed. I. Setlick. Wageningen: Centre for Agricultural Publishing and Documentation.

McCree, K. J. (1974). Equations for the rate of dark respiration of white clover and grain sorghum as functions of dry weight, photosynthetic rate and temperature. *Crop Science* **14**, 509-514.

Penning de Vries, F. W. T. (1972). Respiration and growth. In *Crop Processes in Controlled Environments* pp. 327-348. Eds. A. R. Rees, K. E. Cockshull, D. W. Hand & R. G. Hurd. London: Academic Press.

Penning de Vries, F. W. T. (1974). Substrate utilization and respiration in relation to growth and maintenance in higher plants. *Netherlands Journal of Agricultural Science* **22**, 40–44.

Rhodes, I. (1971). The relationship between productivity and some components of canopy structure in ryegrass (*Lolium* spp.). 2. Yield canopy structure and light interception. *Journal of Agricultural Science, Cambridge* **77**, 283-292.

Rhodes, I. (1973a). Relationship between canopy structure and productivity in herbage grasses and its implications for plant breeding. *Herbage Abstracts* **43**, 129-133.

Rhodes, I. (1973b). The relationship between productivity and some components of canopy structure in ryegrass (*Lolium* spp.). 3. Spaced plant characters their heritabilities and relationship to sward yield. *Journal of Agricultural Science, Cambridge* **80**, 171-176.

Rhodes, I. (1975). The relationship between productivity and some components of canopy structure in ryegrass (*Lolium* spp.). 4. Canopy characters and their relationship with sward yields in some intra-population selections. *Journal of Agricultural Science. Cambridge* **84**, 345-351.

Rhodes, I. & Harris, W. (1979). The nature and bases of differences in sward composition and yield in ryegrass–white clover mixtures. In *Changes in Sward Composition and Productivity*. Eds. A. H. Charles & R. J. Haggar, Occasional Symposium No. 10, British Grassland Society.

Rhodes, I. & Mee, S. S. (1978). Crop morphology–Ryegrass/white clover. *Report of Welsh Plant Breeding Station for 1977,* pp. 142-144.

Rhodes, I. & Mee, S. S. (1979). Changes in dry matter associated with changes for canopy characters in ryegrass. *Grass and Forage Science* (in press).

Rhodes, I., Harris, W. & Mee, S. S. (1979). Crop morphology–Ryegrass/white clover. *Report of the Welsh Plant Breeding Station for 1978*, pp. 156-161.

Robson, M. (1973). The growth and development of simulated swards of perennial ryegrass. II. Carbon assimilation and respiration in a seedling sward. *Annals of Botany* **37**, 501-518.

Scheibe, A. & Meyer zu Drewer, H. (1959). Vergleichende untersuchungen zur Atmungsintensität der wurzeln unterschiedlicher Genotypen bei Getreidearten. *Zeitschrift fur Acker - und Pflanzenbau* **108**, 223-252.

Silcock, R. G. (1978). Water relation in *Festuca. Ph.D. Thesis,* University of Wales.

Stern, W. R. & Donald, G. M. (1962a). Light relationships in grass-clover swards. *Australian Journal of Agricultural Research* **13**, 599-614.

Stern, W. R. & Donald, G. M. (1962b). The influence of leaf area and radiation on the growth of clover in swards. *Australian Journal of Agricultural Research* **13**, 615-623.

Thornley, J. H. M. (1970). Respiration, growth and maintenance in plants. *Nature* **227**, 304-305.

Wilson, D. (1975a). Variation in leaf respiration in relation to growth and photosynthesis of *Lolium. Annals of Applied Biology* **80**, 323-328.

Wilson, D. (1975b). Stomatal diffusion resistances and leaf growth during droughting of *Lolium perenne* plants selected for contrasting epidermal ridging. *Annals of Applied Biology* **79**, 83-94.

Wilson, D. (1976). Physiological and morphological selection criteria in grasses. In *Breeding Methods and Variety Testing in Forage Plants* pp. 9-18. Ed. B. Dennis. Report of meeting of Fodder Crops Section, Eucarpia, September 1976, Roskilde, Denmark.

Wilson, D. (1980). Breeding for morphological and physiological traits. In *Plant Breeding Symposium II.* Ed. K. J. Frey. Iowa State University Press (in press).

Wilson, D., Abdullah, I. B. & Trickey, S. A. (1977). Variation in transpiration rate in *Dactylis.* In *Proceedings of the Thirteenth International Grasslands Congress* 18-27 May 1977, pp. 71-79. Leipzig: GDR.

A physiologist's approach to raising the potential yield of the grass crop through breeding

M. J. Robson
Grassland Research Institute, Hurley, Berkshire, SL6 5LR*

Introduction

This paper examines ways in which an understanding of the physiology of the perennial grass crop can indicate selection criteria which might enable the breeder to increase potential yield. Attention is focussed on a sown crop of a single species grown free from water stress, nutrient deficiency and attack by pests or diseases, and managed to maximize yield. It will be assumed that the crop is already established, and is harvested by infrequent cutting. The quality of the crop, its conservation and utilization, important though they are, will *not* be considered. Nutritional and environmental aspects and, to a lesser extent, cultural practices are dealt with elsewhere, and will only be briefly mentioned here.

Past efforts to increase grass yield by breeding have met with only modest success. Alberda (1971) claims that although actual yields have risen greatly through the increased use of fertilizers, the genetic potential of the crop has changed little in over 100 years. Many of the Aberystwyth 'S' strains, such as S24, introduced over 40 years ago, are still in the NIAB recommended lists (Table 1) and are accredited with yields only 5 per cent below those of the leading varieties – not that these rankings hold in all situations.

Many grass species and varieties are regularly and heavily infected with pests and diseases, for example, ryegrass mosaic virus and the larvae of stem-boring Diptera such as *Oscinella frit* whose largely unseen activities can reduce yields by up to 30 per cent (Clements, Gibson, Henderson and Plumb, 1978). Different genotypes are differently affected, such that ranking orders can be reversed when the pest or disease is eliminated (Table 2). This makes a realistic assessment of the extent of past genetic advance difficult to determine.

* The Grassland Research Institute is financed through the Agricultural Research Council.

Table 1. Dry-matter yields of early and medium early varieties of perennial ryegrass, expressed as a percentage of the total first year yield of S24, from the NIAB Farmers Leaflet No. 16, *Recommended Varieties of Grasses 1978/9*

	Aberystwyth S24	Cropper	Premo	Monta	Reveille (Tet)	Barvestra (Tet)	RvP Hay Pasture	Gremie	Grasslands Manawa	Frances	Grasslands Ruanui
Simulated grazing yield	100	103	97	101	101	100	101	104	100	104	99
Conservation yield	100	98	98	98	100	103	101	99	103	100	92

The source leaflet comments that 'Differences of less than 5 per cent should be treated with reserve.'

Table 2. Total annual (1977) dry-matter yield (kg ha^{-1}) of three varieties of Italian ryegrass treated, or not treated, with a complete insecticide (Clements, unpublished)

	Optima	Aberystwyth S22	Sabalan
Untreated	11 086	10 372	9 146
Treated	12 252	12 810	13 454

Dry-matter partition

One problem the breeder faces is that some selection criteria are not relevant to grasses. During the last thirty years, the cereal breeder has increased the potential yield of winter wheat by 50-60 per cent (Austin, 1978), largely by raising the harvest index – there has been a greater allocation of above-ground dry matter to grain relative to stem. In the forage grasses, there is no discrete energy-rich package, analogous to the cereal grain, into which an increased fraction of the above-ground dry matter can be conveniently diverted. All the above-ground dry matter is harvested, except for stubble necessary to support regrowth.

The only possibility of altering partition might be to increase shoot growth at the expense of roots. An established perennial sward carries, on average, twice the weight of roots (live and dead) as of shoots. But this is misleading. It gives no indication of the gross rate at which the weight of live roots is increasing at any one time, and hence of the rate at which assimilates are being partitioned to them.

The only way to measure such partition is directly, using ^{14}C as a tracer. Indications are that only during seedling establishment does as much as 25 per cent of newly assimilated carbon reach the roots (Fig 1). For most of the time the figure is only 10-15 per cent, falling to 5 per cent or less during rapid stem elongation in the spring (Parsons and Robson, 1978).

Thus, there seems little scope for diverting more assimilate into the harvestable fraction of the crop. The grass breeder is left with the difficult task of increasing total dry-matter yield.

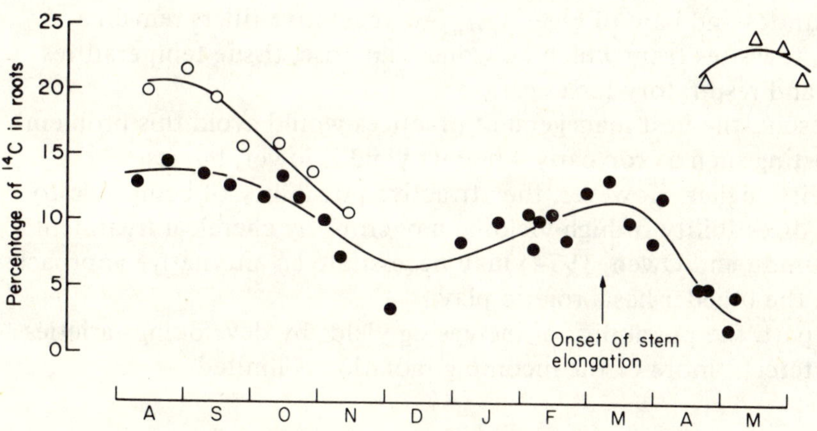

Fig. 1 Seasonal pattern of assimilate partitioning of established (●) and seedling (autumn-sown, ○; spring-sown, △) simulated swards of S24 ryegrass growing under natural conditions; percentage of ^{14}C recovered from plants 48 h after labelling that was present in the roots rather than shoots (Parsons and Robson 1978, and in preparation)

Photosynthetic potential of the leaf canopy

Dry-matter production is limited, ultimately, by the efficiency with which the crop uses light to fix carbon. This, in turn, depends on the proportion of incoming radiation intercepted by the canopy, the way this is distributed over the leaf surface, and the photosynthetic potential of the individual leaves.

Light interception

In annual crops, such as sugar beet, earlier sowing can greatly raise yield, largely because leaf area and light interception are increased (Scott and Allen, 1978). In established perennial crops which maintain a leaf cover throughout the year, no opportunity exists for increasing yields in this way.

Although the grass crop is perennial, harvesting interrupts the leaf cover (Fig. 2(a)) which, in certain circumstances, can be slow to re-establish. For example, this happens after a heavy flowering crop, that has received high levels of nitrogen, is cut during a period of high insolation. In this situation, the residual stubble is often low in reserve carbohydrates and bare of green leaf, few vegetative tillers remain and there are few sites from which new ones can arise, tissue temperatures are high and respiratory losses great.

At present, the best management practices would avoid this problem by harvesting such a crop early when its yield is lower, but its digestibility higher. However, the attractive possibility of being able to raise the digestibility of high-yielding hay crops by chemical treatment (Mwakatundu and Owen, 1974) may necessitate an alternative approach in which the breeder has a role to play.

This apart, the possibility of increasing yields by developing varieties which intercept more of the incoming radiation is limited.

Light distribution

The hyperbolic response of individual leaves to the level of incoming radiation (in terms of the photosynthetic fixation of carbon) means that a given amount of light will be more efficiently used, the more evenly it is distributed over the leaf surface. Thus, when LAI is high, an erect crop is at an advantage since vertical leaves will be more evenly illuminated. Conversely, when LAI is low, a prostrate crop is at an advantage in that horizontal leaves intercept more light.

Of course, a variety whose leaves are *always* prostrate will be at a disadvantage when LAI is high, as will one that is erect when LAI is low. In practice, the use of varieties with a particular fixed growth habit may be limited to particular management situations. As Rhodes (1969) has

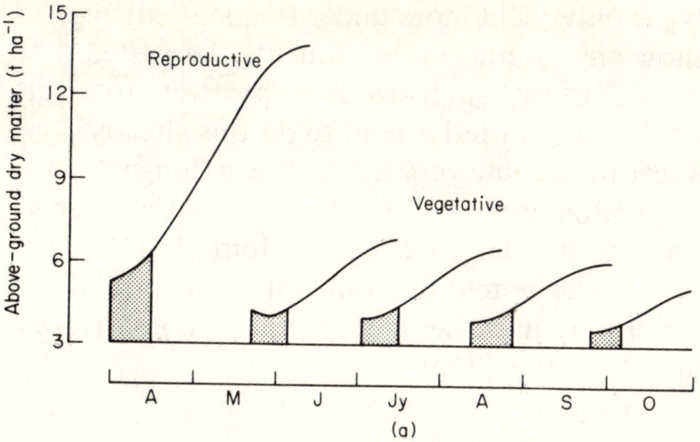

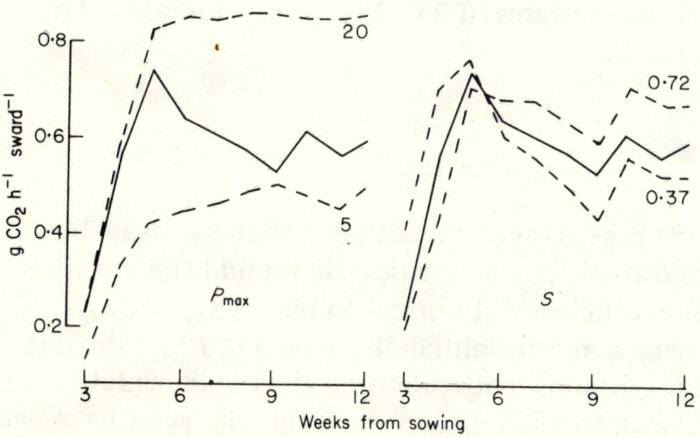

Fig. 2 (a) Accumulated above-ground dry matter production of S24 ryegrass between successive cuts; shaded areas indicate periods of incomplete light interception. Annual total light interception is typically >80 per cent (Leafe, 1978). (b) Rate of canopy gross photosynthesis of swards of S24 ryegrass whose successive newly expanded leaves declined in P_{max} from 20 to 5 mg CO_2 dm^{-2} h^{-1} and whose habit ranged from relatively prostrate (S = 0·37) to relatively erect (S = 0·72) (——), compared with computer predictions which assume that either P_{max} had been constant at 20 or 5, or S values constant at 0·72 or 0·37, throughout the growth period (– – – –) (Robson, 1973b); note, S is the fraction of light passing through unit leaf layer without interception (Monteith, 1965).

shown, prostrate genotypes only yield more under frequent cutting, while erect genotypes show an advantage if infrequently harvested.

For general use, we need grasses that change from prostrate to erect as the crop grows denser. Existing varieties tend to do this already (although, if the crop is left to become very dense, it can 'lodge'). Thus, here too, the scope for increasing yield by genetic manipulation seems limited. In any case, computer simulation using the Monteith (1965) model suggests that habit is a less potent determinant of the rate of carbon fixation of a grass crop than is the photosynthetic potential of the young well-illuminated leaf (Fig. 2(b)).

Photosynthetic potential of individual leaves

If developed under, and subsequently exposed to, favourable environmental conditions, young expanded leaves of north temperate grasses are capable of fixing carbon at rates of 25–35 mg CO_2 dm^{-2} h^{-1}. Can this capability be increased?

Maximum efficiency

Charles-Edwards (1978) has surveyed the genetic variability available in a number of photosynthetic characteristics. He found little variation in α, the light utilization efficiency at normal ambiant CO_2 and O_2 concentrations, but there were substantial differences in P_{max}, the rate of light-saturated photosynthesis, particularly in grasses. Indeed, Cooper and Wilson (1970) report a twofold difference in this character between ecotypes of *Lolium perenne*. Even so, selection based on such differences in single leaf photosynthesis has rarely, if ever, led to greater rates of dry-matter production. One possible reason for this is an inverse correlation between leaf size and photosynthetic rate.

The third character Charles-Edwards examined was the ratio between rates of photorespiration and gross photosynthesis. It has recently become clear that a single enzyme acts on ribulose 1,5-bisphosphate both as a carboxylase and as an oxygenase, and that it is the relative rate of these two reactions that determines the proportion of carbon entering the photorespiratory pathway (Ogren, 1978). It may be that ultimately, either by chemical treatment or gentic manipulation, carboxylation could be increased at the expense of oxygenation. The possible benefits are great and are not limited to grasses, but they are likely to be long in coming.

There would appear to be greater short-term advantage in seeking to ensure that more leaves within a grass crop achieve, and maintain for longer, photosynthetic efficiencies close to the existing maximum, than in trying to raise the maximum. Note, we are *not* here considering leaves which have a low rate of photosynthesis because, at the time of measurement, they are poorly illuminated. We *are* concerned with leaves which, even when illuminated at saturating intensities, fail to make as efficient use of light as we know is possible. The failure is widespread and two main reasons are discussed.

Leaf age

As a grass leaf ages, its photosynthetic potential declines (Fig. 3(a)). Typically, each tiller within an established crop bears three fully expanded leaves, the youngest with the highest photosynthetic potential and the oldest with the lowest. When a new leaf is produced the oldest leaf dies, the rate of turnover being determined by genotype and environment.

If it proved possible, by breeding, to extend the (photosynthetically active) life of a leaf, without delaying the appearance of its successor, then the number of live leaves per tiller could be raised, from say three to four. However, this would not necessarily bring a proportionate increase in the rate of canopy photosynthesis. Leafe (1972) showed that within a closed canopy, about 90 per cent of the incident radiation is intercepted by the two youngest and most efficient fully expanded leaves. The oldest and least efficient leaf intercepts only 10 per cent of the light, so that to delay its senescence would not be a great advantage in these circumstances.

Light environment during development

Not only does photosynthetic potential decline as a leaf ages, but many leaves begin their lives with an already low potential. Controlled environment studies indicate that low light intensities and extremes of temperature lead to leaves with low P_{max}. This, no doubt, explains the values of less than 8 mg CO_2 dm^{-2} h^{-1} shown by leaves expanded in mid-winter (Parsons and Robson, 1978). Fortunately, their ability to make use of the low light intensities they then experience is relatively unimpaired, so that their lower P_{max} matters little.

Very much more serious is the repeated finding (Woledge, 1973; Woledge and Leafe, 1976) that during periods of vegetative regrowth in

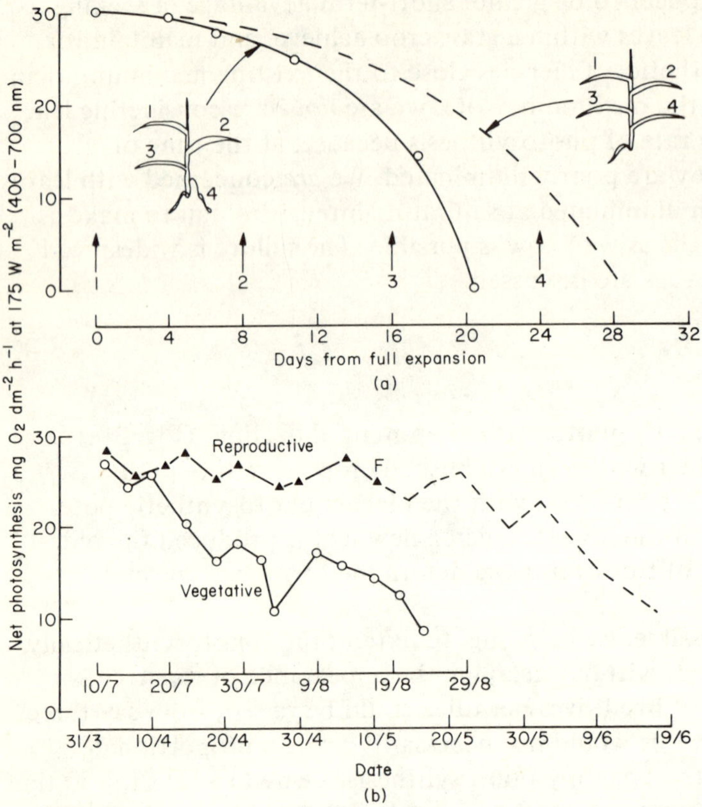

Fig. 3 (a) Decline in P_{max} of an initially efficient leaf of S24 ryegrass as it aged and was superseded by younger leaves (from Woledge, 1979); dashed line indicates possible effect of lengthening leaf life such that each tiller bears four, rather than three live, fully expanded leaves. (b) P_{max} of successive newly expanded leaves of S24 ryegrass developed during July and August within a vegetative sward of steadily increasing density (○) compared with successive leaves developed on flowering tillers within a reproductive sward during April and early May (▲), up to and including the flag-leaf (F) whose subsequent decline through age is indicated by the dashed line (from Woledge and Leafe, 1976).

summer and autumn, successive newly expanded leaves show progressively lower values of P_{max} as the sward in which they develop becomes increasingly dense (Fig. 3(b)). This is *not* an effect of age – the leaves are all newly expanded and at their most efficient. It is a direct effect of density on the development of the leaf's photosynthetic potential. It mimics the effect of growing young spaced plants under increasing levels of artificial shade. It can be prevented by protecting the parent shoot from shading by neighbours and it can be reversed (in that *subsequent* leaves show an increase in potential P_{max}) by removing that shading (Woledge, 1977).

An analogous effect has been reported in tomato, where poorly illuminated leaves develop lower values of P_{max} than leaves in high light, although their light utilization efficiencies are similar (Charles-Edwards and Ludwig, 1975). There is an important practical difference, however, between grass and tomato. In the latter, a leaf whose photosynthetic apparatus has been conditioned by low light normally continues to experience low light. Its inability to make efficient use of high light is irrelevant. In grass, on the other hand, a leaf developed in a dense vegetative canopy emerges into high light (unless the crop 'lodges') where its deficiency is fully exposed. A vegetative sward fixes 30 per cent less carbon that it would if the young leaves at the end of a growth period had used light as efficiently as those at the beginning (Fig. 2(b)) (Robson, 1973b).

This deficiency is not shown by reproductive crops. Once stem elongation begins, successive leaves developed on flowering shoots are as efficient at fixing carbon as their predecessors (Fig. 3(b)). Woledge (1979) suggests that this is because stem elongation keeps young leaves close to the top of the canopy where they can develop in full sunlight, unlike leaves in a vegetative sward which develop in increasing shade at its base; her evidence indicates that in perennial ryegrass, the high requirement of the elongating internodes for substrates exerts no significant 'sink' effect.

These findings do much to explain the high rates of dry-matter production shown by the flowering crop in the spring when compared with any of the subsequent periods of vegetative regrowth (Fig. 2(a)). It might prove possible to circumvent the deficiency of the latter by inducing a high proportion of flowering tillers throughout the growing season. However, the repercussions of such a fundamental change would be so wide-ranging as to make it impossible to predict what the overall effect on dry-matter production would be. It might even endanger the perrenniality of the crop.

An alternative might be to manage the vegetative crop so that it never became dense and the photosynthetic potential of leaves never declined. Continuous and intensive grazing already achieves this (and reduces the loss of carbon through both respiration and leaf detachment). However, the resultant low LAI leads to a loss of light to bare ground, while the removal of leaf laminae before they are fully expanded means that much of the light that is intercepted is wasted on photosynthetically inefficient leaf sheaths. Thus the rate of canopy photosynthesis, per unit of ground area, is actually reduced (Parsons, Leafe, Stiles and Collett, 1979).

The most attractive possibility might be to select genotypes better able to resist the depressant effects of increasing crop density on their capacity to fix carbon. In seeking variability in such a character, it might be possible to substitute artificial shading for the effects of high crop density, although changes in light quality, as well as a reduction in quantity, may be involved.

If appreciable variation were found, it might in part explain past failures to produce higher rates of crop growth from genotypes selected for their great P_{max}. In most cases, screening will have been carried out on single young spaced plants whose ability to fix carbon may bear little relation to that of the same genotype when grown in a dense community.

Loss of assimilated carbon

Even under the most favourable conditions, only a quarter of the carbon fixed in photosynthesis is ultimately harvested (Robson, 1973b). One of the routes by which carbon is lost – partition to non-harvestable fractions of the crop (root and stubble) – has been considered already. Another has been alluded to:

Leaf death and detachment

Leaf death not only reduces the photosynthetic surface of the crop, but detachment and subsequent incorporation into the soil by earthworms represents a direct loss of harvestable tissue. In a grass crop, when the rate of leaf death equals the rate of leaf production, a ceiling yield is reached (Fig. 4(a)). The size of the ceiling is set by the ability of the crop to meet ever-increasing respiratory requirements from a limited supply of current assimilates, while continuing to support leaf growth (Robson, 1973a). Selection for greater leaf longevity would delay detachment and could raise the ceiling yield of the crop – providing, of course, that the retained leaves fixed at least enough carbon to meet their own respiratory requirements.

The flowering crop again has an advantage in that it diverts an increasing fraction of current assimilates into the elongating stem which persists to harvest, unlike the relatively ephemeral leaves. The ultimate in this respect is the tree crop, which year by year accumulates dry matter in its harvestable fraction, the trunk, without a proportionate increase in respiratory burden.

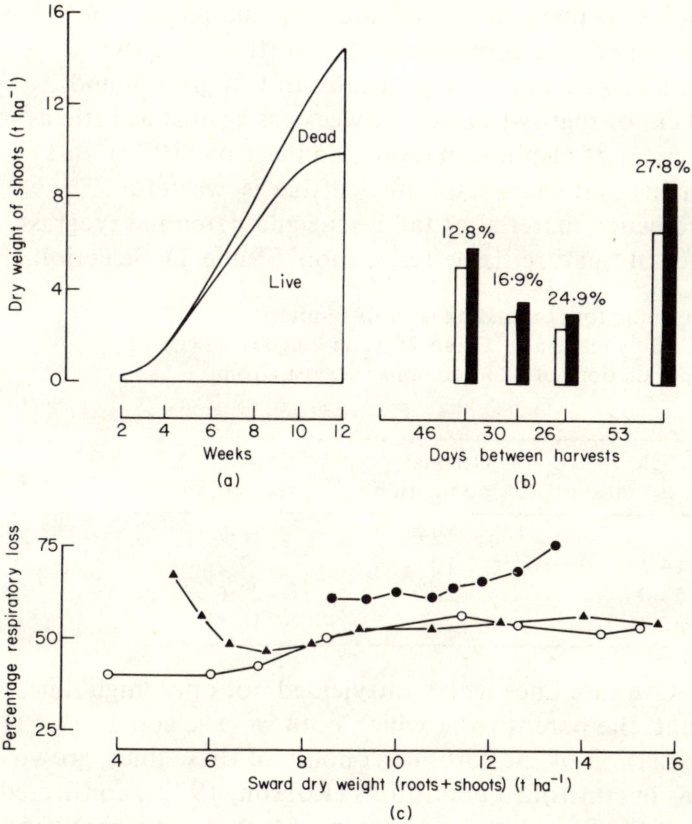

Fig. 4 (a) Shoot dry weight of a seedling sward of S24 ryegrass showing achievement of ceiling yield of live tissue (from Robson, 1973a). (b) Dry weight of harvestable yield (above a 5 cm cutting height) or 'high' (□, GL66) and 'low' (■, GL72) respiration rate lines of S23 ryegrass grown as simulated swards in a growth room at 20°/15°C over four successive regrowth periods. Figures indicate percentage increase in yield of 'low' line. (c) Respiratory loss as a percentage of gross photosynthesis of established swards of S24 ryegrass grown under natural conditions either while vegetative in the autumn (●) or reproductive in the spring (▲), and of seedling swards (○) in a growth room at 22·5/12·5°C (from Robson, 1973b and unpublished).

Respiration

Studies of the fate of $^{14}CO_2$ pulse-fed to single plants (Ryle, Cobby and Powell, 1976) and of the gas exchange of intact communities (Robson, 1973b), indicate that at least half the carbon fixed by illuminated leaves is subsequently lost in the respiratory efflux of CO_2 (Fig. 4(c)). Some of this efflux is associated with vital processes such as the conversion of primary photosynthate into the range of chemical compounds of which new plant tissue is composed, and its rate, no doubt, reflects the rate of that conversion. It is worth noting that genotypes of tall fescue selected for higher rates of leaf growth and yield *per tiller* (30 cm of regrowth after six weeks as against as little as 10 cm) had higher rates of respiration (Jones and Nelson, 1979). But the possibility remains that some respiratory efflux is 'wasteful'.

Wilson (1975) screened material of tall fescue and perennial ryegrass for contrasting rates of 'mature tissue respiration' (Table 3). Selection

Table 3. Initial screening for contrasting rates of respiration ($\mu l\ O_2\ mg^{-1}\ min^{-1} \times 10^{-3}$) made at 8°C and 25°C on long-leaved (L) and short-leaved (S) populations of S23 perennial ryegrass (from Wilson, 1975)

		High respiration	Original population	Low respiration
8°C	(L)	12·4	9·6	6·7
8°C	(S)	14·2	10·1	7·2
25°C	(L)	33·0	25·7	21·6
25°C	(S)	32·3	24·9	18·3

led to 'low' respiration rate lines which outyielded not only 'high' lines but, more important, the parent from which both were selected.

A detailed examination of the carbon economy of these lines, grown as simulated swards in controlled conditions (Robson, 1979), confirmed the superior yield of the 'low' line, and suggested that it came about in two ways: (1) The lower rate of respiration (per unit dry weight) led directly to a progressively greater crop weight (root as well as shoot). (2) This greater weight was expressed in a greater tiller population from which, following cutting, more leaf was expanded (although *not* at a greater rate per tiller) leading to greater light interception, carbon fixation and, again, greater dry-matter production. By the fourth regrowth period the dry-weight advantage, in terms of harvestable shoot, of the 'low' line over the 'high' had risen to a peak of 28 per cent (Fig. 4(b)), equivalent on an annual basis in the field, to a superiority of the 'low' line over its parent of about 10 per cent.

This is a very worthwhile increase in yield. Nevertheless, care is needed. The full biochemical and physiological consequences of selection for 'low' respiration are not yet clear. It may bring with it sensitivity to stress of various sorts. If there is no penalty for having a low rate of respiration, it is hard to see why selection, natural or otherwise, has not already favoured genotypes that possess it to the virtual exclusion of those that do not. Thus, a plant with a greater rate of dry-matter production might be expected to compete strongly with its neighbours while vegetative, and subsequently produce more seed when reproductive. (The same argument applies to photorespiratory losses.) Of course, if there *is* a penalty, it may be that the farmer can readily overcome it by management.

It is important that we establish an understanding of the respiratory metabolism of these lines, not only to forewarn us of any adverse consequences, but to indicate how far selection for low respiration might be usefully pursued.

Use of carbon in leaf growth

The production of harvestable yield is more than the mere collection and retention of carbon. The conversion of assimilate into leaf tissue involves cell division, expansion and a whole range of metabolic processes most of which are temperature dependent. Too low a temperature limits crop growth directly (and leads to an ageing and perhaps photosynthetically less efficient leaf canopy). There has long been an interest in overcoming this limitation and extending the growing season. Priority has gone to achieving earlier spring growth for good agricultural reasons, although there are also good scientific ones.

The seasonal cycle of temperature (most important the temperature just below the soil surface where, in spring, the shoot meristems are sited: Peacock 1975a) lags some weeks behind the cycle of radiation receipt. In mid-April, when the photosynthetic potential of the crop is at its maximum and radiation receipt is high, the ambient temperature is several degrees below that reached in the autumn with indentical radiation receipts. There is a danger of the crop producing assimilates but being limited by the relatively low spring temperatures in its ability to convert them into new leaf tissue. Fortunately, the leaves of many north temperate grasses are able, during flowering in the spring, to expand at a higher rate (at a given temperature) than during the rest of the year (Peacock, 1975b), (Fig. 5(a)).

In S24 ryegrass, the transition from an 'autumn' to a 'spring' mode

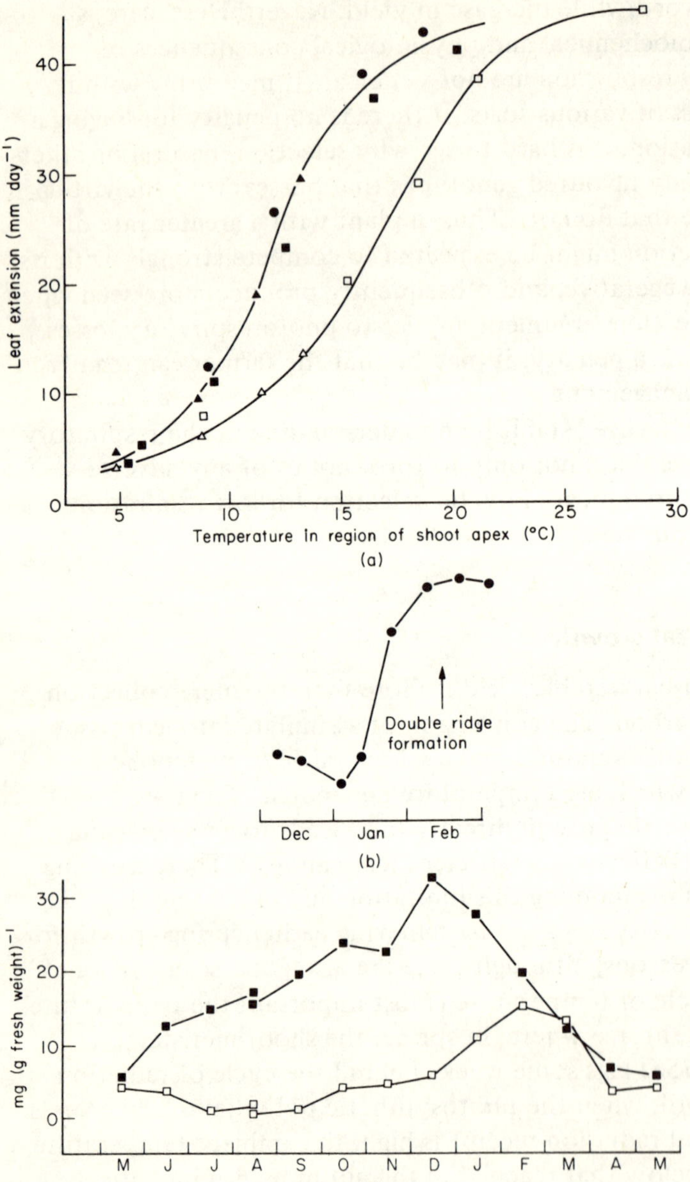

Fig. 5 (a) Rate of extension of leaves of either reproductive (spring, open symbols) or vegetative (autumn, closed symbols) swards of S24 ryegrass at a range of temperatures (from Peacock, 1975a, b and Parsons and Robson, in preperation). (b) Change in the potential leaf extension (at 15°C) of S24 ryegrass in mid-January (from Parsons and Robson, 1978). (c) Seasonal pattern of accumulation and consumption of fructosans (■) and low-molecular-weight sugars (mainly sucrose) (□) in the sheaths and stems of a high water-soluble carbohydrate selection line of S23 ryegrass (from Pollock and Jones 1979).

of growth occurs in early January (Fig. 5(b)), coincidental with the very earliest stages of reproductive development (Parsons and Robson, 1978). Because ambient temperatures are then very low, the change is more in the *potential* to expand leaves than in its realization (that comes later in the spring). Nevertheless, some extra growth is made even though current assimilation is then very low. It is, perhaps, significant that there is a change in the carbohydrate strategy of the crop at this time (Pollock and Jones, 1979). Storage carbohydrates (mainly fructosans) are rapidly mobilized and then utilized (perhaps in leaf growth) (Fig. 5(c)).

A breeder, seeking to increase leaf growth early in the year, may need to know whether the grass with which he is working shows a change from 'autumn' to 'spring' rates and when the transition occurs; he cannot assume that genotypes selected when the plants were vegetative will also show the most rapid leaf growth when they are reproductive. He should also be aware of the apparent association of early spring growth with the consumption of storage carbohydrates, bearing in mind (1) the well-established inverse correlation between leaf expansion at low temperatures and winter hardiness and (2) that the latter has often been related to the level of soluble sugars in the cell cap.

Breeding and management

Until recently, the physiologist has contributed little to breeding beyond explaining why new crop varieties have replaced old. However, as genetic potentials have increased and further improvement becomes more difficult to achieve, the need for selection criteria based on sound physiological principles has grown. This is particularly so with the grass crop where, as we have seen, higher yields must come from greater dry-matter production and not just a change in partition.

Of the criteria examined in this paper, some are already employed in breeding programmes and others are under consideration. Their relative importance varies with location and management. Thus, the ability of a leaf to withstand the depressant effects of increasing crop density on the development of its photosynthetic system may crucially affect its performance in an infrequently cut crop, but is only marginally relevant when the crop is continuously and heavily grazed.

Breeding is but one of several routes to increased yield. The most important, in the past, has been management and there is still immense scope for improvement, particularly in conservation and utilization.

There is a need both to educate and motivate the average farmer into taking up existing techniques and research findings, and to devise even better management practices. Only in this way can the genetic potential of existing varieties be fully exploited – not that this precludes the development of new varieties. Breeding and management go hand in hand and, of course, the better the farmer, the more he has to gain from the advent of improved varieties.

References

Alberda, Th. (1971). Potential production of grassland. In *Potential Crop Production* pp. 159-171. Eds. P. F. Wareing & J. P. Cooper. London: Heinemann.

Austin, R. B. (1978). Maximum yields of cereals as determined by plant type and environment. In *Maximising Yields of Crops,* pp. 18-24. London: HMSO.

Charles-Edwards, D. A. (1978). An analysis of the photosynthetic productivity of vegetative crops in the United Kingdom. *Annals of Botany* **42**, 717-31.

Charles-Edwards, D. A. & Ludwig, L. J. (1975). The basis of expression of leaf photosynthetic activities. In *Environmental & Biochemical Control of Photosynthesis,* pp. 37-44. Ed. R. Marcelle. The Hague: W. Junk.

Clements, R. O., Gibson, R. W., Henderson, I. F. & Plumb, R. T. (1978). Ryegrass: Pest and Virus Problems. *A.R.C. Research Review* **4**, 51-54.

Cooper, J. P. & Wilson, D. (1970). Variation in photosynthetic rate in *Lolium. Proceedings of the 11th International Grassland Congress, 1970,* pp. 522-527.

Jones, R. J. & Nelson, C. J. (1979). Respiration and concentration of water soluble carbohydrates in plant parts of contrasting tall fescue genotypes. *Crop Science* **19**, 367-372.

Leafe, E. L. (1972). Micro-environment, carbon dioxide exchange and growth in grass swards. In *Crop Processes in Controlled Environments,* pp. 157-174. Eds. A. R. Rees, K. E. Cockshull, D. W. Hand & R. G. Hurd. London: Academic Press.

Leafe, E. L. (1978). Physiological, environmental and management factors of importance to maximise yield of the grass crop. In *Maximising Yield of Crops,* pp. 37-49. London: HMSO.

Monteith, J. L. (1965). Light distribution and photosynthesis in field crops. *Annals of Botany* **29**, 17-37.

Mwakatundu, A. G. K. & Owen, E. (1974). *In vitro* digestibility of sodium hydroxide-treated grass harvested at different stages of growth. *East African Agricultural and Forestry Journal* **40**, 1-10.

Ogren, W. L. (1978). Increasing carbon fixation by crop plants. In *Photosynthesis '77,* pp. 721-734. Eds. D. O. Hall, J. Coombs & T. W. Goodwin. London: The Biochemical Society.

Parsons, A. J., Leafe, E. L., Stiles, W. & Collett, B. (1979). The growth of the grass sward under grazing. *Grassland Research Institute Annual Report 1978*, pp. 49-51.

Parsons, A. J. & Robson, M. J. (1978). Entry, partition and utilization of carbon in grass production. *Grassland Research Institute Annual Report 1977,* pp. 66-69.

Parsons, A. J. & Robson, M. J. (1980). Seasonal changes in the physiology of S24 perennial rye grass (*Lolium perenne* L.) I. Response of leaf extension to temperature during the transition from vegetative to reproductive growth. *Annals of Botany* (in press):

Peacock, J. M. (1975a). Temperature and leaf growth in *Lolium perenne.* II. The site of temperature perception. *Journal of Applied Ecology* **12**, 115-123.

Peacock, J. M. (1975b). Temperature and leaf growth in *Lolium perenne.* III. Factors affecting seasonal differences. *Journal of Applied Ecology* **12**, 685-697.

Pollock, C. J. & Jones, T. (1979). Seasonal patterns of fructosan metabolism in forage grasses. *New Phytologist* **83**, 9-15.

Rhodes, I. (1969). The yield, canopy structure and light interception of two ryegrass varieties in mixed culture and mono-culture. *Journal of the British Grassland Society* **24**, 123-127.

Robson, M. J. (1973a). The growth and development of simulated swards of perennial ryegrass. I. Leaf growth and dry weight change as related to the ceiling yield of a seedling sward. *Annals of Botany* **37**, 487-500.

Robson, M. J. (1973b). The growth and development of simulated swards of perennial ryegrass. II. Carbon assimilation and respiration in a seedling sward. *Annals of Botany* **37**, 501-518.

Robson, M. J. (1979). Growth of simulated swards in controlled environments. *Grassland Research Institute Annual Report 1978,* pp. 60-61.

Ryle, G. J. A., Cobby, J. M. & Powell, C. E. (1976). Synthetic and maintenance respiratory losses of $^{14}CO_2$ in uniculm barley and maize. *Annals of Botany* **40**, 571-586.

Scott, R. K. & Allen, E. J. (1978). Crop physiological aspects of importance to maximum yields – potatoes and sugar beet. In *Maximising Yields of Crops,* pp. 25-30. London: HMSO.

Wilson, D. (1975). Variation in leaf respiration in relation to growth and photosynthesis of *Lolium. Annals of Applied Biology* **80**, 323-338.

Woledge, J. (1973). The photosynthesis of ryegrass leaves grown in a simulated sward. *Annals of Applied Biology* **73**, 229-237.

Woledge, J. (1977). The effects of shading and cutting treatments on the photosynthetic rate of ryegrass leaves. *Annals of Botany* **41**, 1279-1286.

Woledge, J. (1979). Effect of flowering on the photosynthetic capacity of ryegrass leaves grown with and without natural shading. *Annals of Botany* **44**, 197-207.

Woledge, J. & Leafe, E. L. (1976). Single leaf and canopy photosynthesis in a ryegrass sward. *Annals of Botany* **40**, 773-783.

The contribution of physiology to breeding for increased yields in grain legume crops

R. J. Summerfield
Plant Environment Laboratory, Department of Agriculture and Horticulture, University of Reading, Shinfield Grange, Shinfield, Reading, Berkshire.

Introduction

In contributions on grain legumes, it is traditional to mention the value and shortage of protein in human diets and/or to refer to Malthusian theory on the need for more food (proteinaceous and calorific) to satisfy the demands of an ever-increasing human population. It will focus objectives better if, accepting that grain legumes are important dietary components, we consider how much of each legume should be produced to satisfy the present and future dietary needs of the producers and market demands. It is well known that national economic yields of grain legumes compared with cereal crops are small, unreliable and, except for groundnuts in the south-eastern USA, have stagnated during the last two decades; the problem is one of scale. Are we seeking to increase yields by, say, an order of magnitude, or is it reasonable to expect that less ambitious increments, perhaps 20-50 per cent over the next twenty years, would meet the effective needs of the increasing population? To meet the first objective, physiologists may consider it essential to restructure these crops completely in space and time. Less dramatic changes utilizing the variability now available in large germplasm collections of many grain legumes would probably contribute significantly to the second objective.

Retrospect and context

It is worth considering some historical aspects of research on grain legume crops which have contributed to preconceptions which are invalid and have prompted the increasingly common, but seemingly unjustified (at least for 'well-managed' farms; e.g. Swanson and Nyankori, 1979), use of terms such as 'the yield barrier' with respect to these crops.

Linguistic, geographical and historical evidence suggest that grain legumes are among man's oldest crops and are coeval with wheat and barley (e.g. Zohary and Hopf, 1973). Why then, agriculturally, has it proved far easier to increase the supply of cereals than of legumes, especially in the recent past? There are probably several contributory factors, not least of which is the limited research effort on leguminous compared with gramineaceous crops. Soya beans have probably been the subject of more research than all other grain legumes combined, yet there were in 1976 no more than 25 soya bean breeders in the USA (the major producing country) compared with more than 400 maize breeders (Pendleton, 1976).

Unfortunately, many farmers regard grain legumes as a secondary crop. For example, chickpea and lentil crops in the Near and Far East are usually grown in marginal, arid areas where their yields are uncertain. They are neglected agronomically, seldom benefiting from fertilizer inputs, pest control or irrigation. Modern crops thus have to compete with weeds and to contend with edaphic conditions not too dissimilar to those encounted by their wild progenitors. Then again, traditional producers in Africa plant cowpeas only when labour becomes available after staple cereal crops such as maize, sorghum or millet have been established and weeded. In the USA, soya beans, which are commonly grown in rotation with maize, cotton or tobacco, receive much of their agronomic attention as a by-product of the methods used for the other crops. Simple alternatives are known to give substantially greater yields, for example narrow (30-50 cm) compared with conventional (75-100 cm) inter-row spacing (Shibles, Anderson and Gibson, 1975).

Cultivars of several grain legumes have a narrow genetic base. For example, commercial soya bean varieties in the USA can be traced back to a few (probably six) plant introductions from the same area of China (Pendleton, 1976) and several cultivars of *Phaseolus vulgaris,* including those most commonly grown in the UK (Hardwick, Hardaker and Innes, 1978), have their origin in the same X-ray mutant. In some cases, e.g. chickpeas and lentils, almost all the few cultivars which have been released originate from selections within small germ-plasm pools and not from hybridization programmes (Hawtin, Singh and Saxena, 1980). Indeed, except perhaps during the last decade, little exotic germ plasm has been released into local populations of any legume. Furthermore, the possibilities of 'hybrid vigour' have been little exploited in legume breeding.

Effectively nodulated grain legumes can utilize atmospheric nitrogen and their value as green manure crops has long been recognized by farmers. However, plentiful supplies of relatively inexpensive nitrogen fertilizer and systems of land management which placed greater emphasis on economic return per unit of cost, probably combined during the 1950s and 1960s to limit drastically research on nitrogen nutrition and, especially, on the inter-relationships between nitrogen nutrition and carbon assimilation. Only relatively recently have economic (and perhaps environmental) pressures and the advent of International Agricultural Research Centres stimulated a dramatic increase in research in these aspects of legume physiology, which offer potential benefits to modern and subsistence farmer alike. In the interim, many grain-legume breeders have neglected nodulation ability completely and, by selecting progeny material from plots heavily fertilized with inorganic nitrogen, may well have failed to detect superior symbiotic combinations (e.g. soya beans in the USA; Dobereiner, 1974).

A final point which has had a bearing on the 'state of the art' in grain-legume physiology is that, among the organisms particularly suitable for biological research (e.g. *Escherichia coli* and *Drosophila melanogaster*), members of the leguminosae, notably peas, soya beans and beans, have often been selected for study by botanists, biochemists and physiologists (see Sutcliffe, 1977). Ironically, little of the resultant information has contributed to the productivity of legume crops in the field. Attributes such as the availability of large, non-dormant seeds which germinate readily; slender stems which are prone to etiolation; trifoliate (a number convenient for manipulation), planophile leaves which are not difficult to accommodate in cuvettes, and a reproductive ontogeny amenable to 'switching on and off' have all contributed to the extensive use of grain legumes in 'pure research'. Of course, the users have commonly expressed the naive hope that their results may ultimately be relevant to growers (see Bunting (1964) for comments on this). However, despite the intensity with which the garden pea has been studied, plant physiologists have not participated to any major degree in the transformation of the species into its modern forms, and are largely ignorant of the factors most likely to limit further increases in productivity (Pate, 1977). Then again, the effects of numerous chemicals with potential 'growth regulating activity' have been evaluated over the past twenty years in an attempt to modify various aspects of soya bean physiology and so to improve yields, and yet there are no compounds commonly used in current commercial soya bean production (Egli, 1976). Moreover, it is possible that the nitrogen nutrition of this most widely studied species, or at least of its most advanced cultivars, bred with ample added nitrogen, is not very typical of grain legumes in general (Sprent, 1979). With *Phaseolus* beans, failure to recognize the interaction effects between factors means that much of the extensive research on the adaptation of this species to different environments has now to be carefully re-evaluated and/or repeated (Wallace, 1980). We have little understanding of the requirements of legume fruits each day (and night), of their growth on the parent plant or why, in all legumes, so many of them absciss (Sinha, 1977).

Adaptation to the environment

Grain legumes share with most other green plants the ability to improve their nutrient uptake by mycorrhizal associations between their roots and soil fungi, but they also have the virtually unique potential for assimilation of dinitrogen through symbiosis with *Rhizobium.* This creates special demands, notably for molybdenum and cobalt, but also for boron, copper, phosphate and zinc (Munns and Mosse, 1980). Nutrient deficiencies, especially of phosphate, can greatly reduce dinitrogen fixation in the field, but how can deficiencies be recognized and ameliorated? Is there a field potential for dual inoculation with mycorrhiza and *Rhizobium,* and can a suitable inoculant be produced?

Since nodulated legumes obtain at least part of their nitrogen from symbiotic fixation, studies of environmental responses of legumes should therefore contain the contribution of the *Rhizobium* genotype, the likelihood of *Rhizobium* × host, *Rhizobium* × environment, or indeed second-order interactions. The more a legume depends upon symbiotically fixed rather than inorganic nitrogen, the more significant these potential sources of variation become. Unfortunately, the symbiotic relationship has all too frequently been ignored in studies of interactions between genotypes and environment in legumes (see Summerfield and Wien, 1980), and in many cases we know little, if anything, about the naturalized population of *Rhizobium* in breeders' nurseries. The yields of all grain legumes are affected in complex and diverse ways by the vagaries of weather and climate. As larger supplies of legume protein are most likely to result from increased *and* stabilized yields of seeds of improved composition, rather than improved seed biochemical efficiency (Boulter, 1980), the search for varieties or populations that are able to make more productive use of the environment(s) for which they are intended is a key area for interaction between legume breeder and physiologist.

No single legume species approaches universal adaptation: some (e.g. cowpeas, pigeon peas and groundnuts) are recognized as distincely tropical, others (e.g. *Phaseolus* beans, field beans, lentils and field peas) are better adapted to sub-tropical or to temperate climates. Some species (e.g. cowpeas) are reported to be better able to withstand short periods of drought than of waterlogging, and *vice versa* (e.g. some cultivars of soya bean); others are acutely sensitive to ephemeral flooding (e.g. lentils). Strains of *Rhizobium* differ dramatically in their ability to infect host roots, and to fix dinitrogen, in different edaphic conditions. We understand little as to why one strain of *Rhizobium* can succeed in circumstances where another fails, or how the infectivity of inoculum strains, in relation to those already in the soil, can be improved. The relative success of different strains of *Rhizobium* is probably the result of many interacting factors which influence the survival of the bacteria when free-living, the colonization of roots and competition for nodule sites, and the susceptibility of the host genotype. To define, clarify and exploit this variation in both the macro- and micro-symbiont is one of the most challenging areas of legume research.

Within genera, major differences in adaptation to photo-thermal regime, to edaphic conditions and to the amount and seasonal distribution of water are based on combinations of physiological adaptations, anatomical variations, morphological patterns, symbiotic associations and on the structure and genetics of populations (see Adams and Pipoly, 1980). Legume breeders have usually selected plants for adaptation to particular sites, chosen to represent particular regions, rather than to specific conditions of, for example, photoperiod and temperature or the concentrations of major and minor nutrients or the common toxic elements in the soil. Even after selections have been grown and tested for a number of seasons at a particular site the results can, at best, be indicative; they can never be decisive (Engledow and Wadham, 1924). Furthermore, the possibility of using statistical procedures to account for the effects of weather on yield (adaptation) trials seems remote (Yates, 1968). Indeed, we should seek to *explain* how environmental variations in time affect physiological and morphological processes, and hence growth, development and yield, rather than merely to *describe* the outcome by statistical procedures (Bunting, 1975). Hitherto, the vogue with legume trials has been to describe the outcome with measurements at crop maturity of economic yield and, traditionally, three of its components, viz. the number of pods that ripen, the average number of seeds in them, and the mean weight of individual seeds. These data fail to provide any direct understanding of why yield or any particular component of it varies; they do not provide any knowledge about the processes by which components of yield are determined nor how these processes are related to plant growth or the environment (and see Gallagher and Biscoe, 1978). Furthermore, since biological yields are recorded only infrequently, the relative investment of dry matter into fruits cannot be quantified. To classify as equal two genotypes which yield 100 units of seed without knowing their respective biomass is unacceptable (see Bunting, 1971), and to test progeny materials at only the current 'standard' density presupposes that large-yielding cultivars at that particular density will outyield all other combinations of plant type and density – which is patently invalid (Donald and Hamblin, 1976). Then again, to neglect root (and nodule) weight in legumes is likely to be a serious omission. Although it is well known that reproductive ontogeny (the initiation and expansion of flower initials and the subsequent development of fruits) is sensitive to photoperiodic regulation (Summerfield and Wien, 1980), there can also be independent, and large photoperiodic effects on root and nodule growth and function. For example, the dramatic increase in root dry weight in long (15 h) compared with short (11-12 h) days could be important for nodulation and cultivar adaptation in chickpeas, but would not be immediately obvious in routine field screens. Indeed, the duration of reproductive growth in chickpea is strongly correlated with, and may depend on, the proportion of roots which senesce during the latter half of the life of the crop. Plants which produced the largest yields when grown in pots in simulated tropical environments in growth cabinets lost relatively small proportions of root dry weight during reproductive growth (Summerfield, Minchin, Roberts and Hadley, 1979). Relationships which involve root and symbiotic longevity are unlikely to be detected by measurements of relatively few shoot components recorded only at maturity.

Principles have been slow to emerge from measurements and analyses of the growth of predominantly non-leguminous crops in terms of plant leaf areas and dry weights (Watson, 1968) and controlled environment facilities have often failed to produce plants which resemble their counterparts in the field – which limits the usefulness of data from these facilities, though perhaps not the frequency or enthusiasm with which they are used. For example, 'most soya bean researchers apparently assumed that they knew how to grow plants [in controlled environments] but the appearance of most chamber-grown soya beans [excessive height, atypical branching and leaf expansion] is obvious testament to the fact that the assumption was wrong' (Tanner and Hume, 1976). However, not only is it possible to produce legume plants in controlled environments which closely resemble spaced plants in the field in terms of morphology and phenology (Summerfield, Minchin and

Roberts, 1978) but also the nitrogen nutrition of fruits can be remarkably similar (Eaglesham *et al.*, 1977). What is required is a prudent selection of pot size, growing medium, nutrient solution, irrigation technique, spectral quality of light and regulation of temperature, vapour-pressure deficit and CO_2 concentration. The work involved is time consuming and expensive; the alternative is to investigate the artefacts of artificial climates, which is unacceptable.

Controlled environments can be a powerful adjunct to field research in screening for adaptation to the aerial environment (e.g. Summerfield, Minchin and Wien, 1979). Legume breeders and physiologists must, however, avoid a pre-occupation with 'single-factor fitness' (e.g. phtoperiodic response) and the assumption that a single combination of factors is optimal for all successive stages of legume development. There is ample evidence that factors which promote the initiation and expansion of leaves, for example, are far from optimal for the length of life of the leaves (and little is known about leaf death in legumes); and this applies also to the initiation and early symbiotic performance of nodules and the persistence of symbiotic activity into the reproductive period. In contrast with cereals, the initiation, expansion and senescence of leaves, flowering, and the setting and ripening of fruits all occur simultaneously in different parts of individual, indeterminate legumes. We know relatively little about the time course of the ratios of fruit to total growth and the effects on them of genotype and environment. Until more is known about these questions in particular, predictive models of crop growth and yield become less and less reliable as flowering and reproductive growth provide more important sinks than vegetative growth. In any event, previous endeavours in crop modelling have revealed much ignorance about the physiology of field crops on which far more work has been done than on grain legumes, and they did not predict growth or yield accurately (De Wit, Brouwer and Penning de Vries, 1970); alternatively, complex dynamic models are becoming so bulky and unwieldy 'that they may suffer the fate of the dinosaurs' (Monteith, 1977). Far more attractive, is the potential to combine micrometeorological measurements throughout crop duration with measurements of crop dry weight, leaf-area expansion (and decline) and reproductive (and symbiotic) development. Such a strategy (Monteith and Biscoe, 1971) has proven reliability in the elegant and detailed physiological analyses of growth and yield of cereals (Biscoe and Gallagher, 1978; Gallagher and Biscoe, 1978). It has exciting potential not only for sole crop legumes but also in mixed crops, provided the crop mixture as a whole is treated as a single commensal physiological system rather than as an area in which different species compete.

Without doubt, physiologists can do much to assist the legume breeder in his choice of parental material, in establishing selection criteria and in developing techniques for screening progeny lines both in the field and in judiciously designed (or specified) controlled environments. It should be feasible to test the effect of morphological and physiological traits on crop productivity through the use of isogenic lines and isogenic populations (e.g. in seeking genotypes adapted to semi-arid environments; Fischer and Turner, 1978), but the research effort required, both in breeding and physiology, would be substantial. However, more frequent harvests on carefully selected ranges of genotypes in *factorial* experiments are required to provide, in addition to yield information, data on the time course of the development of structure and the uptake and distribution of nitrogen and carbon, before breeders can use more critical selection criteria. Such knowledge is unlikely to replace standard plant-breeding procedures but, in guiding the selection of parents, it can permit greater emphasis on specific crosses and so allow the breeder to work with larger populations of fewer crosses (in whose segregating progenies, moreover, the genetics of individual characters can be studied), to select more intensively and hence to become more efficient. Even when the main objective of breeding is to lessen the effects of diseases and pests, it can have a greater effect only if the resistance attributes are incorporated into genotypes or populations which are well adapted to particular environments.

Carbon metabolism and nitrogen nutrition

Several essential events precede active dinitrogen fixation in the *Rhizobium*-legume symbiosis. Failure may occur throughout this complex sequence of physiological processes, from a lack of mutual host-symbiont recognition on the legume rhizoplane through to poor symbiotic fixation when nodules provide insufficient nitrogen for the host plant to achieve maximum growth. Any attempt to improve the nodulation of legumes must involve both bacterium and host (Gibson, 1980).

Strains of *Rhizobium* may compete successfully for nodulation sites but then be ineffective in dinitrogen fixation. It does not follow that inoculation will allow strains with superior fixation capabilities to penetrate the indigenous population. The reasons for differences in competitive ability between strains of *Rhizobium* are unknown and although the host can influence which of the strains in a mixture will form nodules, little is known of the genetic or physiological factors involved. The biochemistry of the infection process is only partially understood, but recognition at infection sites may involve the binding of specific legume lectins to unique carbohydrate structures found exclusively on the surface of the *Rhizobium* (Dazzo, 1980). If this is so, will it prove possible or even desirable to breed rigid strain specificity into a legume? Limitations of space preclude more than a brief summary of these aspects of rhizobiology. Undoubtedly, more effective use could be made of host genetics in order to understand the legume-*Rhizobium* symbiosis better (Gibson, 1980). In some analyses of genetic variations in 'physiological components of yield' neither characteristics of nodulation nor of strains of *Rhizobium* (or indeed any aspect of the symbiotic association) are mentioned (e.g. Wallace, Ozbun and Munger, 1972); and usually the genetics are not calculated quantitatively from segregation ratios.

A common misconception about nitrogen fixation in legume-root nodules is that a great deal of work has been done and that all nodules have the same fixation metabolism. In effect, only a narrow range of species and genotypes has been studied and there is considerable variation in the supporting metabolism. Unfortunately, previous research on 'nodulation' has concentrated almost exclusively on root infection and nodule formation without reference to nodule senescence. Biochemical research hitherto has been strongly biassed towards mechanisms of nitrogenase action and ammonia assimilation whereas the equally important aspects of carbon metabolism remain largely speculative. The nature of carbon compounds involved, their routes of metabolism, the mechanisms of control and the partitioning of metabolites between various sites of utilization are only poorly understood. These aspects (Rawsthorne *et al.*, 1980) and many other facets of the carbon and nitrogen metabolism in legume-root nodules (Beringer *et al.*, 1979; Sprent, 1979) merit intensive study by physiologists. We need to distinguish those traits which are likely to be useful criteria for selection and to determine what proportion of the variation found is due to genetic differences.

For an effectively nodulated legume to fix sufficient nitrogen to meet the demands of the host, several criteria must be satisfied. Adequate photosynthates must be transported to the nodulated root and the nodules must be able to respire and generate, from the photosynthate they receive, the energy and reductant necessary for fixation (as well as for their own growth and maintenance). In addition, the fixation products must be produced, exported and distributed throughout the host plant (which depends on the availability of carbon skeletons to the nodules) at appropriate times and rates and for the duration required to allow maximum growth.

Dinitrogen fixation activity of slender-rooted, annual legumes is paced primarily by the availability to nodules of photosynthate translocated from the shoot (Pate and Minchin, 1980). Thus, changes in the aerial and edaphic environment which influence the rates of initiation and expansion of leaves, canopy photosynthesis and hence crop growth rate are likely to be reflected by changes in symbiotic acitivity. This is in addition to any direct effects that individual or combinations of factors may have on infection, fixation and nodule longevity. The spatial and temporal relationships between carbon metabolism and nitrogen nutrition are complex and involve genetic, physiological and microbiological interactions. It is perhaps not surprising then to find so many apparently contradictory data in the literature; this neither helps the legume breeder nor encourages him to include 'symbiotic traits' among his selection criteria.

It follows that plants dependent on dinitrogen fixation are more sensitive to hot and cold temperatures, excess water and acid substrates than those supplied with inorganic nitrogen (Sprent, 1980). Nevertheless, genetic variability in *Rhizobium* and nodulated legumes in their sensitivity to each of these, and probably to most other environmental factors, is extensive. Conceptually, the ability to manipulate two genomes each of which contribute to the phenotypic expression of legume characters might be seen as an advantage, providing that the breeder is able to identify the more important characters, to screen large populations effectively and then to recombine desirable attributes.

It is probably true that most of the (indirect) selection pressure in grain legumes hitherto has been *against* fixation ability. As discussed earlier, not only has the agronomic management of breeders plots often been inappropriate, but also changes in morphology and crop duration, with a view to mechanized harvesting and to accommodating legumes in established cropping systems, may have been disadvantageous. For example, indeterminate climbing cultivars of *Phaseolus vulgaris* apportion more photosynthate to nodules than most determinate bush cultivars; they also maintain carbohydrate in soluble (available) form rather than as starch stored in stems (Graham, 1978). However, a slight delay of about seven days in flowering in bush types can double the amount of nitrogen fixed. Indeed, because of the exponential relationship in many legumes between nitrogen fixed and plant age through a large part of the life of the crop, an increase of only 10 per cent in the period of nodule activity can double the amount of nitrogen fixed (Hardy and Havelka, 1976). Since considerable variation is found in seasonal fixation profiles, improvements by management and breeding seem distinctly possible. Fixation activity may be first detected after widely different times from emergence, and peak values differ dramatically and occur some time during early reproductive growth.

We have recent evidence that the timing, and perhaps the magnitude, of peak fixation values in soya bean are a consequence of a combination of an inhibitory effect of cotyledons on the fixation efficiency of primary nodules and a marked stimulation of fixation activity per plant associated with the initiation of flower buds and, to a lesser extent, the presence of open flowers and immature fruits (Peat, 1979). Both of these responses suggest that endogenous hormones play a part and so raise the question as to whether or not appropriate chemicals with 'plant-growth regulatory activity' might be used to improve the seasonal fixation profiles of nodule-dependent plants. Alternatively, direct selection for enhanced dinitrogen fixation is a novel endeavour for grain if not for forage legumes (Nutman, 1969) and data are not yet available for comparisons between selections for economic yield *per se,* selection for dinitrogen fixation and selection for related traits such as nodule weight, number, specific activity (efficiency) or ability of hosts to sustain effective nodule populations during reproductive growth. Such studies will require close co-operation between the legume physiologist and breeder.

Fixation during the seed-filling period varies considerably: it may be tailored to demands, decline precipitously or slowly, or not at all. As we have argued before (Summerfield *et al.,* 1978) it is not possible to generalize even with a crop studied in such detail as soya beans. Physiologists can certainly resolve many of these problems but will need to pay greater attention to several factors, including the onset, rate and duration of dinitrogen fixation during the pre-flowering stages; the large effects which can be attributed to crop density; the reliability of techniques which can be used to estimate dinitrogen fixation activity; the inorganic nitrogen status of soils during crop growth, and the relative yield of the crops which are studied. Although grain-legume crops are capable of economic yields far larger than 'average national values', most research has been done on crops which produce relatively meagre yields. Clearly, the type of agricultural system and the time available to the legume crop need to be specified at the outset; in many cases yield may need to be evaluated in both fertilized and dinitrogen fixing conditions (Graham, 1978).

Effectively nodulated, well-adapted legumes can grow and yield as well on dinitrogen fixation as they do on inorganic nitrogen (e.g. Minchin and Summerfield, 1978). The extra carbon and nitrogen assimilated by plants dependent on inorganic nitrogen (or on urea applied to the leaves) does not necessarily increase reproductive potential, rather stems become thicker, branches are heavier and senesced leaves retain larger concentrations of nitrogen than plants which rely on dinitrogen fixation. Thus, if in breeding for increased dinitrogen fixation the objective is to increase seed yield and/or seed protein content, then the additional nitrogen will need to be transported not just to the shoot but to the seeds. It is worth noting here that if the seeds are for human consumption, a small concentration of protein of improved quality may be a better objective than a larger concentration of protein of poorer quality. Of course, whether or not this objective is valid, will depend not only on the magnitude of economic yield and its price but also on factors such as the presence of anti-metabolites, flatus factors and other attributes which determine how much consumers are willing to eat.

Legume seeds derive large proportions of their nitrogen from two principal sources; from assimilation during fruit development and from mobilization and translocation from vegetative organs and pod walls. Mechanisms also exist for the rapid flux of fixed nitrogen through vegetative structures into fruits (Pate *et al.*, 1979). Cultivars which draw on stem and pod-wall reserves rather than nitrogen in leaves might be expected to have larger yield potentials. Illuminated green fruits can play an important role in the recapture of respired CO_2, although fruits exposed above the canopy in the tropics may experience supra-optimal temperatures (but may be easier to harvest and less susceptible to insect depradations than those within the canopy). The relative importance of these diverse factors will depend on the species involved, the location in which the crop is grown and whether it is harvested by hand or mechanically.

Seeking ways to increase total photosynthate production by grain legume crops so that more carbon is available for efficient utilization in dinitrogen fixation by nodules is an attractive research goal (Hardy and Havelka, 1977). Genotypic differences in foliar carbon exchange rates (CER) have been reported for several legumes but the consensus of opinion is that the extent by which net photosynthesis rate, particularly of the crop as a whole, can be advanced by breeding is only speculative. Indeed, there is little evidence of a linear relationship between maximum C-fixation per hectare and economic yield in grain legume crops; attributes such as the rates of leaf initiation and expansion are probably more important during vegetative growth. Moreover, CER values have been measured most frequently at the seedling or early vegetative stage when, in fact, photosynthetic productivity after flowering is the major determinant of economic yield in most grain legumes (Sinha, 1977).

Alternative strategies to maximize total photosynthate production and the efficiency of subsequent utilization seem more attractive. There are at least six possibilities:

(1) Are certain species and cultivars intrinsically superior to others in allocation of assimilatory resources for synthesis of protein-rich seeds during the time available in specific environments?

(2) Can the loss of photosynthate by photorespiration, which occurs in all legumes, be minimized?

(3) Nodulated roots consume during active growth and dinitrogen fixation a substantial fraction (30–50 per cent) of the daily gain of photosynthate by the host shoot, but only a few cultivars of a small number of species have been studied and the effects of environment, strain of *Rhizobium* and genotype on the carbon economy of nodulated roots are largely unknown.

(4) Starchy tap roots (as, for example, in lupin and *Psophocarpus*) may buffer dinitrogen fixation against ephemeral stress and improve activity at night, but what are the implications for infection by *Rhizobium* and for water and nutrient uptake (and perhaps even infestation by nematodes) if the breeder were able to change root form in slender-rooted species?

(5) Since the carbon attached to export products may be the largest item of the nodules' total carbon consumption (about 50 per cent), can we improve the economic usage of carbon in the transport of fixation products? Some species transport mostly amides (e.g. peas and lupins) whereas in others it has been discovered only recently that ureides are the major nitrogenous export product (e.g. soya beans and cowpeas). Ureides have a smaller C:N ratio than amides (1:1 and at least 2:1, respectively) but, if species-specific, this basic biochemical difference may be of little use to the legume breeder and reliable data on the comparative productivity of amide- and ureide-producing species are not available.

(6) The respiratory costs of dinitrogen fixation by nodules in some species is significantly less than in others (e.g. cowpeas and lupins, respectively) which is probably related to the relative activity of carboxylation enzymes (PEP carboxylase) in nodules and/or to the complex relationships between dinitrogen fixation, hydrogen evolution and hydrogenase activity (Pate and Minchin, 1979). The enzyme nitrogenase can waste energy by the reduction of H^+ to H_2 rather than

N_2 to NH_4^+, and H_2 may be lost unless a hydrogenase is present (Schubert and Ryle, 1980). Therefore, until experiments in the field are carried out to assess the significance of nodule efficiency in H_2 metabolism throughout crop duration, both day and night, the potential significance of this process for economic yield cannot be evaluated (Hardy and Havelka, 1977).

Physiological research on grain legume crops has expanded dramatically in the last decade. Attention has focussed on dry-matter production and on carbon metabolism, whereas morphology, phenology and anatomy have been neglected (see Adams and Pipoly, 1980). Furthermore, the complex interactions between carbon metabolism and nitrogen nutrition have received scant attention. Researchers have not been sufficiently prudent in evaluating the reliability of techniques, which are now 'established' and in common use. This deficiency is nowhere more acute than in the use of controlled environments (see Summerfield and Wien, 1980) and in the establishment of an effective symbiosis and assessment of dinitrogen fixation activity. It is difficult to estimate dinitrogen fixation reliably in the field and non-nodulating isolines are available only in soya beans. If useful genetic variation does exist for even some of the many attributes of the nodulated legume described earlier, the breeder will probably require assay techniques which are reliable, rapid, inexpensive, applicable to large numbers of plants and to the field, require a minimum of labour and, perhaps, are non-destructive. This can be achieved. For example, the need to inoculate soya beans could be circumvented by genetically incorporating the promiscuity of some lines into larger-yielding, improved varieties. The success of such a breeding scheme depends on the ability to select plants possessing promiscuity from a segregating population. This can now be done by determining the amount of allantoin in the last fully developed leaf: shoots of soya-bean plants fixing dinitrogen contain large quantities of ureide compounds (allantoin and allantoic acid) whereas those relying on inorganic nitrogen contain only small amounts of these compounds. The method is non-destructive and the assay system has been automated (IITA, 1979).

Concluding remarks

It is generally accepted that the world's most economically important grain legumes are soya beans, *Phaseolus* beans, field beans, field peas, groundnuts, cowpeas, pigeon peas and chickpeas. Species of increasing importance include lentils, mung beans and lupins. Collectively, these species differ in genetic composition (diploid and polyploid); some are predominantly self pollinated, others have appreciable amounts of out-crossing; their growth may be determinate or indeterminate, annual or perennial; roots may be large and starchy or predominantly fibrous and may be nodulated only by specific strains of *Rhizobium* or have far less exacting requirements; and seeds may or may not contain large quantities of oil as well as various amounts of protein. Very few of the most economically important grain legumes are grown principally in temperate regions, many are cultivated predominantly in the semi-arid tropics. Irrespective of location, estimates of the nitrogen fixed by these species vary from trivial amounts to about 600 kg N ha^{-1} $annum^{-1}$. If the amount of nitrogen removed when the crop is harvested is greater than the amount fixed during growth, then the legume crop will deplete soil reserves which, of course, negates their assumed beneficial role in crop rotations. Clearly, the climatic back-cloth, the agricultural system and the time available to produce a given crop of legume seeds need to be carefully specified at the outset. It is difficult and dangerous to extrapolate to the tropics the results of research in the temperate environment, or from individual species.

Close collaboration between legume physiologist, bacteriologist, agronomist and breeder in programmes devoted to each of the major legumes is essential if the variations which undoubtedly exist for many physiological traits are to be exploited to result in larger amounts of dinitrogen fixation and greater economic yields in future legume crops. Those concerned with crop protection are also likely to benefit from the input of physiologists in a legume-breeding programme. For example, the lectins are not only involved with nodulation, these glycoproteins are intimately involved with cell surface recognition phenomena, so they may also function in host-pathogen interactions (Smith and Ingham, 1980). A second example concerns the often extended period of flowering and seed set which is a critical stage in reproductive development. It is characterized in grain legumes by marked internal competition leading (*inter alia*) to the profligate loss of flower buds and/or open flowers. Such losses can be much more severe as a result of mechanical damage caused by insects (e.g. thrips infestation on cowpea flowers). Physiologists reasoned that the wound response of the plant probably led to the evolution of ethylene and were able to mimic the damage by spray application of ethephon. The frequency and optimum stage of development at which to apply the chemical were evaluated and a reliable screening technique for susceptibility to thrips damage, which satisfied many of the criteria mentioned earlier, became available (IITA, 1979).

Clearly, there are ample opportunities for physiologists to point the way to further advances in legume productivity. It is not surprising that much remains to be done: hitherto, 'in no area [other than research on biological nitrogen fixation] has the global scientific community been so derelict in its responsibility' (Wittwer, 1976).

Acknowledgements. The publications of Professor J. S. Pate, Dr J. I. Sprent and Professor J. L. Monteith, discussions with Dr P. J. Dart, Dr H. C. Wien, Dr F. R. Minchin, Dr D. S. Drennan, Professor J. Elston, Dr B. Jeffcoat and Dr P. Hadley, and stimulating exchanges with Professor E. H. Roberts and Professor A. H. Bunting, as mentors, have all contributed in no small way to the views expressed in this chapter. The generous financial support from the UK Ministry of Overseas Development is gratefully acknowledged.

References

Adams, M. W. & Pipoly, J. J. (1980). Biological structure, classification and distribution of economic legumes. In *Advances in Legume Science*, pp. 1-16. Eds. R. J. Summerfield & A. H. Bunting. London: HMSO.

Beringer, J. E., Brewin, N., Johnston, A. W. B., Schulman, H. M. & Hopwood, D. A. (1979). The *Rhizobium*-legume symbiosis. *Proceedings of the Royal Society, London, Series B* **204**, 219-233.

Biscoe, P. V. & Gallagher, J. N. (1978). A physiological analysis of cereal yield. I. Production of dry matter. *Agricultural Progress* **53**, 34-50.

Boulter, D. (1980). Ontogeny and development of biochemical and nutritional attributes in legume seeds. In *Advances in Legume Science*, pp. 127-134. Eds. R. J. Summerfield & A. H. Bunting. London: HMSO.

Bunting, A. H. (1964). Master Gaster: or the nature and prospects of applied biology. *Annals of Applied Biology* **53**, 351-362.

Bunting, A. H. (1971). Productivity and profit or is your vegetative phase really necessary? *Annals of Applied Biology* **67**, 265-272.

Bunting, A. H. (1975). Time, phenology and the yield of crops. *Weather* **40**, 312-325.

Dazzo, F. B. (1979). Infection processes in the *Rhizobium*-legume symbiosis. In *Advances in Legume Science*, pp. 49-59. Eds. R. J. Summerfield & A. H. Bunting. London: HMSO.

Dobereiner, J. (1974). Legumes in Latin American agriculture. *Proceedings of the Indian National Science Academy, Series B* **40**, 768-794.

Donald, C. M. & Hamblin, J. (1976). The biological yield and harvest index of cereals as agronomic and plant breeding criteria. *Advances in Agronomy* **28**, 361-405.

Eaglesham, A. R. J., Minchin, F. R., Summerfield, R. J., Dart, P. J., Huxley, P. A. & Day, J. M. (1977). Nitrogen nutrition of cowpea (*Vigna unguiculata*). III. Distribution of nitrogen within effectively nodulated plants. *Experimental Agriculture* **13**, 369-380.

Egli, D. B. (1976). Planting date, row width, population, growth regulators. In *World Soybean Research.* pp. 56-62. Ed. L. D. Hill. Illinois: Interstate Printers & Publishers.

Engledow, F. L. & Wadham, S. M. (1924). Investigation on yield in cereals. *Journal of Agricultural Science, Cambridge* **24**, 325-345.

Fischer, R. A. & Turner, N. C. (1978). Plant productivity in the arid and semi-arid zones. *Annual Review of Plant Physiology* **29**, 277-317.

Gallagher, J. N. & Biscoe, P. V. (1978). A physiological analysis of cereal yield. II. Partitioning of dry matter. *Agricultural Progress* **53**, 51-70.

Gibson, A. A. (1980). Host determinants in nodulation and nitrogen fixation. In *Advances in Legume Science*, pp. 69-75. Eds. R. J. Summerfield & A. H. Bunting. London: HMSO.

Graham, P. H. (1978). Some problems and potentials of field beans (*Phaseolus vulgaris* L.) in Latin America. *Field Crops Research* **1**, 295-317.

Hardwick, R. C., Hardaker, J. M. & Innes, N. L. (1978). Yields and components of yield of dry beans (*Phaseolus vulgaris* L.) in the United Kingdom. *Journal of Agricultural Science, Cambridge* **90**, 291-297.

Hardy, R. W. F. & Havelka, U. D. (1976). Photosynthate as a major factor limiting nitrogen fixation by field-grown legumes with emphasis on soybeans. In *Symbiotic Nitrogen Fixation in Plants.* IBP Synthesis Vol. 7, pp. 421-439. Ed. P. S. Nutman. London: Cambridge University Press.

Hardy, R. W. F. & Havelka, U. D. (1977). Possible routes to increase the conversion of solar energy to food and feed by grain legumes and cereal grains (crop production): CO_2 and N_2 fixation, foliar fertilization, and assimilate partitioning. In *Biological Solar Energy Conversion*, pp. 299-322. Ed. A. Mitsui. New York & London: Academic Press.

Hawtin, G. C., Singh, K. B. & Saxena, M. C. (1980). Some recent developments in the understanding and improvement of *Cicer* and *Lens.* In *Advances in Legume Science*, pp. 613-624. Eds. R. J. Summerfield & A. H. Bunting. London: HMSO.

IITA (1979). Research Highlights 1978. International Institute of Tropical Agriculture, Ibadan, Nigeria.

Minchin, F. R. & Summerfield, R. J. (1978). Potential yield improvement in cowpea (*Vigna unguiculata*): The role of nitrogen nutrition. *Annals of Applied Biology* **88**, 468-473.

Monteith, J. L. (1977). Climate. In *Ecophysiology of Tropical Crops,* pp. 1-27. Ed. P. Alvim. London: Academic Press.

Monteith, J. L. & Biscoe, P. V. (1971). Meteorological measurements of photosynthesis and transpiration. *Report of the School of Agriculture University of Nottingham,* pp. 66-75.

Munns, D. N. & Mosse, B. (1980). Mineral nutrition of legume crops. In *Advances in Legume Science*, pp. 115-125. Eds. R. J. Summerfield & A. H. Bunting. London: HMSO.

Nutman, P. S. (1969). Genetics of Symbiosis and nitrogen fixation in legumes. *Proceedings of the Royal Society, London, Series B* **172**, 417-437.

Pate, J. S. (1977). The pea as a crop plant. In *The Physiology of the Garden Pea*, pp. 469-484. Eds. J. F. Sutcliffe & J. S. Pate. London: Academic Press.

Pate, J. S. & Minchin, F. R. (1980). Comparative studies of carbon and nitrogen nutrition of selected grain legumes. In *Advances in Legume Science*, pp. 105-114. Eds. R. J. Summerfield & A. H. Bunting. London: HMSO.

Pate, J. S. Layzell, D. B. & McNeil, D. L. (1979). Modelling the transport and utilisation of carbon and nitrogen in a nodulated legume. *Plant Physiology* **63**, 730-737.

Peat, J. R. (1979). Physiological and practical problems associated with the search for plant growth regulators for soya beans. *PhD Thesis*, University of Reading.

Pendleton, J. W. (1976). The key to maximum soybean production. *Extension Bulletin Asian and Pacific Council Food and Fertilizer Technology Centre* no. 82. Taipei, Taiwan.

Rawsthorne, S., Minchin, F. R., Summerfield, R. J., Cookson, C. & Coombs, J. (1980). Carbon and nitrogen metabolism in legume root nodules. *Phytochemistry* **19**, 341-355.

Schubert, K. R. & Ryle, G. J. A. (1980). The energy requirements for nitrogen fixation in nodulated legumes. In *Advances in Legume Science*, pp. 85-96. Eds. R. J. Summerfield & A. H. Bunting. London: HMSO.

Shibles, R. M., Anderson, I. C. & Gibson, A. H. (1975). Soybean. In *Crop Physiology: some case histories*, pp. 151-189. Ed. L. T. Evans. London: Cambridge University Press.

Sinha, S. K. (1977). Food legumes: distribution, adaptability and biology of yield. *FAO Plant Production and Protection Paper* no. 3. Rome: FAO.

Smith, D. A. & Ingham, J. L. (1980). Legumes, fungal pathogens and phyto-alexins. In *Advances in Legume Science*, pp. 207-223. Eds. R. J. Summerfield & A. H. Bunting. London: HMSO.

Sprent, J. I. (1979). The Biology of Nitrogen-fixing Organisms. London: McGraw-Hill.

Summerfield, R. J. & Wien, H. C. (1980). Effects of photoperiod and air temperature on growth and yield in economic legumes. In *Advances in Legume Science*, pp. 17-36. Eds. R. J. Summerfield & A. H. Bunting. London: HMSO.

Summerfield, R. J., Minchin, F. R. & Roberts, E. H. (1978). Realisation of yield potential in soya bean (*Glycine max* L. Merr.) and cowpea (*Vigna unguiculata* L. Walp.). In *Proceedings BCPC/BPGRG Symposium on Opportunities for Chemical Plant Growth Regulation*, pp. 125-134. *British Crop Protection Council Monograph* no. 21.

Summerfield, R. J., Minchin, F. R. & Wien, H. C. (1979). Screening for environmental adaptability in soya bean and cowpea: an improved strategy. *World Crops* **31**, 21-23.

Summerfield, R. J., Minchin, F. R., Roberts, E. H. & Hadley, P. (1979). *Reading University-International Crops Research Institute for the Semi-Arid Tropics Internal Communication* no. 2.

Sutcliffe, J. F. (1977). History and use of the pea in plant physiology research. In *The Physiology of the Garden Pea,* pp. 1-19. Eds. J. F. Sutcliffe & J. S. Pate. London: Academic Press.

Swanson, E. R. & Nyankori, J. C. (1979). Influence of weather and technology on corn and soybean yield trends. *Agricultural Meteorology* **20**, 327-342.

Tanner, J. W. & Hume, D. J. (1976). The use of growth chambers in soybean research. In *World Soybean Research,* pp. 342-351. Ed. L. D. Hill. Illinois: Interstate Printers & Publishers.

Wallace, D. H. (1980). Adaptation of *Phaseolus* to different environments. In *Advances in Legume Science*, pp. 349-357. Eds. R. J. Summerfield & A. H. Bunting. London: HMSO.

Wallace, D. H., Oxbun, J. L. & Munger, H. M. (1972). Physiological genetics of crop yield. *Advances in Agronomy* **24**, 97-146.

Watson, D. J. (1968). A prospect for crop physiology. *Annals of Applied Biology* **62**, 1-9.

Wit, C. T. de, Brouwer, R. & Penning de Vries, F. W. T. (1970). Simulation of photosynthesis systems. In *Prediction and Measurement of Plant Photosynthetic Productivity,* pp. 47-70. Ed. I. Setlik. Wageningen: Centre for Agricultural Publishing and Documentation.

Wittwer, S. H. (1976). Biological nitrogen fixation. In *Proceedings of the First Annual Review Meeting of the Fertilizer INPUTS Project.* Honolulu, Hawaii: East-West Centre.

Yates, F. (1968). Investigations into the effects of weather on yields. *Report Rothamsted Experimental Station for 1968,* pp. 46-49.

Zohary, D. & Hopf, M. (1973). Domestication of pulses in the old world. *Science, New York* **182**, 887-894.

Physiological potential for increasing yields of sugar beet

G. F. J. Milford, P. V. Biscoe, K. W. Jaggard, R. K. Scott and A. P. Draycott
Broom's Barn Experimental Station, Bury St. Edmunds, Suffolk.

Introduction

Before considering how physiology can help improve yields from sugar beet it is worth comparing past yield trends with those of other arable crops in the UK, to examine how far yields have been restrained by genome and husbandry. Between 1930 and 1960, national average yields of sugar-beet roots increased, in relative terms, as fast as yields of cereal grain and faster than yields of potato tubers (Fig. 1). However, from 1960 to 1970, yields of sugar beet remained almost constant at about 35 t ha^{-1}, 1·7 times the average yield between 1926 and 1930, whilst the yields of cereals and potatoes continued to increase. The substantial increase in the national average yield of sugar beet between 1945 and 1960 was the consequence of better varieties, improved nutrition, earlier sowing and the partial control of virus yellows disease. During this time the crop was grown from multigerm seed and hoed by hand to remove weeds and establish the required population of plants.

During the 1960s major changes occurred in the way sugar beet was grown. Increasingly the national crop became precision drilled at wider spacings using pelleted, monogerm seed and treated with pre- and post-emergence herbicides (Fig. 2). All of these modifications aimed at reducing the need for manual labour and containing costs, rather than increasing yields. Indeed, their overall effect was to slow the trend of increase in yield. This was partly because yield losses were incurred whenever seedling establishment was poor or plants were damaged by herbicides. Previously, the hoer was able to select the largest seedlings and could partly compensate for inadequate or irregular plant population. Also, pressure from the industry to eradicate the need for expensive hand labour led to the introduction of monogerm varieties at a stage when their performance was poorer than the best multigerm material. The need for the breeder to combine monogermity with bolting and disease resistance, whilst reconciling the conflicting requirements of high root yield for the grower and high quality for the processor, meant that new varieties were, and still are, compromises and genetic improvement in yield has been slow relative, for example, to cereals.

Successive severe infections of sugar beet with virus yellows in 1974, 1975 and 1976, and years of drought in 1975 and 1976 depressed national sugar-beet yields more than those of cereals and potatoes. In the last three years the crop has been free from disease but yields have still not exceeded the plateau levels of the 1960s.

The challenge to the sugar-beet industry is to achieve another period of increasing yields. The response to this challenge must come from physiologists, nutritionists and agronomists working in close collaboration with breeders and pathologists. Hundreds of experiments have been done to optimize the agronomy and nutrition of sugar beet (Draycott, 1972). In most, only final yields were measured, with the result that they provided little information on the processes governing

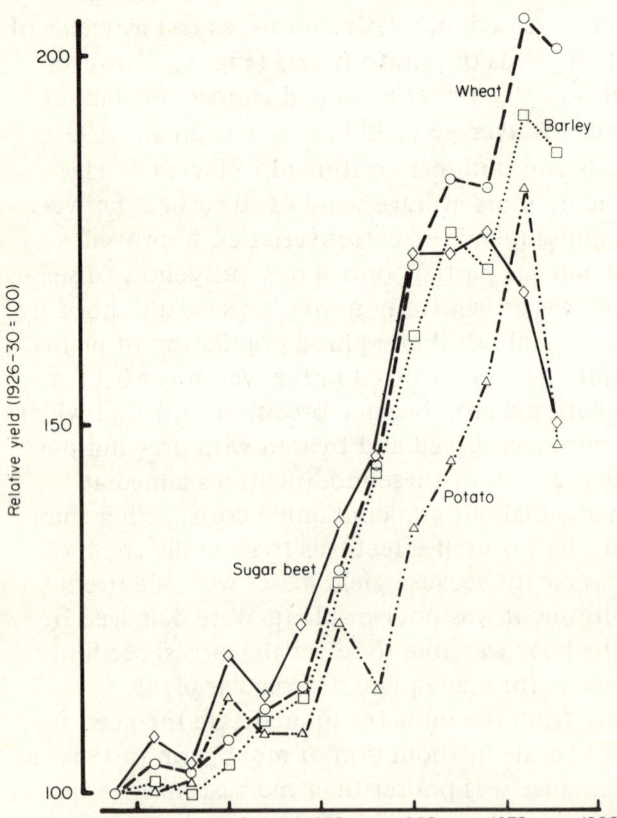

Fig. 1 Trends in national average yields of arable crops in the UK relative to average yields for 1926-30 = 100. (Actual average yields 1926-30; wheat, 2·20 t grain ha^{-1} ; barley, 2·07 t grain ha^{-1} ; sugar beet, 20·3 t roots ha^{-1} ; potatoes, 16·3 t tubers ha^{-1})

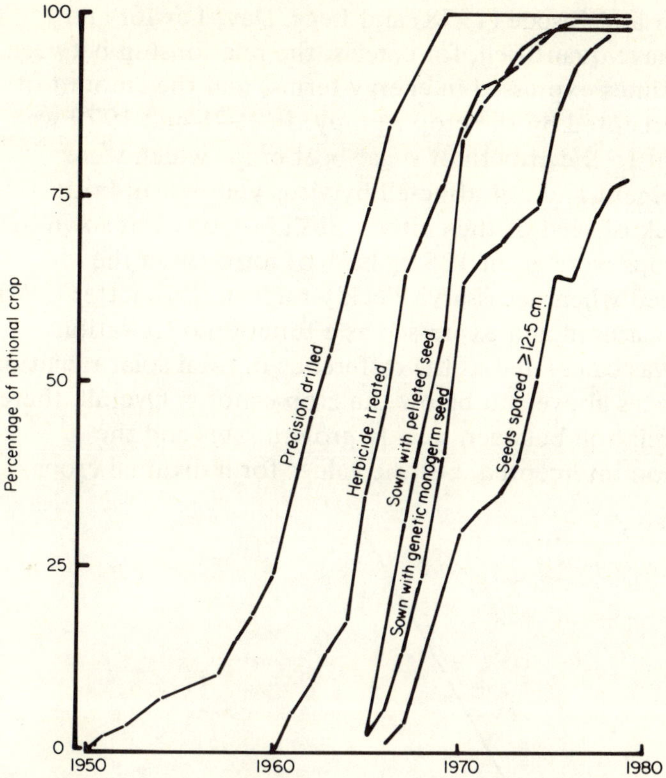

Fig. 2. Changes in husbandry practices for sugar beet grown in the UK since 1950.

the production of dry matter and its partition to stored sugar in the root. Recently a research programme has been started to understand these processes and to identify those that restrict growth, with the ultimate aim of discovering how they might be manipulated to increase yield. This paper describes the approach that has been adopted and the measurements being made by the multidisciplinary research group.

Dry-matter production

In the literature the most detailed analyses of the growth of sugar-beet crops are those of Watson (1952) and Goodman (1968), both of which relied on conventional growth analysis. These studies emphasized the importance of rapid, early growth of leaves and consequently this established the important relationships between leaf-area growth and the seasonal pattern of light-energy receipts for maximizing yields.

Recently, Gallagher and Biscoe (1978) and Legg, Day, Lawlor and Parkinson (1979) have quantified, for cereals, the relationship between growth rate (sometimes expressed in energy terms) and the amount of solar radiation intercepted by the crop canopy. In 1978 and 1979 we applied this concept to the growth of sugar-beet crops which were irrigated to requirement and not affected by virus yellows. In both years the same stock of seed of the cultivar Bush Mono G was sown in mid-April. Both crops were given 125 kg ha^{-1} of nitrogen in the seedbed and irrigated when necessary. Weekly rates of dry-matter (DM) production were measured and expressed as a function of ratiation intercepted. This was calculated as the difference in total solar radiation on horizontal surfaces above and below the crop canopy. Overall, there was a good linear relation between weekly growth rates and the amounts of radiation intercepted, but the values for individual crops

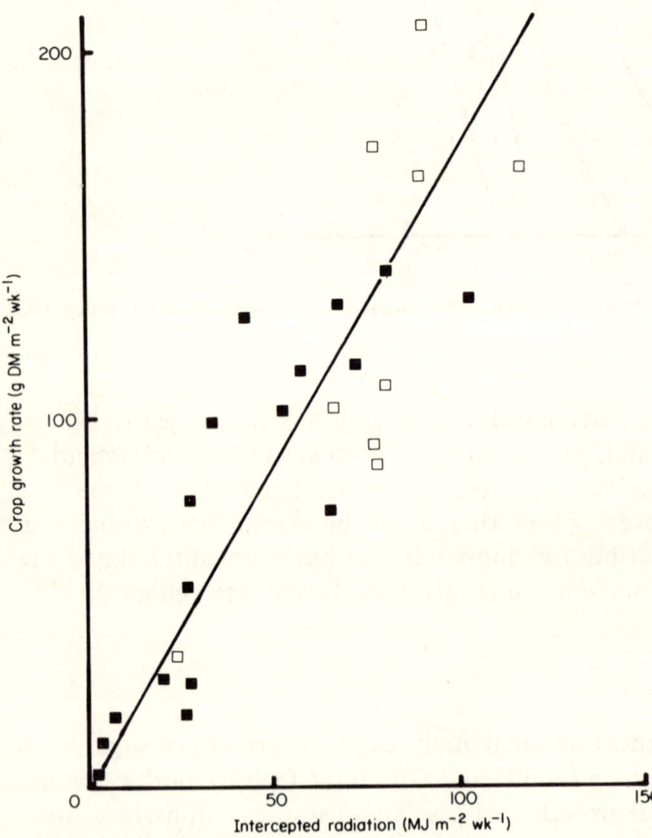

Fig. 3 Weekly crop growth rates of sugar beet as a function of intercepted solar radiation: 1978 (■), 1979 (□).

were too scattered to discern whether their regressions had different slopes and whether the two crops had different efficiencies of converting intercepted radiation to total DM (Fig. 3). The common linear regression, constrained through the origin, has a slope of 1·71 g MJ^{-1}. This is similar, although a little higher than the efficiencies of conversion measured in wheat (Gallagher and Biscoe, 1978), barley (Legg *et al.*, 1979) and apples (Monteith, 1977) and estimated for potatoes (Scott and Allen, 1978).

Alternatively, the efficiency of conversion can be estimated from the slopes of lines relating accumulated dry weight to accumulated intercepted radiation (Littleton, Dennett, Monteith and Elston, 1979). This type of graph is shown for the 1978 and 1979 crops in Fig. 4(a) and the derived efficiencies of conversion of intercepted radiation to total DM (e_T) are given in Table 1. In both years the regressions were linear till crops had intercepted 1 GJ m^{-2}; this was all the 1978 crop intercepted by the time it was harvested in the third week of October. The 1979 crop had intercepted a further 400 MJ m^{-2} by the time it was harvested at around the same date. However, in this case a departure from a linear relationship toward the end of growth suggests that this

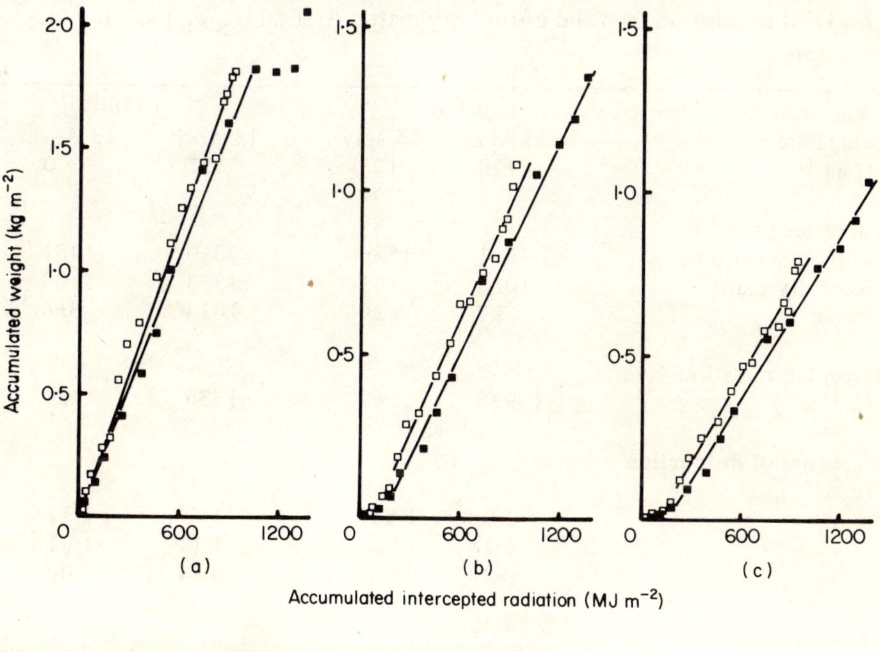

Fig. 4 The accumulation of (a) total dry matter, (b) root dry matter, and (c) sugar by sugar-beet crops as a function of intercepted solar radiation: (1978 (■), 1979 (□)). (—— represents the slope and extent of fitted regressions; the regressions in (a) are constrained through the origin.)

crop either used radiation less effectively during autumn than in summer or that its total DM yield was seriously underestimated because large amounts of leaf material died and were not harvested. Therefore, in estimating the values of e_T for the two crops, only the linear portion of the 1979 curve up to 1 GJ m^{-2} was used. The values of e_T were similar, i.e. 1·98 g MJ^{-1} in 1978 and 1·77 g MJ^{-1} in 1979, and similar to the efficiency of conversion derived from the analysis of weekly crop growth rates (Fig. 3).

In 1978 the growth of the standard crop, sown on 24 April, was compared with that of one sown three weeks later on 18 May; in 1979 the standard crop was compared with one sown on the same day, but given no nitrogen fertilizer. The values of e_T for these crops are given in Table 1. The e_T value for the May-sown crop (1·88 g MJ^{-1}) was similar to that of the April sowing, but omitting nitrogen fertilizer decreased

Table 1. The yields, amounts of radiation intercepted and the efficiencies of production for total dry matter (e_T), root dry matter (e_R) and sugar (e_S) for sugar-beet crops sown on different dates in 1978 and given various amounts of nitrogen fertilizer in 1979. The partition ratios of total dry matter to roots (e_R/e_T) and to sugar (e_S/e_T) and of root dry matter to sugar (e_S/e_R) for the same crops

Season	1978		1979	
Sowing date	24 April	15 May	13 April	13 April
kg N ha^{-1}	125	125	125	0
Final yields (g m^{-2}):				
Total dry matter	1811	1546	2060	1481
Root dry matter	1083	893	1353	1108
Sugar	797	639	1034	866
Intercepted radiation ($MJ\ m^{-2}$)	935	801	1336	935
Efficiencies of Production (g MJ^{-1})				
e_T	1·98	1·88	1·77	1·52
e_R	1·18	1·25	1·11	1·31
e_S	0·89	0·90	0·84	1·00
Partition Ratios				
e_R/e_T	0·60	0·67	0·63	0·86
e_S/e_R	0·75	0·72	0·76	0·76
e_S/e_T	0·45	0·48	0·48	0·66

e_T to 1·52 g MJ^{-1}. On this evidence, change in nitrogen status might also be the cause of the lower efficiency of the 1979 crop during the autumn (Fig. 4(a)). At this time the larger leaves that intercept most of the incoming radiation are ageing and their nitrogen reserves are moving elsewhere in the plant (Bürkey and Biscoe, 1979). It is important, therefore, to optimize nitrogen nutrition to ensure that crops convert intercepted radiation to DM as efficiently as possible and to maintain that efficiency throughout their growth.

Radiation interception

The key to increasing DM production by sugar-beet crops is to maximize the amount of radiation they intercept. Radiation receipts are not within the grower's control but he can manipulate the husbandry of his crops to increase the proportion of the radiation intercepted. The fraction of incident radiation intercepted by crops increases hyperbolically with leaf-area index (LAI); in sugar beet 80-85 per cent is intercepted when LAI reaches about 3·0 and 95 per cent when LAI reaches 5·0. Leaf-area index is the product of plant population and leaf area per plant; the latter is the product of leaf number and size. Apart from sowing the crop earlier, for which there is little further scope with current genotypes because of the limits imposed by low temperatures, the main inputs that affect leaf area per plant and which the grower can control are nitrogen and water. These affect the sizes of individual leaves rather than the rate of leaf production (Morton and Watson, 1948). An extreme example of the effect of nitrogen on radiation interception is given by the two crops grown in 1979 with 0 and 125 kg ha^{-1} of nitrogen. The crop given no nitrogen fertilizer intercepted only 973 MJ m^{-2} during its growth compared with 1361 MJ m^{-2} for the crop given the recommended amount (Table 1). When a wider range of nitrogen rates were tested on several soil types, it was found that an early response in leaf DM yield to nitrogen applications was a prerequisite for any substantial responses in total and root DM production (Last and Draycott, 1975).

The factor which has an over-riding influence on the early growth of leaves is temperature. Controlled-environment experiments have shown that both leaf production and expansion increase linearly with increase in temperature between 7 and 20 °C and cease at temperatures close to 2 °C. Low temperatures experienced early in a leaf's development can markedly reduce its final size (Milford and Riley, 1980). Scott (1980) used this information to predict how different spring temperatures would influence the time taken by a crop to reach LAI of 3·0, so altering the amount of radiation intercepted and DM yield. He showed that, in a year of average radiation receipts, temperatures 1 °C warmer than the long-term averages during the early months of crop growth would be expected to increase both radiation interception and sugar yield by about 10 per cent. By analogy, improving the rate of leaf expansion with temperature by 1 °C should increase sugar yields to the same extent so long as there were no adverse effects on leaf persistence.

The grower can do little to modify the temperature environment of his crop, so these improvements must come through plant breeding. Table 2 shows that certain commercial varieties expand their leaf area twice as fast as others when grown at a constant 7 °C, the mean air temperature in April. There is often greater plant-to-plant variability within than between varieties, so sufficient variability in leaf expansion

Table 2. Rates of expansion of leaf area per plant at 7°C in commercial varieties of sugar beet.

	Rate of expansion (cm^2 day^{-1})	
	Mean	Range
Anglo Maribo Polybeet	10·4	6 - 15
Amono	8·6	5 - 23
Sharpe's Klein Megabeet	7·9	2 - 17
Hilleshog Monotri	7·5	3 - 16
Bush Mono	6·7	3 - 13
Nomo	6·1	4 - 10
Vytomo	5·9	3 - 19
Sharpe's Klein Monobeet	5·2	2 - 15
Klein AA	5·3	2 - 14
S.E.D.	2·10	

in low temperatures clearly exists for improvement by selection. Of the varieties tested, Klein AA (included because it was the oldest genotype available) had the slowest rate of leaf-area expansion at 7 °C. This illustrates the advance that has been made in increasing early leaf growth at cool temperatures. However, such improvements are only useful if they are not at the expense of other temperature-dependent attributes, i.e. bolting resistance. Unfortunately, there is evidence that this might not be so. In Table 2 the varieties Klein AA and Nomo are the most resistant to bolting, but have relatively poor leaf-expansion rates.

The rate at which the leaf canopy of sugar beet grows and the size that it eventually reaches vary with both site and season (Goodman,

1968). The effect of this on LAI and radiation interception are illustrated for two crops, grown at Broom's Barn in 1973 and 1978 (Fig. 5); both were free from drought and diseases. The leaf canopy grew faster in 1973 than in 1978, largely because mean air temperatures between May and July were 1·5 °C warmer. Consequently, the 1973 crop reached a bigger maximum LAI (4·7 cf. 3·3) and achieved LAI of 3·0 towards the end of June rather than late in July. Therefore, it achieved almost complete interception of incoming radiation whilst radiation receipts were still near their maximum values. Fig. 5 also

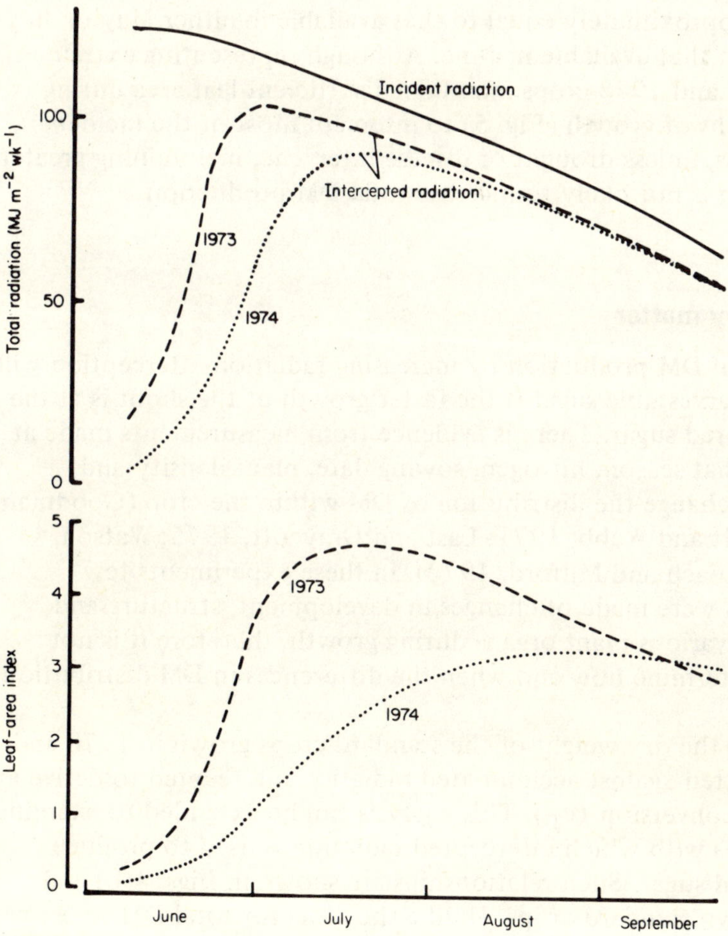

Fig. 5 Patterns of leaf-area index in sugar-beet crops at Broom's Barn in 1973 and 1978 and the estimated amounts of radiation intercepted by such crops in a year of average radiation receipts.

shows the long-term averages of the weekly amounts of solar radiation received at Broom's Barn. In 1973 the crop would have failed to intercept about 300 MJ m^{-2} of the 1500 MJ m^{-2} available and the 1978 crop failed to intercept 500 MJ m^{-2}. Neither crop would have intercepted much of the 420 MJ m^{-2} of radiation incident in May. On the basis of the data in Figs. 3 and 4(a), the 200 MJ m^{-2} difference in radiation intercepted by the two crops is equivalent to about 3 t ha^{-1} of extra DM at harvest. If the crop could be manipulated to intercept even half of May's radiation it would add 3·5 t DM ha^{-1}.

The average amount of radiation available in September and October combined is approximately equal to that available in either May or July, and is less than that available in June. Although representing extremes, both the 1973 and 1978 crops maintained sufficient leaf area during the later months of growth (Fig. 5) to intercept most of the incident radiation. Thus, unless drought or disease intervene, maintaining greater LAI in autumn is not likely to increase total DM production.

Partition of dry matter

Increasing total DM production by increasing radiation interception will not increase harvestable sugar if the faster growth of the shoot is at the expense of stored sugar. There is evidence from measurements made at final harvest that season, nitrogen, sowing date, plant density and irrigation can change the distribution of DM within the crop (Goodman, 1968; Draycott and Webb, 1971; Last and Draycott, 1975; Watson, Motomatsu, Loach and Milford, 1972). In these experiments few measurements were made of changes in development, structure and weight of the various plant organs during growth, therefore it is not possible to determine how and when the differences in DM distribution occurred.

In Fig. 4(a) the dry weight of the standard crops grown in 1978 and 1979 was plotted against accumulated radiation intercepted to derive an efficiency of conversion (e_T). This analysis can be extended to examine the efficiencies with which intercepted radiation is used to produce root dry matter and sugar. Such relationships are shown in Figs. 4(b) and 4(c) for the two standard crops. Unlike the plots for total DM, these are not completely linear. Initially, the crops partitioned more of their DM to the shoot and the amounts of root dry matter and sugar accumulated per MJ of radiation intercepted was relatively small, but increasing. In both crops this stage lasted until they had intercepted about 200 MJ m^{-2} of radiation, a small proportion of the total intercepted during the growing season. However, to intercept this 200 MJ m^{-2} required approximately half the life of the crop, thirteen weeks from sowing in 1978 and ten weeks in 1979.

The noticeable feature of both graphs is that beyond 200 MJ m^{-2}, both root DM and sugar increased linearly with intercepted radiation. This linearity lasted until harvest when radiation interception had reached 1400 MJ m^{-2} and total DM had increased sevenfold, i.e. from 0·25 to 1·75 kg m^{-2}. This implies that over this stage of growth the efficiencies, given by the slopes of the regressions, with which the crops produced root DM (e_R) and sugar (e_S) were constant (Figs. 4(b) and 4(c); Table 1). Both of the standard crops produced root DM and sugar with similar efficiencies; e_R was 1·11-1·18 g MJ^{-1} and e_S 0·84-0·89 MJ^{-1}. Table 1 also includes estimates of e_R and e_S for the other crops grown for comparison in the two years. That sown in May 1978 produced root DM and sugar from intercepted radiation with slightly higher efficiencies than the standard crop sown in April; the crop given no nitrogen fertilizer in 1979 produced both with a much higher efficiency than the standard fertlized crop. From August to October, when most growth occurred in these crops, root DM and sugar production were directly proportional to intercepted radiation and total DM production.

The ratios between e_R and e_T, between e_S and e_T and between e_S and e_R during this period give the DM partitioning between these components of the crop. The use of these ratios has the advantage that they are independent of time and are, therefore, more realistic than estimates of partitioning derived from single harvests which can be distorted, particularly by losses through death of shoot material late in the season. These partition ratios for the two standard and the two comparison crops grown in 1978 and 1979 are given in Table 1 to show the origin of differences in yields of DM and sugar. The standard crop grown in 1979 partitioned more of its total DM to the root than the standard crop of 1978 and, therefore, partitioned more to sugar. The late-sown crop in 1978 partitioned much more of its total DM to the roots, but less of its roots DM to sugar. The crop given no nitrogen fertilizer in 1979 partitioned much more of its total DM to root DM than the fertilized crop but the partition of root DM to sugar was the same for both.

The time at which the partitioning of total DM to roots and sugar became constant varied between seasons; it occurred earlier when the crop intercepted more radiation early in its growth and did not greatly affect e_R/e_T or e_S/e_R (Table 1). This suggests that maximizing the interception of radiation and production of DM early in the life of the crop should not decrease the partitioning of that DM to sugar and, therefore, might be expected to increase sugar yields. The partitioning ratios in Table 1 indicate that there could be scope for increasing yields of sugar, independently of increasing total DM production, by manipulating its partition to root DM and sugar. On the basis of the values for e_S/e_T in Table 1, we conservatively estimate that by optimizing the various husbandry options available to the grower, it should be possible to alter the fraction of total DM partitioned to sugar from 0·45 to 0·50. If this can be done consistently, it will represent, at the rates of DM production shown in Fig. 4(a), an additional 0·6 t ha^{-1} of sugar at final harvest.

The contribution of physiological studies to increased yields

Expressing DM production as a function of intercepted radiation has allowed us to analyse the productivity of different sugar-beet crops relative to the potential of their environment. An extension of the analysis to include root DM and sugar has provided a comparable basis for evaluating the partition of DM to storage roots and sugar. In this way, the extent of variations in radiation interception arising from different patterns of leaf growth, the efficiency with which intercepted radiation is converted to DM, and the ratios of partitioning of DM to sugar were examined in four crops grown in different ways in two seasons. Consideration has been given to the potential contribution that alterations to each process might make to final sugar yield. These initial studies have shown the value of this approach for sugar beet, but more comprehensive measurements are needed to determine how far constraints upon the production and distribution of DM originate in the crop's genome, nutrition or environment. This will entail further compatible measurements of growth, development, nutrient and water uptake and use in response to soil and aerial environments. This can only be done when agronomists, nutritionists and physiologists all contribute their expertise in common experiments. In this paper we have presented the first steps in such a programme.

References

Bürkey, K. & Biscoe, P. V. (1979). The effect of nitrogen on leaf persistence. *Rothamsted Experimental Station Annual Report for 1978,* Part 1, p. 65.
Draycott, A. P. (1972). *Sugar-beet nutrition.* London: Applied Science Publications.

Draycott, A. P. & Webb, D. J. (1971). Effects of nitrogen fertilizer, plant population and irrigation on sugar beet. I. Yields. *Journal of Agricultural Science, Cambridge* **76**, 261-267.

Gallagher, J. N. & Biscoe, P. V. (1978). Radiation absorption, growth and yield of cereals. *Journal of Agricultural Science, Cambridge* **91**, 47-60.

Goodman, P. J. (1968). Physiological analysis of the effects of different soils on sugar beet crops in different years. *Journal of Applied Ecology* **5**, 339-357.

Last, P. J. & Draycott, A. P. (1975). Growth and yield of sugar beet on contrasting soils in relation to nitrogen supply. II. Growth, uptake and leaching of nitrogen. *Journal of Agricultural Science, Cambridge* **55**, 27-37.

Legg, B. J., Day, W., Lawlor, D. W. & Parkinson, K. J. (1979). The effects of drought on barley growth: models and measurements showing the relative importance of leaf area and photosynthetic rate. *Journal of Agricultural Science, Cambridge* **92**, 703-716.

Littleton, E. J., Dennett, M. D., Monteith, J. L. & Elston, J. (1979). The growth and development of cowpeas (*Vigna unquiculata*) under tropical field conditions. 2. Accumulation and partition of dry weight. *Journal of Agricultural Science, Cambridge* **93**, 309-320.

Milford, G. F. J. & Riley, J. (1980). The effects of temperature on leaf growth of sugar-beet varieties. *Annals of Applied Biology* **94**, pp. 431-443.

Monteith, J. L. (1977). Climate and the efficiency of crop production in Britain. *Philosophical Transactions of the Royal Society, London* B **281**, 277-297.

Morton, A. G. & Watson, D. J. (1948). A physiological study of leaf growth. *Annals of Botany* **47**, 281-310.

Scott, R. K. (1980). In *Proceedings, Climatic change and European Agriculture.* Centre for European Agricultural Studies, Wye College, Kent (In preparation).

Scott, R. K. & Allen, E. J. (1978). Crop physiological aspects of importance to maximum yields – potatoes and sugar beet. In *Maximising crop yields, ADAS-ARC Symposium*, pp. 25-30. London: HMSO

Watson, D. J. (1952). The physiological basis of variation in yield. *Advances in Agronomy* **4**, 101-145.

Watson, D. J., Motomatsu, T., Loach, K. & Milford, G. F. J. (1972). Effects of shading and seasonal differences in weather on the growth, sugar content and sugar yield of sugar-beet crops. *Annals of Applied Biology* **71**, 159-185.

Section 2 **Nutritional and Environmental Aspects**

Long-term trends in the aerial and edaphic environment

J. F. Elston, D. J. Greenland and M. D. Dennett
Department of Plant Sciences, The University, Leeds, UK;
International Rice Research Institute, Los Banos, The Philippines and
Department of Agricultural Botany, The University, Reading, UK.

'In a free country no man can be compelled to prophesy against his will'
W. H. Beveridge (1922).

Introduction

It is now conventional to state that agriculture is a complex and interacting system. The first crude division of the system recognizes edaphic, atmospheric and biotic components, and alterations in one component will produce concomitant alterations in other parts of the system. The concomitant alterations are rarely simple and linear and are often entirely unexpected in nature, scale and direction.

Environmental factors, in soil and atmosphere, may vary in some apparently random way with time, or may show some more-or-less regular increase or decrease. That is to say that in addition to apparently random variation there may be fluctuations or there may be trends. The distinction is not always clear, a part of a long fluctuation may appear to be a trend if the period of observation is much shorter than the period of the fluctuation.

Certainly crop yields appear to show trends with time. Wheat yields in England and Wales have increased from about 2 t ha^{-1} in 1880 to about 5 t ha^{-1} now (Fig. 1). For a first approximation, a pair of straight lines can be fitted to these yields (Dennett, 1980). The first line, from 1885-1947, shows an increase of 0·004 t ha^{-1} yr^{-1} ($P < 0{\cdot}001$) and the second line shows one of 0·074 t ha^{-1} yr^{-1} ($P < 0{\cdot}001$). Similar trends can be found for other crops in many different agricultural systems (e.g. Elston and Dennett, in press). These increases in yield have been produced by changes in several parts of the agricultural system.

There is now quite a lot of information about changes with time in some variables of the soil and atmosphere that affect crop and animal production. There is, for example, a formidable amount of information from Rothamsted and other research stations. Manley (1974) has organized an impressive time series for temperatures for England from

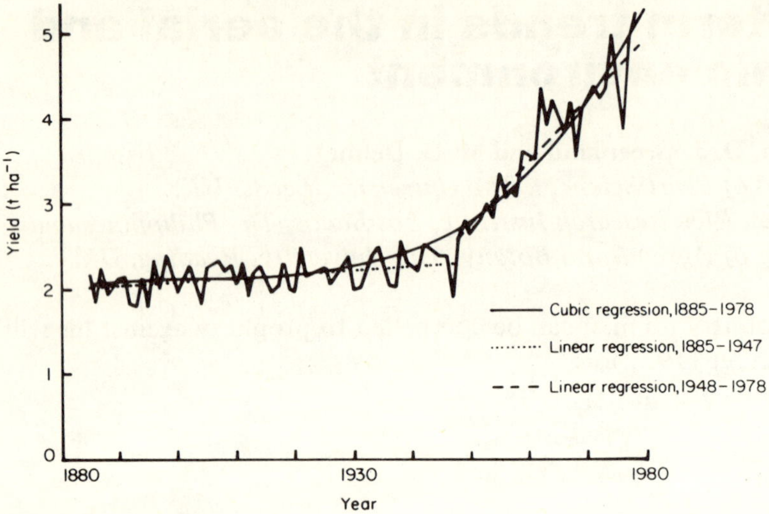

Fig. 1 The average yield of wheat in England and Wales in t ha^{-1} from 1885 to 1978. The increase in yield can be represented by different fitted regressions.

1659. However there are large gaps in our knowledge, for example of solar irradiance, and the available information is not conveniently organized or even freely available.

The development of numerical models, stochastic or heuristic, begins to let us assess the quantitative consequences of some of these changes. These models are more effective for predicting crop than animal yields at the moment and tend, perhaps from the institutional organization of science, to reflect a special interest in some particular compartment of the agricultural system rather than an attempt to pull together all the relevant components.

There are some practical differences between the trends in the soil and atmospheric compartments. Additions of materials to the soil, or CO_2 to the atmosphere, are often cumulative, the amount of the material in the soil appearing to increase more or less steadily. Climatic factors on the other hand vary continuously through the day, the season and the year. However, both matter and energy obey the laws of physics and are usually intimately associated in our systems. Both move into a field, may be transformed there and eventually may move away from the field. We can measure the fluxes into and out of an area, and think of the field as a reservoir; perhaps gaining in, say, heat energy or soil N at one time and losing them at another. We can think of the system as a simple compartmental model. If the rates of influx or efflux change then the quantity of a soil or atmospheric variable in the field will change, and move towards a new equilibrium. This adjustment is often much slower for matter added to soil than for energy or matter in the atmosphere.

There is also a difference in scale of variation, as well as in the rate of change, between soil and atmosphere. In general, there is much more variation horizontally in soil than in climate, though both may vary vertically. The measurements from a meteorological station can represent a substantial area, while measurements from a field may only represent a small area.

Analysis

The analysis of the components of the soil and atmospheric compartments is more straightforward than that of other parts. The variables are more easily defined and measured and continuous sets of measurements, time series, of varying reliability, are often available. Statistical techniques can be used to assess variance, reliability and stability. Trends can be estimated from regressions with time and so rates of change can be assessed. Estimates of the probability of extreme events can be derived from series consisting of the extreme values experienced in each year.

The extrapolation of a regression beyond the set of measurements upon which it is based is often both interesting and unreasonable. For example, sugar beet yields have been systematically estimated for England and Wales since 1926 (Fig. 2). Observers in 1941 and 1956 might have estimated that yields were increasing at a rate of about 0·3 t ha^{-1} yr^{-1} and in 1971 at 0·4 t ha^{-1} yr^{-1}. Their projections for 1975 would be between 35 and 40 t ha^{-1}. That yield was only 24·8 t ha^{-1} and appeared to be declining at a rate of about 1 t ha^{-1} yr^{-1}.

Trends in climate

Temperature, photosynthetically-active irradiance and the water balance are generally believed to be the most important weather variables agriculturally in Britain. Variation in CO_2 concentration may affect climate but is less likely to affect photosynthesis directly (Monteith, 1970). We know little of the variation in irradiance. Manley (1974) has organized a set of monthly temperatures for central England from 1659. This has been analysed by Dyer (1976) and Mason (1976). The annual means vary substantially, and have a tendency to oscillate with periods at

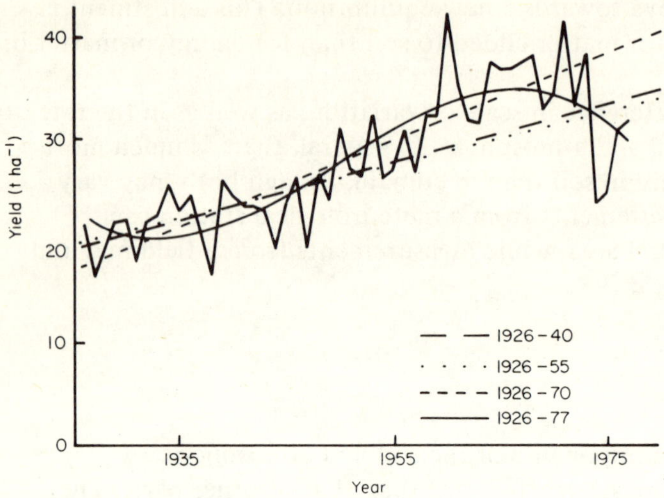

Fig. 2 The average yield of sugar beet in England and Wales in t ha^{-1} from 1926 to 1977. The fitted lines represent regressions calculated for different periods.

about 2, 5, 8, 14, 23 and 76 years (Mason, 1976). The period from 1900-1960 was warmer than average but there is no convincing evidence for a continuous trend, as opposed to continuing fluctuations, in temperature.

In Britain the length of the growing season is determined by temperature. Lamb (1973), after masterly analyses of many sorts of evidence, suggests that the average length of the growing season in north-western Europe shortened from the early Middle Ages until the seventeenth century, then lengthened until about 1950 and is now shortening again; largely as springs become later and cooler. There is now a slower rise of soil and air temperatures in spring.

Data for rainfall have been organized by Craddock (1976) for England since 1725, by Gray (1976) for south-east England from 1840 and by Thom and Ledger (1976) for Edinburgh since 1785. Again, the annual means vary substantially and much of the variation appears to be random (Gray, 1976). There do seem to be a number of oscillations with periods of 2·5 years or less, and one of about 50 years. Gray has used her data to predict the mean rainfall for the decades beginning 1980 and 1990. The mean monthly rainfall for the period 1840-1970 is 57·9 mm $month^{-1}$. Her predictions for the two decades to come are $56{\cdot}1 \pm 0{\cdot}51$ and $55{\cdot}4 \pm 0{\cdot}76$ mm $month^{-1}$ respectively. The systematic variation in average rainfall, as such, is not substantial.

However, rainfall matters primarily because it is a component of the water balance. Analyses have been made of potential evaporation (Wales-Smith, 1977), and soil water deficit (Wigley and Atkinson, 1977) for south-east England and run-off for Edinburgh (Thom and Ledger, 1976). Variation in summer rainfall accounts for about 65 per cent of the variation in estimated soil-water deficit (Wigley and Atkinson, 1977), fortunately for the conventional belief that potential evaporation is less variable than rainfall! Again the data are very variable. Annual potential evaporation at Kew ranged from 430 to 660 mm during the period 1698-1970 (Wales-Smith, 1977). Rainfall varied from 308 to 970 mm over the same period. The period 1960-1976 was substantially drier than any other period in south-east England since 1700. The scale of the apparently random components of the variation is much greater than any systematic trend.

Trends in soil physical conditions

Structural determination in soils with continued use is often asserted to lead to reduced yields (see, e.g. Strutt, 1970). This may arise from the development of plough pans which restrict the root volume, from the decrease in the amount of water which can be extracted by plant roots or in other ways. Mostly this damage can be corrected by proper soil management (Greenland, 1977) and there are few indications of declining yield trends in the UK to support the assertion that serious structural damage is occurring to soils, although a slowing of the rate of increase in cereal yields over the past ten years could be partly due to poorer soil physical conditions. The current decline in sugar beet yields in East Anglia may also be due to the adverse effects of large machinery used for harvesting sugar beet in the late autumn when soils are wet and easily damaged. On easily compacted clay soils the shift to minimum tillage methods of crop production appears to be counteracting the adverse effects which have occurred in the past.

Trends in soil nutrients

The flows and transformations of nutrients in the soil are amazingly complex, especially when contrasted with the relative simplicity of the statistical analysis of weather variables at one place. Elston and Snaydon (1976) set out a simplified compartmental model for the exchanges of nitrogen in an ecosystem. Even in this model, there are 10 compartments and 28 different fluxes. A similar one for potassium has 10 compartments and 24 fluxes and a rather more detailed one for the

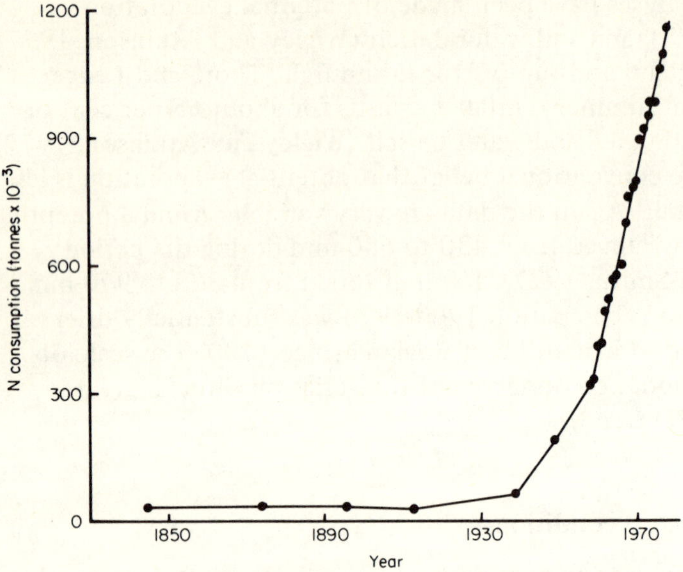

Fig. 3 The amount of nitrogen used each year in the UK in thousands of tonnes, since 1845, based in part upon Cooke (1972).

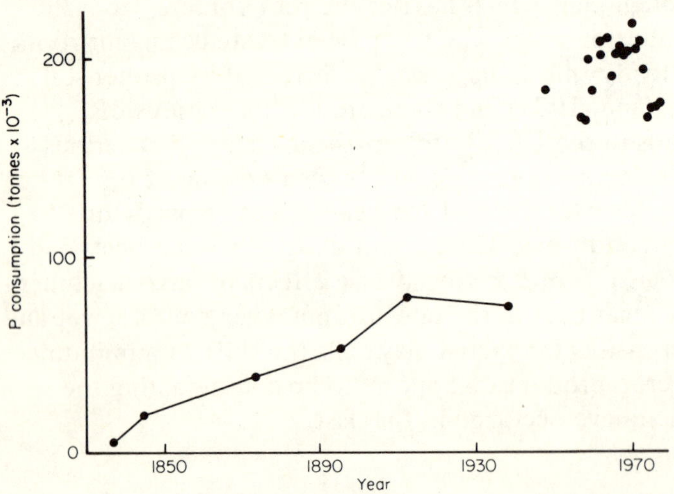

Fig. 4 The amount of phosphorus used each year in the UK in thousands of tonnes, since 1837, based in part upon Cooke (1972).

flow of phosphorus in UK agriculture (Centre for Agricultural Strategy, 1978) has 13 compartments and 25 fluxes.

Cooke (1972) has organized some estimates of the quantities of N, P and K fertilizers used in the UK since 1837. This represents the influx of these nutrients into an 'average' soil. We have extended these to 1977 (Figs. 3, 4 and 5). Fertilizer use has increased enormously since 1930,

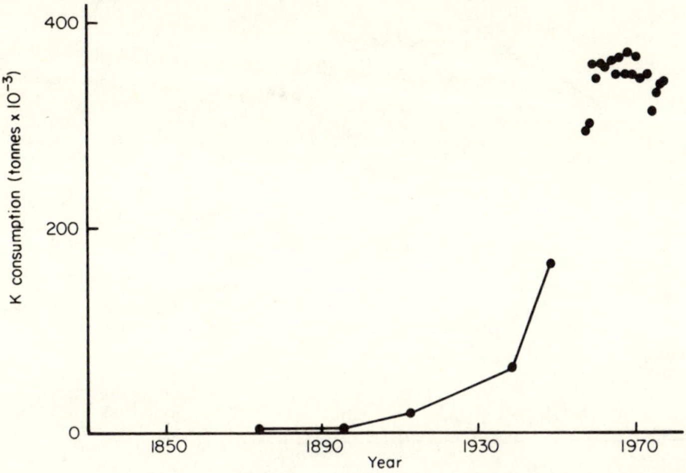

Fig. 5 The amount of potassium used each year in the UK in thousands of tonnes, since 1874, based in part upon Cooke (1972).

and the use of N is still increasing (Fig. 3). However, that of P and K (Figs. 6 and 7) is not. Over the period 1958-1977, the use of N increased from an average for all arable and grassland of about 25 kg N ha^{-1} yr^{-1} to about 90 kg N ha^{-1} yr^{-1}. P and K increased a little from 1958-1965 and then decreased, and averaged about 15 kg P ha^{-1} yr^{-1} and about 30 kg K ha^{-1} yr^{-1} over the whole period. This levelling in fertilizer inputs was caused by the considerable rise in the price of P and K fertilizers. If the price of these fertilizers falls relative to the value of the extra produce, then there may be a further increase in use.

There are no comparable national figures for the efflux of these nutrients from an 'average' soil. It is clear from measurements on individual soils that mineral N is very mobile and easily leached from most soils. The forms of P and the fates of P and K in the soil are complex, but, in general, the rates of leaching are very much slower.

Johnston (1969) gives time series of the percentage of N and soluble P and K in the soil from various plots of Broadbalk field, Rothamsted, from 1865-1966. In general, the data give a clear picture of the soil as a reservoir moving relatively slowly to a new equilibrium when the influx is altered. For example, the soil N content has been steady at about 0·1 per cent where no additional N has been given and has risen to about 0·25 per cent N on the farmyard manure plot. This rise took about 50 years, from the start of the experiment in 1843.

Both Johnston (1969) and Draycott, Durrant and Webb (1978) give admirable analyses of the nutrient balances in the soil under different fertilizer regimes. Draycott *et al.* show that the change in soil content

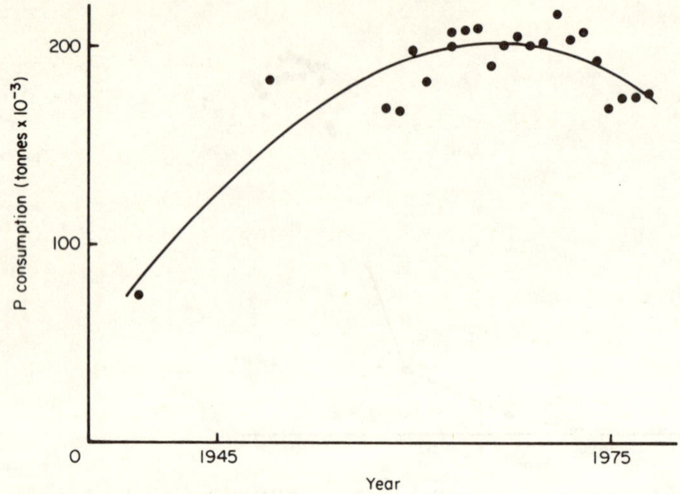

Fig. 6 The recent pattern of consumption of phosphorus for the UK, in thousands of tonnes, largely based upon data from the Fertilizer Manufacturer's Association.

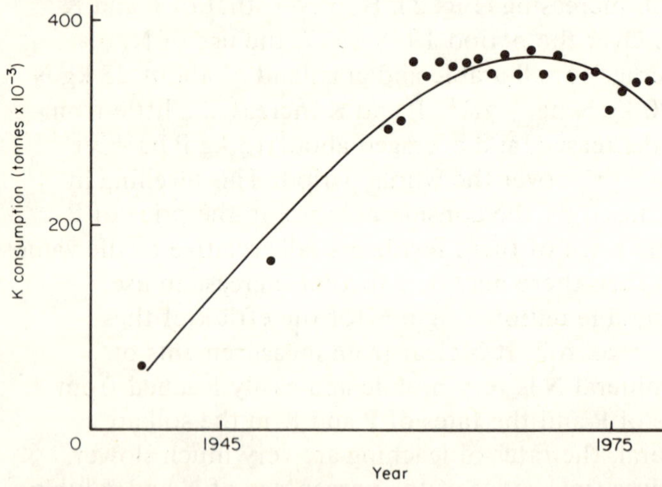

Fig. 7 The recent pattern of consumption of potassium for the UK, in thousands of tonnes, largely based upon data from the Fertilizer Manufacturer's Association.

depends, as expected from the simple compartmental models, upon the balance between nutrient applied and nutrient removed. When the amount removed exceeds the amount supplied, then the extractable soil P content does not decrease much. If yield depends upon the extractable P content rather than the total P applied, then we may expect P fertilizers to become relatively less important on farms where a lot has been applied in the past. The soil K content did decrease, but the rate of change was not as great as when the balance of K was positive.

The future

Prophecy is the most gratuitous form of error, and prophets are rightly without honour in their own countries. If the present trend of increase in wheat yields of about 0·75 t ha^{-1} in ten years continues, then there will be an average yield of about 12·5 t ha^{-1} in 2080 and one of about 45 t ha^{-1} in 2500. Commonsense fortunately limits such extrapolation.

Random variation and fluctuations are the notable feature of the climatic variables that influence agriculture. These features are likely to continue and the variation may be greater than in the first half of this century, which was a time of relatively constant climate.

The costs of fertilizers have risen substantially over the last seven years. The return to the farmer from sales of produce has also risen, but there is increasing pressure on the economic margins of production systems. Consumption of P and K has steadied and is likely to decrease a little, since these nutrients accumulate in the soil. Consumption of N is still increasing (Fig. 8). Our yields depend upon the effective interception of solar radiation and that depends upon a large leaf-area index. We produce this by adding N to replace the substantial losses from the soil each year. The increase in the price of N in Britain has been limited by our access to cheap supplies of natural gas, but the price is likely to rise more steeply in the future so that biological fixation will become more important. The general pattern of N use will be strongly influenced by the cost to benefit ratio and that will depend upon the value of the extra yield produced. That is even more difficult to predict!

Acknowledgement. The authors wish to acknowledge the help given by Kevin Birkby in this work.

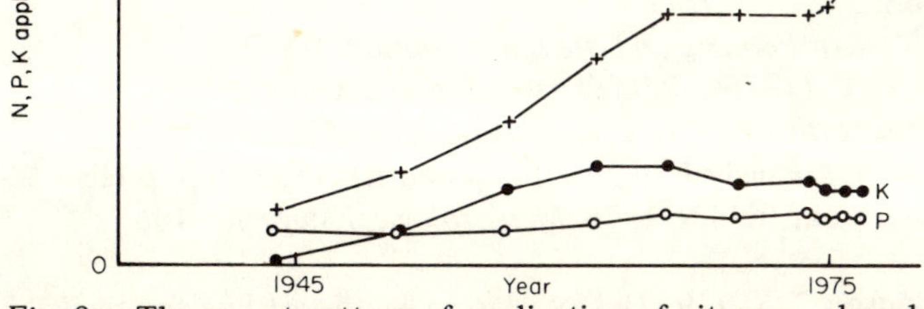

Fig. 8 The recent pattern of application of nitrogen, phosphorus and potassium, in kg ha^{-1}, to winter wheat in the UK.

References

Beveridge, W. H. (1922). Wheat prices and rainfall in Western Europe. *Journal of the Royal Statistical Society* **85**, 412-478.

Centre for Agricultural Strategy, (1978). Phosphorus: a resource for UK agriculture. *CAS Report*, Vol. 2.

Cooke, G. W. (1972). *Fertilizing for Maximum Yield,* 296 pp. London: Crosby Lockwood.

Craddock, J. M. (1976). Annual rainfall in England since 1725. *Quarterly Journal of the Royal Meteorological Society* **102**, 823-840.

Dennett, M. D. (1980). Variability of annual wheat yields in England and Wales. *Agricultural Meteorology*. (in press).

Draycott, A. P., Durrant, M. J. & Webb, D. J. (1978). Long-term effects of fertilizers at Broom's Barn, 1971-76. *Rep. Rothamsted exp. Stn. for 1977,* Part 2, pp.15-30.

Dyer, T. G. J. (1976). An analysis of Manley's central England temperature data: 1. *Quarterly Journal of the Royal Meteorological Society* **102,** 871-888.

Elston, J. & Dennett, M. D. (in press). Climates of the tropics in relation to crop productivity. In *Characterisation of Soils of the Humid Tropics*. Ed. D. J. Greenland. London: Oxford University Press.

Elston, J. & Snaydon, R. W. (1976). Flows and cycles of natural resources. In *Food Production and Consumption*, pp. 27-41. Eds. A. N. Duckham, J. G. W. Jones & E. H. Roberts. Amsterdam: North Holland.

Gray, B. M. (1976). Medium term fluctuations of rainfall in south-eastern England. *Quarterly Journal of the Royal Meteorological Society* **102**, 627-638.

Greenland, D. J. (1977). Soil damage by intensive arable cultivation: temporary or permanent? *Philosophical Transactions of the Royal Society, London* B **281**, 193-208.

Johnston, A. E. (1969). Plant nutrients in Broadbalk soils. *Rep. Rothamsted exp. Stn. for 1968,* Part 2, pp. 93-115.

Lamb, H. H. (1973). Climatic change and foresight in agriculture: the possibilities of long-term weather advice. *Outlook on Agriculture* **7**, 203-210.

Manley, G. (1974). Central England temperatures: monthly means 1659 to 1973. *Quarterly Journal of the Royal Meteorological Society* **100**, 389-405.

Mason, B. J. (1976). Towards the understanding and prediction of climatic variations. *Quarterly Journal of the Royal Meteorological Society* **102**, 473-498.

Monteith, J. L. (1970). Prospects for photosynthesis from A.D. 1970 to A.D. 2000. *Weather* **25**, 456-462.

Strutt, H. W. (1970). *Modern Farming and the Soil*. London: HMSO.

Thom, A. S. & Ledger, D. C. (1976). Rainfall, run-off and climatic change. *Proceedings of the Institution of Civil Engineers* **61**, 633-652.

Wales-Smith, B. G. (1977). An analysis of monthly potential evaporation totals representative of Kew from 1698 to 1976. *Meteorological Magazine* **106**, 297-313.

Wigley, T. M. L. & Atkinson, T. C. (1977). Dry years in south-east England since 1698. *Nature* **265**, 431-434.

Grain production of wheat in relation to nitrogen, weather and diseases

J. H. J. Spiertz
Centre for Agrobiological Research (CABO), Wageningen, the Netherlands.

Introduction

The growth and yield of crops depends strongly on the response of the plants to specific growing conditions, which are mainly determined by weather and soil characteristics. Crop response to weather and soil factors is affected by the occurrence of pests and diseases (Fig. 1) and

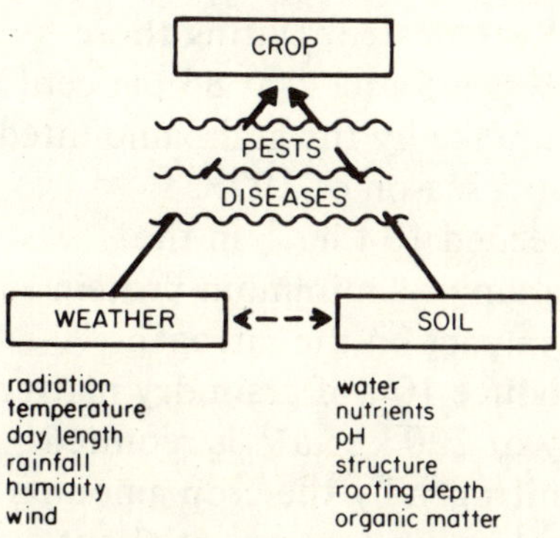

Fig. 1 Soil-weather-crop relations and interference by pests and diseases.

by management practices. The importance of weather to grain yield in wheat has been recognized in agricultural research for nearly a century (Lawes and Gilbert, 1880). In 1926, Kiesselbach and Sprague analysed the variations in grain yield caused by changes in soil fertility and environment in terms of the grain-yield components. A very high correlation between grain yield and grains per unit land area was subsequently reported by Locke, Rauchschwalbe and Mathews, (1942). The number of ears per m^2 usually appears to be more important than the number of grains per ear in adjustments of a crop to environmental conditions. This matter was reviewed thoroughly by Biscoe and Gallagher (1977). They concluded that the influence of weather on the physiological and developmental processes which determine the number of grains is more important than on the processes determining grain size.

Considerable emphasis has been given to deriving maximum grain yields from the photosynthetic potentials for growing conditions in various climatic regions. Gallagher and Biscoe (1978) provided one illustration with a calculation of the maximum yield of grain for English growing conditions. This assumed that 756 MJ m^{-2} of photosynthetically active radiation was absorbed with an efficiency of 4·5 per cent and a harvest index of 0·53, and showed that the grain yield could be as great as 10·3 t ha^{-1}.

However, grain yield of wheat generally is restricted not only by radiation but also by other growth factors, including nitrogen supply and the occurrence of diseases, which have an important influence on crop growth, dry-matter distribution and grain yield (Pearman, Thomas and Thorne, 1978; Spiertz, 1978).

Grain yield, nitrogen concentration of the grain and nitrogen uptake vary widely, as was shown by Benzian and Lane (1979) in their analyses of production between 1954 and 1973 at Rothamsted. During those years the maximum yield of winter wheat was 9·5 t ha^{-1} at 85 per cent dry matter (DM), and the largest nitrogen uptake by the grain amounted to 170 kg ha^{-1}. However, in the cool growing season of 1978, maximum grain yields of winter wheat exceeded 10 t ha^{-1} in the Netherlands. Further yield increases with grain of a minimum protein content of 12 per cent would have a great impact on the nitrogen economy of the whole wheat plant. To produce 10 t of grain dry matter with 12 per cent protein, a nitrogen supply of 200 kg ha^{-1} is required for the grain alone. The total demand for nitrogen by the crop amounts then to about 250 kg ha^{-1}. In modern varieties, we can expect about 100–120 kg ha^{-1} of nitrogen to be available for relocation from the vegetative parts to the grains. So the uptake and relocation of nitrogen in the wheat plant after anthesis becomes a very important feature of a high-yielding wheat crop.

In this paper some aspects of the carbon and nitrogen economy of the wheat plant will be discussed in relation to nitrogen supply, climatic factors and disease control.

Uptake and use of nitrogen

The total uptake of nitrogen by a wheat crop depends on the amount of available soil nitrogen and the recovery of fertilizer nitrogen (Fig. 2). The relationship between total nitrogen uptake and nitrogen fertilizer can be presented by

$$N_t = N_0 + (r \times N_g),$$

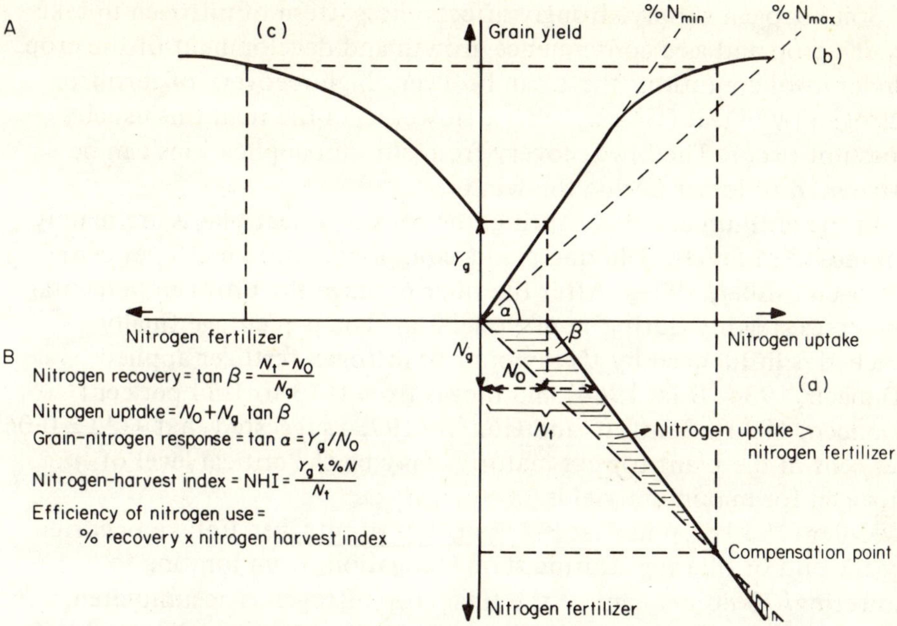

Fig. 2 A. Relations between (a) nitrogen fertilizer and nitrogen uptake, (b) nitrogen uptake and grain yield, (c) nitrogen fertilizer and grain yield. B. Characteristics for the nitrogen economy of a wheat crop.

where N_t is the total nitrogen uptake in the above-ground crop, N_0 is the nitrogen uptake without nitrogen fertilizer, N_g is the amount of nitrogen fertilizer and *r* is the recovery fraction of applied nitrogen. The availability of nitrogen in the soil is determined by the balance between nitrogen mineralization and fertilizer on the one hand, and nitrogen losses (leaching, volatilization, denitrification) and fixation by micro-organisms on the other. The amount of nitrogen uptake without nitrogen fertilizer depends mainly on the inorganic nitrogen made available from soil organic matter, and on rooting characteristics (e.g. root depth, duration of root activity). Generally the rate of nitrogen uptake is higher than the rate of mineralization. As a consequence, the mineral nitrogen reserve in the soil is depleted very quickly under a crop without added fertilizer (Table 1).

Table 1. Mineral nitrogen reserve (kg ha^{-1}) in the soil to 1m depth at Easter Flevopolder in 1978

Dates:	8/2	17/4	15/5	15/6	17/7	14/8
Bare soil	77	78	92	130	154	214
Under crop	–	73	24	33	21	26
Difference	–	5	68	97	133	188

Soil-nitrogen supply strongly affects the pattern of nitrogen uptake by the crop and as a consequence growth and development of the crop. Under ideal conditions there can be a very high recovery of fertilizer nitrogen by wheat (Gasser, 1961). However, in the field this usually does not occur. The low recovery from autumn applications can be attributed to losses during the winter.

In the autumn and early spring, the tops of wheat plants are mainly composed of leaves. Tillering plants commonly contain 4-5 per cent nitrogen (Miller, 1939). After the tillering stage the nitrogen percentage decreases steadily during stem elongation. The percentage finally reached is influenced by the amount of nitrogen fertlizer applied (Doneen, 1934; Dilz, 1964) and ranges from 0·73 to 1·54 per cent (Doneen, 1934). Stanford and Hunter (1973) suggested that 1·22 ± 0·06 per cent in the plant tops at maturity may be the critical level of nitrogen for maximum yields.

Miller (1939) found that 8-22 per cent of nitrogen uptake occurred by the end of tillering. During stem elongation, from jointing to flowering, 70-80 per cent of the final crop nitrogen is accumulated. With low rates of nitrogen fertilizer, nitrogen uptake usually reaches a maximum around heading, but with high rates nitrogen uptake may continue until soft dough stage (Carpenter, Haas and Miles, 1952). The rate of nitrogen uptake by a wheat crop growing at a rate of 200 kg DM ha^{-1} d^{-1} during the phase of leaf growth is about 6 kg ha^{-1} d^{-1}. During the later stages of growth, the rate of nitrogen uptake may be restricted by an inadequate supply of nitrogen or a reduced capacity of the roots to accumulate nitrogen.

Large differences in the amount of nitrogen uptake from soil reserves (N_0) and in nitrogen recovery (r) exist between years and soils (Fig. 3), but only minor differences occur between varieties. The former differences are mainly related to the nitrogen residues of the preceding crop, the amount of winter rainfall and the organic matter content of the soil. The organic matter contains an organic bound nitrogen pool of 3000-15 000 kg ha^{-1} of nitrogen of which 1·5-3 per cent mineralizes each year (Vlassak, 1978). The rate of mineralization depends on the C/N ratio, the soil moisture content and temperature. When fungicide treatments were applied against ear and foliar diseases, it turned out that N_0 and r were higher in the healthy than in the diseased crop (Fig. 4). This improved nitrogen uptake was probably brought about by a longer duration of root activity. It seems likely that the roots were maintained longer because more assimilates were available to the roots.. Hence the nitrogen mineralized in the soil during the grain-filling phase was utilized more effectively.

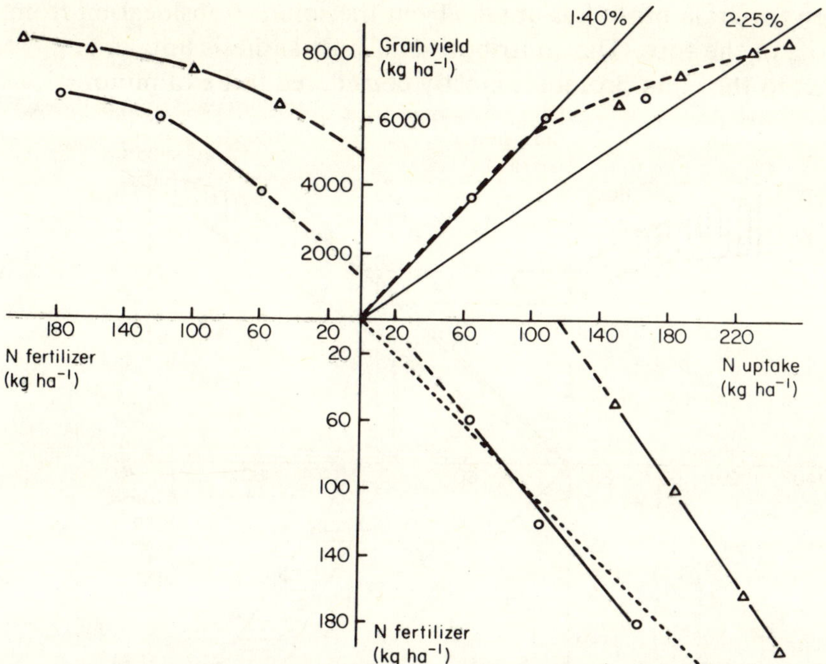

Fig. 3 Seasonal differences in the relations between nitrogen fertilizer, nitrogen uptake and grain yield for Lely winter wheat grown at Flevopolder in 1975 (○) and 1976 (△). In both years fungicides were used.

Growth of the grain creates a demand for nitrogen which is determined by the genetically defined chemical composition of the grain and by temperature (Spiertz, 1977; Vos, 1980). The extent to which this demand is satisfied depends on the supply of nitrogen to the grains.

The nitrogen in the grain is derived both from leaf nitrogen and from nitrogen taken up by roots during grain maturation. When the grain nitrogen increases rapidly 60-85 per cent of the nitrogen in the leaves is relocated to the grains (van de Sande Bakhuyzen, 1937; Neales, Anderson and Wardlaw, 1963; Spiertz and Ellen, 1978). Thus the main source of nitrogen for seed protein is the organic nitrogen stored in the vegetative parts (leaves, stem and chaff). When nitrogen uptake by the plant ceases at anthesis, the translocation of nitrogen from the vegetative tissues can proceed to such an extent that the nitrogen concentration of the straw is depleted to very low levels. For example, values of 0·3 per cent nitrogen in the dry matter of depleted tissues are possible (Dilz, 1964; Spiertz and Ellen, 1978). Another source of nitrogen for grain protein is uptake from the soil or translocation from the roots to the tops. The contribution of post-anthesis uptake of nitrogen to the grain protein is mostly considered to be of minor

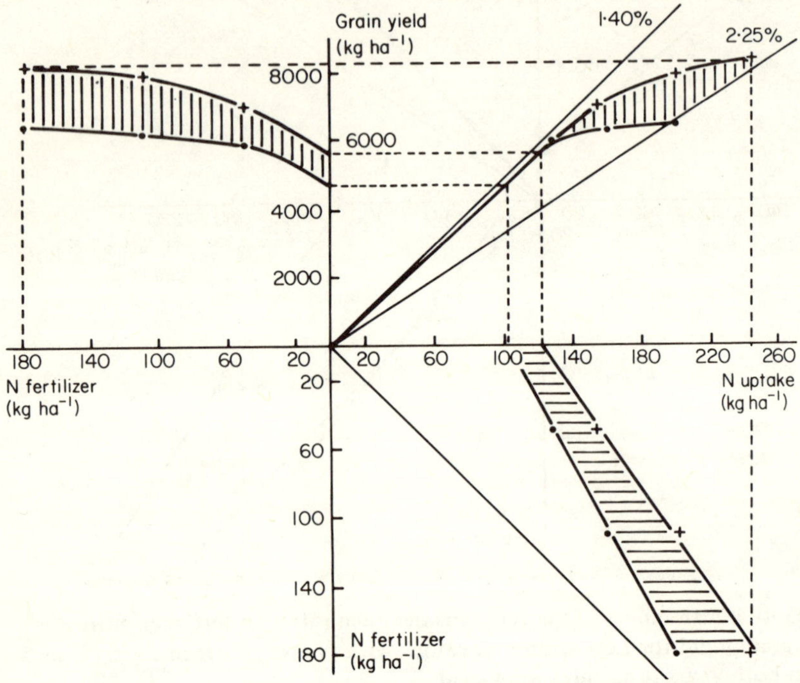

Fig. 4 Effects of disease control on the relations between nitrogen fertilizer, nitrogen uptake and grain yield for Arminda winter wheat grown at Flevopolder in 1978 with (+) and without (•) fungicides.

importance, amounting to only 10-20 per cent of the final grain nitrogen (Austin, Ford, Edrich and Blackwell, 1976). However, with high-yielding wheat crops, that source becomes very important since the nitrogen available from vegetative tissues is only 100-120 kg N ha^{-1} and does not meet the demands of the grain. High rates of nitrogen uptake after flowering have been found in healthy crops under conditions of high solar radiation, and without water stress (Hucklesby, Brown, Howell and Hagemann, 1971; Spiertz and Ellen, 1978).

The moisture supply appears to be a major factor affecting the recovery of fertilizer nitrogen by wheat. Evidence on the effects of soil moisture on nitrogen uptake can be obtained from irrigation experiments. Robins and Domingo (1962) found that low water supply before heading and from boot to milk stage decreased nitrogen uptake. The lack of response of grain yield to larger amounts of nitrogen fertilizer can be associated with a more rapid depletion of soil water. In many situations there will be a shortage of water in the root zone during the grain-filling period and that will limit the influence of nitrogen. Therefore, yield-nitrogen relationships are confused as other growth-limiting factors become dominant (de Wit, van Laar and van Keulen, 1979; Terman, 1979).

Late nitrogen supply and diseases

Nitrogen fertilizer can affect crop growth and grain yield in various ways. Early top dressings with nitrogen stimulate tiller formation and leaf growth, which results in an increase of shoot dry weight and leaf-area index (Thorne, Thomas and Spearman, 1979). Especially in a warm spring, single dressings of large amounts of nitrogen strongly promote leaf growth. This increases mutual competition for light in the crop and the susceptibility of the crop to several foliage and ear diseases (*Septoria* spp., *Puccinia* spp., *Fusarium* spp.,). But the main result is a less favourable harvest index (HI) since extra nitrogen increases straw weight relatively more than grain yield (Pearman, Thomas and Thorne, 1977 and 1978). The reduction in HI reflects the adverse effect of increased competition between tillers on ear formation and ultimately on the number of grains per fertile tiller, that is on sink capacity.

The adverse effects of high rates of nitrogen dressing on crop structure may be prevented by applying the total amount of nitrogenous fertilizer in several, smaller amounts. There is still much discussion about the most favourable amount and timing of the nitrogen fertilizer. The varying recommendations for nitrogen dressing among the north-western European countries reflects more the historical differences in farming and research approaches than they reflect differences in crop response due to specific weather and soil conditions (Laloux, Poelaert and Falisse, 1975; Dilz, 1978; Penny, Widdowson and Jenkyn, 1978). In farm practice the optimum amount of nitrogen fertilizer for a variety will also depend to a large extent on its physiological response, resistance to lodging and susceptibility to diseases (Blackman, Bingham and Davidson, 1978).

The beneficial effect of split-dressings under conditions of a high nitrogen supply to the crop is shown in Table 2 and Fig. 5. With single dressings, there was only a minor response to increased nitrogen fertilizer unless chlormequat was applied. However, when the nitrogen was split between an early and a late top dressing of 35 and 140 kg ha^{-1}, respectively, grain yield was much larger than with a single dressing of 175 kg ha^{-1}. This yield increase was brought about by an increase of

Table 2. Effects of various nitrogen dressings and chlormequat (CCC) application on yield components and crop charcteristics of Arminda winter wheat grown on the north-east polder in 1978.

Nitrogen dressing (kg N ha^{-1})	Chlormequat	DM yield (g m^{-2})		Dry-Matter	Nitrogen yield (g m^{-2})		Nitrogen	Yield components	
		Grain	Straw	HI	Grain	Straw	HI	Grain weight mg	Number of grains per m^2 ($\times 10^3$)
35	–	741	967	0·43	14·7	4·4	0·77	36·3	20·4
	+	739	993	0·43	13·8	3·8	0·78	35·9	20·6
105	–	776	1068	0·42	17·2	6·1	0·74	34·5	22·5
	+	806	1008	0·44	17·6	5·7	0·76	34·2	23·6
175	–	777	1026	0·43	18·0	7·0	0·72	34·5	22·5
	+	830	1013	0·45	18·9	6·8	0·74	34·3	24·2
105 + 70	–	805	1033	0·44	18·4	7·3	0·72	35·8	22·5
	+	841	1039	0·45	18·7	6·1	0·75	35·5	23·7
35 + 140	–	871	1049	0·45	20·2	6·5	0·76	38·1	22·9
	+	873	1060	0·45	19·4	6·0	0·76	38·0	23·0

grain weight from 41·5 to 45·8 mg while grain number remained constant at about 20 000 grains m^{-2}. The largest grain yield, 8·73 t DM ha^{-1}, was associated with an above-ground total dry-matter yield of 19·3 t ha^{-1} and a nitrogen yield of 254 kg ha^{-1}. An increase in nitrogen fertilizer did not affect the dry-matter HI, but decreased the nitrogen

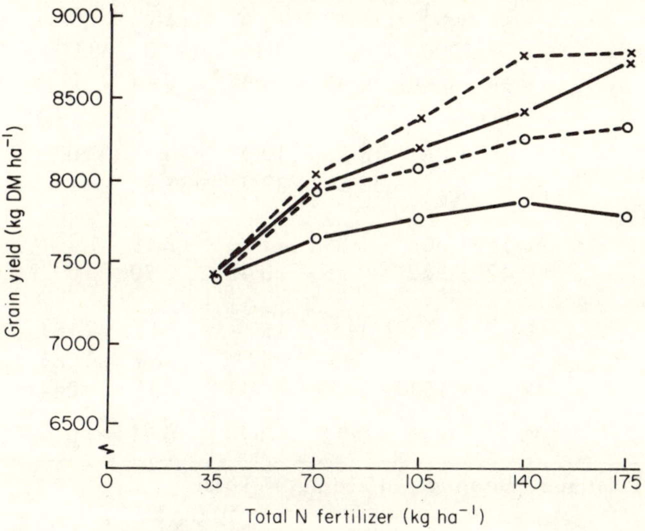

Fig. 5 Response of winter wheat to nitrogen fertilizer applied as a single dressing (○) or a split dressing (X) either with (– –) or without (—) the application of CCC.

HI from 0·78 to 0·73, while split-dressings slightly increased total dry matter as well as nitrogen HI. Thus in healthy crops with ample water supplies, the dry matter HI did not decrease with larger amounts of nitrogen fertilizer.

The effect of late nitrogen top dressings strongly interacts with the occurrence of diseases in the crop. A field experiment in 1978 with the cultivar Arminda showed a poor response to late nitrogen dressings when the wheat crop was not protected with fungicides and insecticides (Table 3). An increase in the amount of nitrogen applied from 50 to 180 kg ha^{-1} gave an additional grain yield of 430 and 1270 kg ha^{-1} for the control and the disease-protected crops, respectively. In various experiments with split dressings of nitrogen and disease control, it was found that an additional nitrogen dressing at the boot stage raised both the number of grains per m^2 and grain weight (Table 3 and Spiertz and Ellen, 1978). Then the yield level increases to such an extent that no compensation occurs between grain number and individual grain weight. Also the nitrogen uptake and the relocation of nitrogen to the grains was promoted by effective disease control (Table 3). In the presence of

Table 3. Effects of nitrogen fertilizer and disease control on grain yield, yield components and nitrogen economy of Arminda winter wheat grown at East Flevopolder in 1978.

Nitrogen fertilizer	N_1		$N_1 + N_2$		$N_1 + N_2 + N_3$	
Fungicides + insecticides	—	+	—	+	—	+
Total DM yield (t ha^{-1})	13·6	15·2	14·3	16·9	14·8	17·9
Grain yield (kgDM ha^{-1})	6040	7000	6250	7930	6470	8270
Dry-matter HI	0·44	0·46	0·44	0·47	0·44	0·46
Number of grains per m^2 (×10^3)	18·2	18·6	18·8	19·9	19·2	20·0
Grain weight (mg grain^{-1})	33·0	37·7	33·3	39·7	33·8	41·4
Nitrogen concentration (%)						
grain	1·54	1·68	1·84	1·94	2·18	2·19
straw	0·42	0·42	0·55	0·48	0·70	0·64
Nitrogen yield (kg ha^{-1})						
grain	93	118	115	157	141	181
straw	35	35	45	44	60	63
total	128	153	160	201	201	244
Nitrogen HI	0·72	0·78	0·72	0·78	0·71	0·74

Nitrogen fertilizer	Tillering	Booting	Heading	Total
N_1:	50			50 (kg ha^{-1})
N_2:	50	+ 60		110 (kg ha^{-1})
N_3:	50	+ 60	+ 70	180 (kg ha^{-1})

diseases, nitrogen uptake by the grains was less and the nitrogen retention in the vegetative tissues was greater. The magnitude of this positive interaction between nitrogen and disease control is of course determined not only by the type and degree of disease infection but also by environmental conditions (temperature, water stress, etc.).

Effects of temperature and nitrogen on rate and duration of grain growth

The rate and duration of grain growth determine grain yield. Both rate and duration are governed by temperature, warmth enhances the rate of grain growth and shortens the duration (Sofield, Evans, Cook and Wardlaw, 1977; Spiertz, 1977). A fast rate of grain growth can only be maintained with an ample supply of assimilate. Carbohydrate supply is supported mainly by current photosynthesis and only a small amount (about 20 per cent) by relocation of stored stem reserves (Austin, *et al.* 1977, Bidinger, Musgrave and Fischer, 1977). At higher temperatures, stem reserves will deplete very fast, while at lower temperatures, grain growth is promoted more than current photosynthesis (Spiertz, 1977).

Higher temperatures also increase the rate of nitrogen uptake by the grains more than it increases the rate of carbohydrate accumulation in the grains. So warmth during the grain-filling period increases the nitrogen content of the grain, but it also accelerates the depletion of the nitrogen reserves in the vegetative parts (Spiertz, 1977; Vos, 1980).

Large requirements for nitrogen by the grains might accelerate leaf senescence and thus reduce photosynthesis at the end of the growing season (Sinclair and de Wit, 1975). It has been shown that late top dressings of nitrogen can increase appreciably the nitrogen concentrations of the leaves and delay their senescence, providing the crop is healthy and sufficient water is available (Spiertz and Ellen, 1978). In this case, the grain may mature before the vegetative organs senesce. It was found that additional nitrogen dressings at the boot stage did not affect the total duration of grain growth (Spiertz, 1979), which is governed generally by a given number of degree-days after flowering. From experiments in controlled environments and data from the literature it was found that there exists a linear relationship between the logarithm of the duration of grain growth and mean daily temperature (Vos, 1980).

The duration of the period with *linear* grain growth depends on the availability of assimilates. The relation between supply and utilization of assimilates, available reserves and duration of grain filling can be expressed as follows:

$$R_g \times D_g = R_a \times D_a + W_t, \qquad (1)$$

where R_g is the average rate of grain growth, D_g the duration of grain growth, R_a the average rate of current assimilation, D_a the duration of current assimilation and W_t the amount of reserves relocated to the grain. Using e.g.(1) and representative values from experiments, a grain yield of 9 t DM ha^{-1} can be achieved in both a cool and a warm growing season as follows:

(a) based on the *carbohydrate* economy,

Cool season:

225 kg ha^{-1} d^{-1} × 40 davs = 200 kg ha^{-1} d^{-1} × 40 days + 1000 kg ha^{-1}

Warm season:

300 kg ha^{-1} d^{-1} × 30 days = 250 kg ha^{-1} d^{-1} × 30 days + 1500 kg ha^{-1}

(b) based on the *nitrogen* economy,

Cool season:

4·5 kg ha^{-1} d^{-1} × 40 days = 2·5 kg ha^{-1} d^{-1} × 40 days + 80 kg ha^{-1}

Warm season:

6·5 kg ha^{-1} d^{-1} × 30 days = 3 kg ha^{-1} d^{-1} × 30 days + 105 kg ha^{-1}.

Generally the amount of reserves is already determined at the beginning of grain growth and the rate of grain growth will depend on daily temperature during grain filling. The only variable affecting the effective duration of grain growth remains the current rates of net photosynthesis and of nitrogen uptake. Especially high temperatures promote the rate of grain growth more than current assimilation by the whole plant and so may reduce the duration of the effective grain-filling period. Temperature also effects transpiration and as a consequence to a certain extent the nitrogen uptake by the plant. Thus uptake rates of nitrogen, rates of relocation from vegetative parts to grains and duration are different between warm and cool growing seasons.

Source-sink relations

The rationale for considering solar radiation as the main factor limiting yield is that the absorption and conversion of light energy to chemical energy, with the associated dark processes, provides the energy and the organic substrates required for growth and yield. The production of carbohydrates depends on the rate of canopy photosynthesis, which appears to increase with increasing leaf-area index (LAI), reaching a maximum at LAI values of about 4 (Puckridge and Ratkowsky, 1971; Evans, Wardlaw and Fischer, 1975). The application of nitrogen and the control of diseases usually increases the production rate per unit land area, because the time taken to achieve complete ground cover is reduced and leaf longevity is extended (de Vos, 1975; Pearman *et al.* 1979). Thorne *et al.* (1979) found that in spring and winter wheat the total dry-matter yield was enhanced more than grain yield when the nitrogen fertilizer was increased at an early stage. However, the reverse may occur with split or late nitrogen dressings (Table 1 and 2 in Spiertz and van de Haar, 1978).

In cereals, grain yield is the ultimate goal and therefore the partitioning of dry matter between grain and vegetative parts is of great importance (Donald and Hamblin, 1976). The partitioning of dry matter in the wheat plant depends on the distribution between leaves, stem and ear before anthesis and on the relocation of stem and leaf reserves after anthesis. Photosynthesis up to anthesis affects kernel number. Storage capacity might continue to be affected by photosynthesis until two to three weeks after anthesis when the endosperm cell number is determined (Fischer, 1975; Brocklehurst, Moss and Williams, 1978; Evans, 1978). A stronger mutual competition for light in a more leafy crop reduces the number of grains per ear. To avoid excess leaf growth, which may occur in soil with high fertility, the crop should be kept under slightly sub-optimal conditions for nitrogen during tillering and the period of leaf growth. This requires the nitrogen fertilizer to be applied two or three times during the season. Using this approach, additional nitrogen applied at the boot stage increased the number of grains per ear (Table 3 in Spiertz and Ellen, 1978) and total dry matter production. Split-dressing of nitrogen under conditions of high soil fertility is a tool for the farm to improve the storage, or sink capacity of the wheat crop.

Under favourable growing conditions, the grain yield appears to depend more on the storage capacity than on assimilate supply. The storage capacity is a function of grain size and especially number of grains. Under Dutch growing conditions at least 18 000 grains per m² are required for achieving maximum grain yields (Darwinkel, 1978). Quite often, grain yield is restricted by the number of grains set as well as by the degree of grain filling (Fig. 6). The factors limiting grain

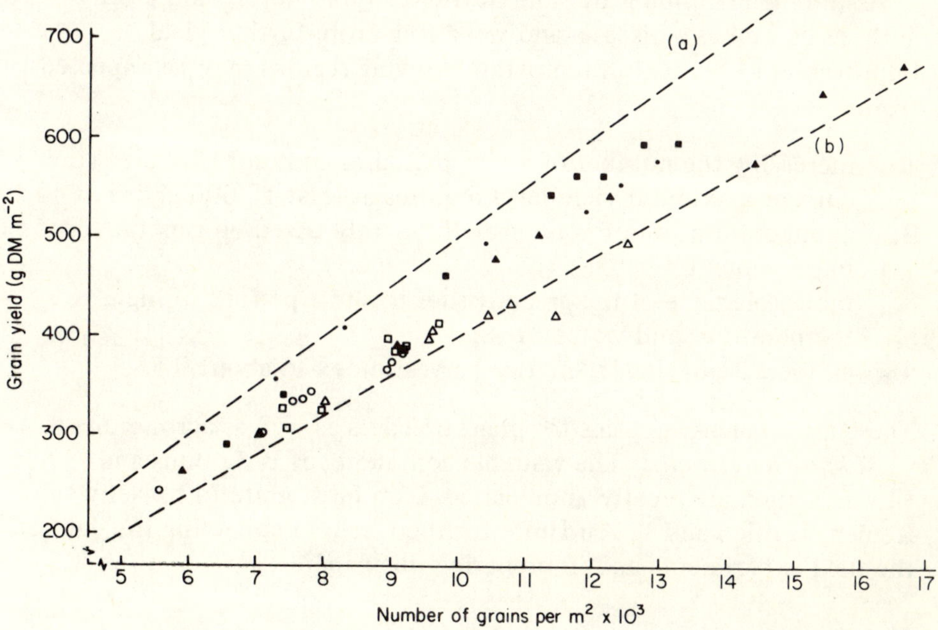

Fig. 6 Relation between number of grains per m² and grain yield for crops of winter wheat. Capelle-Desprez in 1975 (●) and 1976 (○), Fundin in 1975 (▲), Hobbit in 1976 (△), Huntsman in 1975 (■) and 1976 (□). The dotted lines represent the relation if the mean weight per grain is constant at (a) 50 mg or (b) 40 mg (after Pearman *et al.*, 1978 and data from Gillian N. Thorne, Rothamsted).

setting are incompletely understood. Generally it was found that there exists a linear relationship between radiation during ear development and number of grains per unit land area (Fischer, 1975; Evans, 1978;

Ritchie, personal communication). Further research will be required to investigate how this relationship can be manipulated genetically or by crop management.

Opportunities for increasing grain yields

Fiddian (1973) stated that nitrogen fertlizer was being used at near the level of maximum economic response and that further gains must come from breeding varieties able to make better use of nitrogen. However, it has been shown since that with disease control large supplies of nitrogen can further increase the yield of winter wheat with a considerable improvement in the efficiency of nitrogen use. Preventing diseases and other limiting factors (e.g. water stress) has given more scope for yield responses to nitrogen than breeding for a higher metabolic efficiency.

Assuming an ample water and nutrient supply before and after anthesis as well as a disease- and weed-free crop, further yield improvement of wheat in temperate growing regions may be expected from:

1. increasing the number of grains per culm and unit land area (the current maximum yield level requires at least 18 000 grains m^{-2});
2. a longer duration of grain growth, mainly based on genetic improvement;
3. more tolerance of the green tissues to short periods of high temperatures and water stress;
4. an increase in the dry-matter harvest index to about 0·55.

These are challenging goals for plant breeders as well as agronomists.

Acknowledgements. The valuable comments of B. Loomis and H. van Keulen are greatly appreciated. I am most grateful to Nelly van Keulen, J. Ellen and L. ten Holte for their help in collecting the data of the field experiments and in preparing this paper.

References

Austin, R. B., Edrich, J. A., Ford, M. A. & Blackwell, R. D. (1977). The fate of the dry matter, carbohydrates and ^{14}C lost from the leaves and stems of wheat during grain-filling. *Annals of Botany* **41**, 1309-1321.

Austin, R. B., Ford, M. A. Edrich, J. A. & Blackwell, R. D. (1976). The nitrogen economy of winter wheat. *Journal of Agricultural Science, Cambridge* **88**, 159-167.

Benzian, B. & Lane, P. (1979). Some relationships between grain yield and grain protein of wheat experiments in South-East England and comparisons with such relationships elsewhere. *Journal of the Science of Food and Agriculture,* **30**, 59-70.

Bidinger, F., Musgrave, R. B. & Fischer, R. A. (1977). Contribution of stored pre-anthesis assimilate to grain yield in wheat and barley. *Nature* **270**, 431-433.

Biscoe, P. V. & Gallagher, J. N. (1977). Weather, dry matter production and yield. In *Environmental Effects on Crop Physiology*, pp. 75-100. Eds. J. J. Landsberg & C. V. Cutting. London: Academic Press.

Blackman, J. A., Bingham, J. & Davidson, J. L. (1978). Response of semi-dwarf and conventional winter wheat varieties to the application of nitrogen fertilizer. *Journal of Agricultural Science, Cambridge* **90**, 543-550.

Brocklehurst, P. A., Moss, J. P. & Williams, W. (1978). Effects of irradiance and water supply on grain development in wheat. *Annals of Applied Biology* **90**, 265-276.

Carpenter, R. W., Haas, H. J. & Miles, E. F. (1952). Nitrogen uptake by wheat in relation to nitrogen content of soil. *Agronomy Journal* **44**, 420-423.

Darwinkel, A. (1978). Patterns of tillering and grain production of winter wheat at a wide range of plant densities. *Netherlands Journal of Agricultural Science* **26**, 383-398.

Dilz, K. (1964). On the optimum nitrogen nutrition of cereals. *Verslagen landbouwkundig Onderzoek* (Agricultural Research Reports) **641**.

Dilz. K. (1978). [Stikstofbemestingsproblemen in de nederlandse landbouw]. *Landbouwkundig tijdschrift* **90**, 82-88.

Donald, C. M. & Hamblin, J. (1976). The biological yield and harvest index of cereals as agronomic and plant breeding criteria. *Advances in Agronomy* **28**, 361-405.

Doneen, L. D. (1934). Nitrogen in relation to composition, growth and yield of wheat. *Washington Agricultural Experimental Station Bulletin* **296**.

Evans, L. T. (1978). The influence of irradiance before and after anthesis on grain yield and its components in microcrops of wheat grown in a constant daylength and temperature regime. *Field Crops Research* **1**, 5-19.

Evans, L. T., Wardlaw, I. F. & Fischer, R. A. (1975). Wheat. In *Crop Physiology: some case histories,* pp. 101-149. Ed. L. T. Evans. Cambridge: University Press.

Fiddian, W. E. H. (1973). Proceedings of the Association of Applied Biologists. *Annals of Applied Biology* **75**, 123-149.

Fischer, R. A. (1975). Yield potential in a dwarf spring wheat and the effect of shading. *Crop Science* **15**, 607-613.

Gallagher, J. N. & Biscoe, P. V. (1978). Radiation absorption, growth and yield of cereals. *Journal of Agricultural Science, Cambridge* **91**, 47-60.

Gasser, J. K. R. (1961). Transformation, leaching and uptake of fertlizer nitrogen applied in autumn and spring on a heavy soil. *Journal of the Science of Food and Agriculture* **12**, 375-380.

Hucklesby, D. P., Brown, C. M., Howell, S. E. & Hagemann, R. H. (1971). Late spring applications of nitrogen for efficient utilization and enhanced production of grain and grain protein of wheat. *Agronomy Journal* **63**, 274-276.

Kiesselbach, T. A. & Sprague, H. B. (1926). Relation of the development of the wheat spike to environmental factors. *Journal of the American Society of Agronomy* **18**, 40-60.

Laloux, R., Poelaert, J. & Falisse, A. (1975). [Stikstofbemesting bij graangewassen]. Nitrogen dressing of cereal crops. *Landbouwtijdschrift* **5**, 1155-1184.

Lawes, J. B. & Gilbert, J. H. (1880). Our climate and our wheat crops. *Journal of the Royal Agricultural Society of England* **16**, 173-210.

Locke, L. F., Rauchschwalbe, O. E. & Mathews, O. R. (1942). The relations to yield of certain plant characters of winter wheat as influenced by different tillage and sequence treatments. *Journal of the American Society of Agronomy* **34**, 628-645.

Miller, E. C. (1939). A physiological study of the winter wheat plant at different stages of development. *Kansas Experimental Station Technical Bulletin* **47**.

Neales, T. F., Anderson, M. J. & Wardlaw, I. F. (1963). The role of the leaves in the accumulation of nitrogen by wheat during ear development. *Australian Journal of Agricultural Research* **14**, 725-736.

Pearman, I., Thomas, S. M. & Thorne, G. N. (1977). Effects of nitrogen fertilizer on growth and yield of spring wheat. *Annals of Botany* **41**, 93-108.

Pearman, I., Thomas, S. M. & Thorne, G. N. (1978). Effect of nitrogen fertilizer on growth and yield of semi-dwarf and tall varieties of winter wheat. *Journal of Agricultural Science, Cambridge* **91**, 31-45.

Pearman, I., Thomas, S. M. & Thorne, G. N. (1979). Effect of nitrogen fertilizer on photosynthesis of several varieties of winter wheat. *Annals of Botany* **43**, 613-621.

Penny, A., Widdowson, F. V. & Jenkyn, J. F. (1978). Spring top-dressings of 'Nitro-Chalk' and late sprays of a liquid N-fertilizer and a broad spectrum fungicide for consecutive crops of winter wheat at Saxmundham, Suffolk. *Journal of Agricultural Science, Cambridge* **90**, 509-516.

Puckridge, D. W. & Ratkowsky, D. A. (1971). Photosynthesis of wheat under field conditions. IV. The influence of density and leaf area index on the response to radiation. *Australian Journal of Agricultural Research* **22**, 11-20.

Robins, J. S. & Domingo, C. E. (1962). Moisture and nitrogen effects on irrigated spring wheat. *Agronomy Journal* **54**, 135-138.

Sande Bakhuyzen, H. L. van de (1937). *Studies on Wheat Grown under Constant Conditions.* California: Stanford University Press.

Sinclair, T. R. & Wit, C. T. de (1975). Photosynthate and nitrogen requirements for seed production by various crops. *Science* **189**, 565-567.

Sofield, I., Evans, L. T., Cook, M. G. & Wardlaw, I. F. (1977). Factors influencing the rate and duration of grain filling in wheat. *Australian Journal of Plant Physiology* **4**, 785-797.

Spiertz, J. H. J. (1977). The influence of temperature and light intensity on grain growth in relation to the carbohydrate and nitrogen economy of the wheat plant. *Netherlands Journal of Agricultural Science* **25**, 182-197
Spiertz, J. H. J. (1978). Grain production and assimilate utilization of wheat in relation to cultivar characteristics, climatic factors and nitrogen supply. *Verslagen landbouwkundig Onderzoek* (Agricultural Research Reports) **881**, 1-35.
Spiertz, J. H. J. (1979). Weather and nitrogen effects on rate and duration of grain growth and on grain yield of wheat cultivars. In *Crop Physiology and Cereal Breeding*, pp. 60-64. Eds. J. H. J. Spiertz & Th. Kramer. Wageningen: Pudoc.
Spiertz, J. H. J. & Ellen, J. (1978). Effects of nitrogen on crop development and grain growth of winter wheat in relation to assimilation and utilization of assimilates and nutrients. *Netherlands Journal of Agricultural Science* **26**, 210-231.
Spiertz, J. H. J. & Haar, H. van de (1978). Differences in grain growth, crop photosynthesis and distribution of assimilates between a semi-dwarf and a standard cultivar of winter wheat. *Netherlands Journal of Agricultural Science* **26**, 233-249.
Stanford, G. & Hunter, A. S. (1973). Nitrogen requirements of winter wheat (*Triticum aestivum* L.) varieties 'Blueboy' and 'Redcoat'. *Agronomy Journal* **65**, 442-447.
Terman, G. L. (1979). Yields and protein content of wheat grain as affected by cultivar, N, and environmental growth factors. *Agronomy Journal* **71**, 437-440.
Thorne, G. N., Thomas, S. M. & Pearman, I. (1979). Effects of nitrogen nutrition on physiological factors that control the yield of carbohydrate in the grain. In *Crop Physiology and Cereal Breeding*, pp. 90-95. Eds. J. H. J. Spiertz & Th. Kramer. Wageningen: Pudoc.
Vlassak, K. (1978). Biologische stikstoffixatie en stikstofmineralisatie. In *Stikstof en plantenproduktie* **5**, 1-16, Genootschap Plantenproduktie en Ekosfeer, Leuven.
Vos, J. (1980). Effects of temperature and nitrogen on carbon-exchange rates and on grain growth of wheat. *Ph. D thesis* (in preparation).
Vos, N. M. de (1975). Field photosynthesis of winter wheat during the grain-filling phase under highly fertile conditions. *Proceedings of the Second International Winter Wheat Conference, Zagreb, Yugoslavia*, pp. 251-255.
Wit, C. T. de, Laar, H. H. van & Keulen, H. van (1979). Physiological potential of crop production. In *Plant Breeding Perspectives*, pp. 47-82. Eds. J. Sneep & A. J. T. Hendriksen. Wageningen: Pudoc.

Recent progress in the nutrition of vegetable crops

D. J. Greenwood and G. F. Collier
National Vegetable Research Station, Wellesbourne, Warwick, UK

Introduction

Many important developments in plant nutrition have taken place over the past ten years. We review three, of very different types, that concern vegetable production. One is a 'short-cut' approach to forecasting the optimum levels of fertilizer for different vegetable crops. Another is the improvement in yield that can be achieved by modifying the spatial distributions of nutrients and roots within the soil profile. The third emphasizes the importance of calcium in the maintenance of good crop quality.

Forecasting the effective of level of fertilizer on the yield of vegetable crops

Problem

The development of a basis for forecasting the fertilizer requirements of vegetable crops is made difficult by the diversity of crops. More than twenty of them are grown in the UK on widely different soils. It is, therefore, quite impracticable to develop fertilizer advice in the usual way by carrying out numerous trials on each crop on all the soils where it is grown. Somer alternative 'short-cut' approach was clearly essential.

Approach

One possibility was suggested by studying the results from nationwide fertilizer trials that had been carried out with agricultural crops. This indicated that if one crop was more responsive than another to an application of a given nutrient on one site then it was generally more responsive on other sites. For example, potatoes were more responsive to potassium fertilizer than cereals wherever the crops were grown. It seemed, therefore, that it might be possible to deduce response curves of the various vegetable crops on a wide range of soils from the results of fertilizer experiments with all the crops on one site and with only one of the crops on different sites.

Model

To develop this idea further some form of model for fertilizer response was needed. We were attracted to a modification of one first introduced more than 50 years ago by Balmukand (1928), but since largely neglected. It is:

$$\frac{1}{Y} = \frac{1}{1-(N_S + N_F)/\alpha_N} \times \left[\frac{1}{A} + \frac{1}{B_N\,(N_S + N_F)} + \frac{1}{B_P\,(P_S + P_F)} + \frac{1}{B_K\,(K_S + K_F)}\right], \tag{1}$$

where Y is yield, A is the maximum yield, α_N is a term for the harmful effects of high nitrogen fertilizer levels on yield, N_F is the amount of nitrogen applied as fertilizer, N_S is the amount of nitrogen that is made available to the crop in unfertilized soil, B_N is the increase in yield per unit of available nitrogen at the origin, B_P, P_S, P_F, B_K, K_S, and K_F have meanings that parallel those of the corresponding symbols for nitrogen but refer to phosphorus (P) and potassium fertilizers (K).

The equation can be derived from existing principles for important soil/plant processes (Greenwood, *et al.*, 1971) and takes account of the interactions between the effects of nutrients on crop growth in a way that is consistent with most experimental evidence (Wood *et al.*, 1972). Moreover, the equation fitted the yield data from multi-level N, P and K fertilizer experiments very satisfactorily, and the values of the parameters obtained by fitting the equation to the yield data were of the same order of magnitude as those derived from chemical analyses of soils and plants (Greenwood, Wood, Cleaver and Hunt, 1971). For these reasons the equation was used in further work.

When the levels of N and P fertilizer are high, eq. (1) simplifies to:

$$\frac{A^1}{Y} = 1 + \frac{A^1}{B_K\,(K_S + K_F)}, \tag{2}$$

where A^1 is the maximum yield that can be obtained with any level of K fertilizer. This equation gives equally good fits to responses with different combinations of values of K_S and A^1/B_K. In fact, unless responses are enormous, the relation between A^1/Y and K_F is almost entirely defined by the single composite parameter $\dfrac{A^1}{B_K K_S}$.

It has also been deduced that for any two crops:

$$\left(\frac{A^1}{B_K K_S}\right)_{\text{crop 1}} \div \left(\frac{A^1}{B_K K_S}\right)_{\text{crop 2}} \triangleq \text{Constant.} \qquad (3)$$

on many soils (Greenwood, Cleaver and Turner, 1974).

Procedure

These considerations suggested that it should be possible to predict the response curves for the 20 different crops on a range of soils in the following way:

(1) Measure the response curves of yield against level of K fertilizer for the 20 different vegetable crops under comparable conditions. Fit eq. (2) separately to the data from each experiment and thereby obtain a value for $\frac{A^1}{B_K K_S}$ for each crop.

(2) Measure the response curves of one of the crops to K fertilizer on different sites. Fit eq. (2) to the data from each experiment and thereby obtain a value of $\frac{A^1}{B_K K_S}$ for each site.

Relate $\frac{A^1}{B_K K_S}$ to the soil-exchangeable K.

(3) From $\frac{A^1}{B_K K_S}$ for each crop, from the relation between $\frac{A^1}{B_K K_S}$ for one of the crops and exchangeable K, and from eq. (3), calculate the relations between $\frac{A^1}{B_K K_S}$ for all the other crops and exchangeable K. Then by substitution in eq. (2) calculate the response curves for each crop for soils with different values of exchangeable K.

Experiments

The main experimental problem was finding some way of measuring the response curves of the many different crops when grown under comparable conditions. Our approach was to maintain a field at low nutrient status and then carry out experiments, over a number of years, on different sites within it. Special compact designs of multi-treatment experiment were used in which there were three strips of plots (Fig. 1). In one strip the level of K fertilizer increased progressively from one end to the other; all plots, however, received the same high levels of N and P fertilizer. There were also strips of plots in which the levels of N and P fertilizer varied. No guard rows were present between adjacent plots but there were guard rows round each strip of plots. There were three replicates and the order of the strips and the direction of increase of fertilizer were chosen at random. Plot size was small, less than $0{\cdot}5\ m^2$,

Fig. 1 A block of plots laid out systematically. Unharvested guard area (shaded). In one strip of plots the levels of N fertilizer increased in proportion to the number on the plot, but all plots in the strip received the same heavy application of P and K fertilizer. Likewise there was a strip of plots where the P fertilizer varied with a heavy application of N and K and a strip of plots where the K fertilizer varied with a heavy application of N and P.

and the whole experimental area was generally less than 150 m^2. The area occupied by each experiment was, therefore very small indeed.

On average, three separate experiments were carried out per crop, and although inter-year variability in response was substantial it was insufficient to mask the considerable differences between crops. Examples of the average responses to K fertilizer are given in Fig. 2. The uptake of K, as well as other nutrients, was measured and the amount

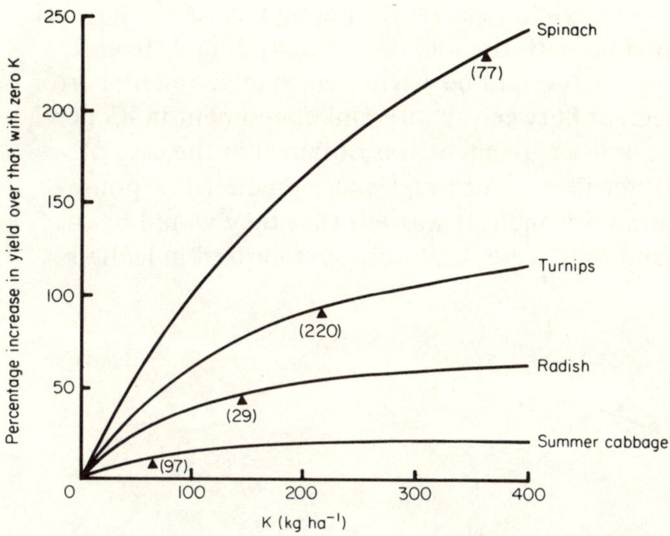

Fig. 2 Percentage increases in yield from applying K fertilizer to four crops. ▲ indicates the optimum level of K fertilizer (defined as that level where a further 10 kg ha^{-1} of K gave a 1 per cent increase in yield) and the figure beneath is the amount of K (kg ha^{-1}) taken up by the crop with that level of fertilizer.

taken up with the optimum levels of K fertilizer (defined as those where a further 10 kg ha^{-1} of K gave a 1 per cent increase in yield) are also given in Fig. 2. Paradoxically, crops that took up most K were often those that were best able to satisfy their needs for the element from unfertilized soil and required least fertilizer.

Predictions

From the response curves of the different crops grown on one site and from the results of nationwide fertilizer experiments on one of the crops (potatoes) that were kindly made available to us by the Ministry of Agriculture, Fisheries and Food, we predicted curves of percentage increase in yield against level of applied K fertilizer for all crops on soils containing widely different levels of exchangeable K. Their validity was checked against the results of experiments that were entirely independent of those used for making the predictions. They had been carried out by ourselves, by commercial firms and by the Agricultural Development and Advisory Service (ADAS) on growers' holdings throughout the country. For each experiment a forecast was made of the percentage increase in yield as a result of applying 100 kg ha^{-1} of K, and this was compared with the increase actually found. If differences of ± 10 per cent could be attributed to experimental error, then there was agreement between theory and experiment in 83 per cent of cases. Rather worse agreement was obtained in the case of responses to N and P fertilizers. Although these predicted response curves left room for improvement it was felt that they would be of interest to growers and they were, therefore summarized in leaflets

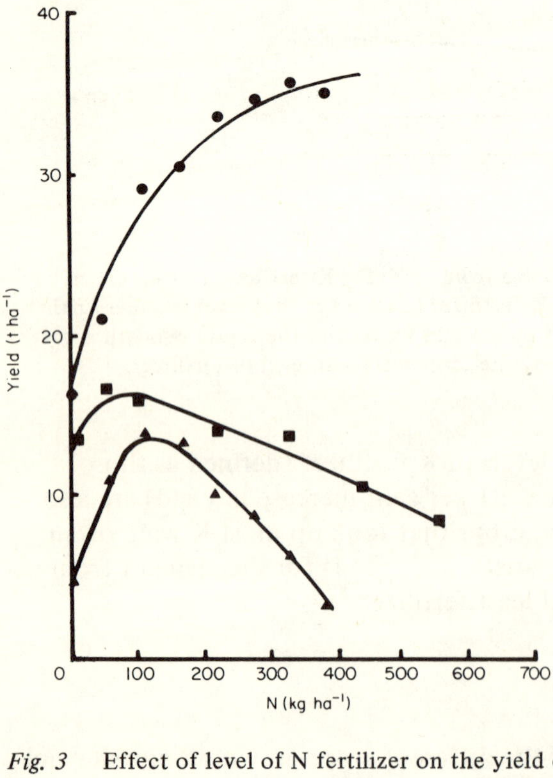

Fig. 3 Effect of level of N fertilizer on the yield of lettuce in experiments carried out in different years within the same field.

which have been much in demand. But perhaps the most important conclusion is that few vegetable crops would be expected to respond appreciably to fresh applications of K and P fertilizers on soils containing more than 200 ppm of exchangeable K or 50 ppm of bicarbonate-extractable phosphate.

Effect of weather conditions

Inevitably, there must be uncertainty in predictions based on the foregoing methods because they can only take account of some of the many factors influencing fertilizer response. They are, for example, unable to predict the influence of weather conditions on crop response. This can be a serious drawback, especially for responses to N fertilizer, as is illustrated by Fig. 3, which gives the yields of lettuce grown on adjacent sites of the same soil in different years. On some occasions, response curves were of a 'diminishing returns' type and in others an 'overturning' type. Yet the N status of the soil was always similar and the seedlings had been thinned to a constant stand shortly after emergence. Many of the differences in response were attributed to the extent to which rain and irrigation had moved the ions down the soil profile and the time when this occurred (Greenwood, Cleaver and Niendorf, 1974).

Computer simulation

Equations have been developed for calculating the extent of leaching of ions down the soil profile from standard meteorological data and easily measurable properties of the soil. They have given very reliable estimates of leaching in unstructured soils (Burns, 1977). Many other soil/plant processes have also been defined by comparatively simple equations. These different types of equation have been combined into dynamic models that attempt to calculate the daily increments in growth and in nutrient uptake of crops from daily values of rainfall and evapotranspiration and from values of parameters characterizing the soil and crop (Barnes, Greenwood and Cleaver, 1976).

The degree of agreement between yields predicted by the models and those obtained in experiments is illustrated by the data in Fig. 4. In addition, the models have so far correctly predicted the general effects of weather conditions on response to fertilizer at different stages of growth (Barnes *et al.*, 1976).

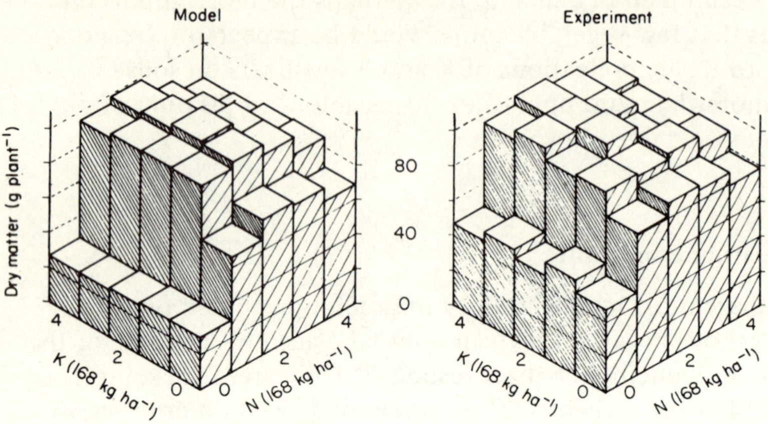

Fig. 4 Effect of five different levels of N fertilizer (each increment was 168 kg ha^{-1}) in all combinations with five different levels of K fertilizer (each increment was 168 kg ha^{-1}) on the yield of cabbage, found in an experiment and calculated by means of a model (see text).

Although this model-building approach is in its infancy, it shows much promise of providing a simple, effective way of forecasting how to adjust nitrogen fertilizer practice for differences in past weather conditions. It may be that mechanistic models will play an increasing role in interpreting experimental data and extending it to different situations.

Effect of modifying the spatial distributions of nutrients and roots in soil on crop growth

Yields vary greatly from soil to soil even when heavy applications of fertilizer are made. One of the causes of this phenomenon is that many of the nutrients may be rendered unavailable to the crop by the way they are distributed in soil.

Seedling growth

The point can be well illustrated by considering the responses of seedlings to applications of potassium fertilizer. Seeds seldom contain more than 0·7 per cent K in their dry matter (Haworth and Cleaver, 1964; Freeman, 1967), but immediately after emergence the young seedling must contain at least 10 times that amount of K (i.e. 7 per cent if photosynthesis is to be maximal (Haworth and Cleaver, 1964). In practice, the plant often cannot extract potassium quickly from soil so there is a lag between the time of emergence and the time when the required concentration of potassium in the seedling is reached. The delay can be serious even when large quantities of K fertilizer have been incorporated, immediately before drilling, in a soil of low K status (Greenwood, 1978). Potassium only moves slowly from the fertilizer granules into the bulk of soil. Thus, when K fertilizer is applied as KCl in granules it may increase potassium concentrations in such a small volume of soil that the seedling may be unable to derive any benefit for a considerable time. Assuming that potassium moves in soil by diffusion, then applying 100 kg ha^{-1} of K as 2 mm radius granules in the top 25 cm of soil would increase the K concentration of only 1 per cent of wet soil within a week of incorporation. If the soil was dry the proportion of soil enriched would be even less. There could, therefore, be a substantial delay before the root of a young seedling intersected a region of potassium-enriched soil. These processes could explain why once the soil potassium status has been allowed to fall to a deficiency level yields are often low in the following year, irrespective of how much fertilizer is applied immediately before drilling (Cooke, 1967). Likewise, it may be possible to explain why residual soil phosphate (phosphate remaining from past applications) is often regarded as being of great importance in maintaining high yields (Cooke, 1967).

Subsoil improvement

Another aspect of spatial distribution that may often have a decisive influence on yields is the physical and chemical conditions of the subsoil. Although it has long been recognized that many crops obtain most of their water from the subsoil, there has been little recent interest in improving it, at least in the UK. This may have occurred partly because of the results from the numerous experiments carried out with stationary tine subsoilers between 1945 and 1965 which showed that the benefits from subsoiling were generally small. Thus, in one set of 47 experiments on potatoes, the average increase in yield was only 2 per cent and there were only three experiments where the increases exceeded 15 per cent (Russell, 1957). Similar patterns of results were obtained by other workers on other crops with worthwhile improvements being confined to soils having distinctive hard pans (Swain, Thomas and Little, 1969).

The lack of response on soils which had no distinctive pans but which were generally compact below the plough layer was surprising. The possibility existed that the tines did not effectively loosen the soil, as the degree of loosening in the experiments was seldom measured. Supporting evidence for this possibility was provided by Stone (1978) who found that *thorough* loosening of the soil to a depth of 1 m almost doubled yields of broad beans on a soil where there was no pan and where roots penetrated the unloosened soil to at least 70 cm. McEwen and Johnston (1979) found that double digging by hand increased yields of cereals by an average 23 per cent on a site where yields had been almost unaffected by fixed-tine subsoiling.

Nutrients in the subsoil

For generations almost all chemical analyses of soils have been on samples taken from the top 20 cm. Recent work, however, has emphasized the importance of the subsoil. Jungk and Wehrmann (1978) found that the inorganic nitrogen in the top metre, and not just the top 20 cm, was available to many crops. They assessed the value of determining the inorganic nitrogen in soil (in spring) as a guide to the nitrogen fertilizer requirements of cereals. The total inorganic N in the top 25 cm was a poor guide, whereas the total amount in the top metre was a good one. Other workers found that deep incorporation of K, P and lime could result in higher yields than could be obtained by incorporating the same amounts of nutrients in the surface layers of soil (Kohnke and Bertrand, 1956; Rowse, Stone and Goodman, 1978; McEwen and Johnston, 1979). In some cases, there was no benefit from thorough loosening unless this was associated with the deep incorporation of nutrients (Engelbert & Truog, 1956).

The duration of the benefits from subsoil improvement have been studied to only a small extent. McEwen and Johnston (1979) and Stone (personal communication) found that the benefits from thorough loosening of a sandy loam soil persisted for at least four years and the former workers also showed that the benefits from deep incorporation of nutrients could last for a similar period. It may well be, therefore, that improvements in the subsoil need to be made only once every 5 or 10 years to get the best results.

Machines for subsoil improvement

Progress has been made in the development of machines to bring about effective loosening and nutrient enrichment of the subsoil. The degree of soil loosening by pulling tines through the subsoil has been much increased by causing them to vibrate (Schulte-Karring, 1975) and by modifying their shape (Spoor, 1975). Even after modifications, however, effective loosening occurs only if the soil is reasonably dry, which is seldom the case in many parts of the country. One way of overcoming the difficulty is to rotavate the subsoil and a machine has been designed to do this without altering the relative positions of sub- and surface soils (Warboys, Wilkes, Gooderham and Wilkins, 1976). Use of this machine resulted in an average increase in yield of 21 per cent for 4 vegetable crops. The machine also enabled fertilizer to be incorporated at depth and this resulted in further increases in yield of up to 10 per cent (Rowse *et al.*, 1978).

Future work

It appears, therefore, that practical methods exist for improving the physical and chemical conditions of the subsoil which can give substantial improvements in yield. The benfits, however, vary greatly from soil to soil and crop to crop, and some means is need of predicting their effectiveness in different situations.

Crop quality and calcium deficiency

Quality is becoming increasingly important in vegetable production. This arises partly from the development of pre-packed vegetables sold in supermarkets and partly because consumers are becoming increasingly exacting in their demands. Crops that in the past would have been accepted by the retail trade are now rejected because of blemishes or disorders such as blossom-end rot of tomatoes, tipburn of lettuce, blackheart of celery and internal rust spot of potatoes (Collier, Wurr and Huntington, 1978). These, and at least 17 other disorders, are caused by a localized deficiency of calcium in the affected tissues (Shear, 1975). The disorders occur in crops grown on widely different soils, including those that contain large quantities of calcium (Marschner, 1974).

Causes of calcium deficiency

Localized deficiencies of calcium may occur in plants grown on calcium-rich soils because, unlike most elements, calcium is not transported in the phloem. The phloem has a high pH and contains much phosphate, so that any calcium entering it would be immediately precipitated as calcium phosphate (van Goor and Wiersma, 1974). Little calcium, however, enters the phloem because plants have evolved a mechanism which prevents calcium moving across membranes separating the xylem from the phloem (Marschner, 1974). The net result is that, unlike other major nutrients, calcium cannot be redistributed within the plant from an area of sufficiency to one of deficiency even if this is the growing point. Calcium must, therefore, be absorbed and transported in at least trace amounts (Wallace, Frolich and Lunt 1966) in the xylem to all plant parts at all times if membrane integrity is to be preserved and necrosis prevented (Simon, 1978).

Thus, although calcium is more abundant in the soil than most other elements, the plant's demand for it is usually more exacting.

Uptake of calcium by plant roots

The uptake of calcium ions by plant roots is confined to a small segment of unsuberized root near its tip. This is in contrast to other nutrients which are absorbed over a much greater length of root (Drew and Clarkson, 1978). Consequently, any process such as mechanical impedance or poor soil aeration which restricts root growth and reduces the proportion of unsuberized root might be expected to suppress the uptake of calcium more than that of other ions (Scaife and Clarkson, 1978).

The rate of calcium uptake increases with an increase in calcium concentration until this reaches a higher value than is the case with the major nutrients (Bar-Yosef, 1971; Loneragan and Snowball, 1969). In soil solution the calcium concentration increases as the concentration of anions rises, because calcium is invariably the predominant cation and there must always be equal concentrations of positively and negatively charged ions. Thus, an increase in, say, nitrate or sulphate concentration will raise the calcium concentration and tend to enhance uptake by the roots. At the same time, however, the rate of uptake of calcium is also affected by the presence of other cations, with hydrogen, ammonium, potassium and magnesium being particularly effective in impeding calcium uptake (Geraldson, 1967; Bedi and Sekhon, 1977). It may be, therefore, that regions of soil of, say, low pH within the rooting zone could severely restrict uptake by part of the root system and thus the supplies to some of the plant tissues.

Translocation of calcium within the plant

Calcium moves into the roots and through the xylem in association with inorganic and organic anions. For plants which rely on ammonium and not nitrate ions as their source of nitrogen, the uptake of inorganic cations is greater than that of inorganic anions. Electroneutrality is maintained by the synthesis of organic anions within the roots. This process can be so slow that both the uptake of all cations, including calcium, and their subsequent translocation to the leaves is severely impeded (Dijkshoorn, Lathwell and De Wit, 1968). Most vegetable crops, however, obtain almost all of their nitrogen as nitrate so that calcium uptake is seldom likely to be restricted because of slow rates of organic-acid synthesis within the roots.

Once the calcium is absorbed by the plant roots, its subsequent movement to the various plant parts is much affected by the transpiration stream. Calcium moves to those parts of the plant that are losing water most rapidly (Wiersum, 1966). For a lettuce or cabbage plant this means that the outer leaves, which transpire freely, contain ample quantities of calcium but the growing point, which is enclosed and, therefore, loses water very slowly, is often deficient in calcium. Leaf morphology might, therefore, be expected to have a decisive influence on whether all tissues get an adequate supply of calcium or not.

The supply of calcium to low-transpiring tissues is enhanced greatly by root pressure flow (Palzkill, Tibbitts and Williams, 1976; Bradfield and Guttridge, 1979). This phenomenon is most likely to occur at night when evapotranspiration is low. Essentially, a water stress within the plant greater than that in the soil results in the absorption of water from the soil, which is then forced throughout the entire plant, simultaneously transporting calcium with it. Clearly, the effects are greatest when the soil is wet and least when it is subjected to severe moisture stress. This could explain why calcium disorders have been induced by high salt concentrations in the root zone (Robbins, 1937; Bradfield and Guttridge, 1979).

Prevention of deficiency of calcium

The supply of calcium to plant tissues is thus affected by a number of interacting processes which are often antagonistic, and the best means of eliminating calcium deficiencies depends on circumstance.

Occasionally, the susceptible tissues are exposed (e.g. celery) and in these cases the disorders can be eliminated by frequent spraying with calcium salts (Cox and Dearman, 1978). However, with other crops, (e.g. lettuce) calcium sprays are of no value because the susceptible tissue is protected by the outer wrapper leaves (Thibodeau and Minotti, 1969). Alternative approaches are, therefore, needed.

One possibility is breeding. For example, in lettuce, there are undoubtedly big genetic differences in the susceptibility of cultivars to tipburn but a cultivar that is tipburn-resistant in one environment is often susceptible in another (Cox and McKee, 1976). Greater understanding of the causes of localized deficiencies of calcium has suggested ways of selecting plants that should possess resistance under a wider range of conditions (Collier, Huntington and Cox, 1979) and it is to be hoped that there will be further progress along these lines.

Finally, it seems likely that the incidence of the disorders could be alleviated by agronomic practices such as modifying the composition of the soil cation-exchange complex, improving soil physical conditions and altering plant water regimes.

References

Balmukand, Bh. (1928). Studies in crop variation. V. The relation between yield and soil nutrients. *Journal of Agricultural Science, Cambridge* **18**, 602-627.

Barnes, A., Greenwood, D. J. & Cleaver, T. J. (1976). A dynamic model for the effects of potassium and nitrogen fertilizers on the growth and nutrient uptake of crops. *Journal of Agricultural Science, Cambridge* **86**, 225-244.

Bar-Yosef, B. (1971). Fluxes of P and Ca into intact corn roots and their dependence on solution concentration and root age. *Plant and Soil* **35**, 589-600.

Bedi, A. S. & Sekhon, G. S. (1977). Effect of potassium and magnesium application to soils on the dry-matter yield and cation composition of maize. *Journal of Agricultural Science, Cambridge* **88**, 753-758.

Bradfield, E. G. & Guttridge, C. G. (1979). The dependence of calcium transport and leaf tipburn in strawberry on the relative humidity and nutrient solution concentration. *Annals of Botany* **43**, 363-372.

Burns, I. G. (1977). Nitrate movement in soil and its agricultural significance. *Outlook on Agriculture* **9**, 144-148.

Collier, G. F., Huntington, Valerie C. & Cox, E. F. (1979). A possible role for chlorogenic acid in calcium-related disorders of vegetable crops with particular reference to lettuce tipburn. *Communications in Soil Science and Plant Analysis* **10**, 481-490.

Collier, G. F., Wurr, D. C. E. & Huntington, Valerie C. (1978). The effect of calcium nutrition on the incidence of internal rust spot in the potato. *Journal of Agricultural Science, Cambridge* **91**, 241-243.

Cooke, G. W. (1967). *The Control of Soil Fertility*, p. 258, London: Crosby Lockwood.

Cox, E. F. & Dearman, A. S. (1978). The control of blackheart of celery with calcium sprays. *Experimental Horticulture* **30**, 1-6.

Cox, E. F. & McKee, J. M. T. (1976). A comparison of tipburn susceptibility in lettuce under field and glasshouse conditions. *Journal of Horticultural Science* **51**, 117-122.

Dijkshoorn, W., Lathwell, D. J. & de Wit, C. T. (1968). Temporal changes in carboxylate content of ryegrass with stepwise change in nutrition. *Plant and Soil* **29**, 369-390.

Drew, M. C. & Clarkson, D. T. (1978). The growth and function of the root system in relation to maximum yield. In *Maximising Yields of Crops*, pp. 50-56. Eds. J. K. R. Gasser & B. Wilkinson. London: HMSO.

Engelbert, L. E. & Truog, E. (1956). Crop response to deep tillage with lime and fertilizer. *Soil Science Society of America Proceedings* **20**, 50-54.

Freeman, G. G. (1967). Studies on potassium nutrition of plants. I. Effects of potassium concentration on growth and mineral composition of vegetable seedlings in sand culture. *Journal of the Science of Food and Agriculture* **18**, 171-176.

Geraldson, C. M. (1967). Evaluation of the nutrient intensity and balance system of soil testing. *Proceedings of the Soil and Crop Science Society of Florida* **27**, 59-67.

Goor, B. J. van & Wiersma, D. (1974). Redistribution of potassium, calcium, magnesium and manganese in the plant. *Physiologia Plantarum* **31**, 163-168.

Greenwood, D. J. (1978). Factors influencing dry matter production of vegetable crops. In *Maximising Yields of Crops,* pp. 92-99. Eds. J. K. R. Gasser & B. Wilkinson. London: HMSO.

Greenwood, D. J., Cleaver, T. J. & Niendorf, K. B. (1974). Effects of weather conditions on the response of lettuce to applied fertilizers. *Journal of Agricultural Science, Cambridge* **82**, 217-232.

Greenwood, D. J., Cleaver, T. J. & Turner, M. K. (1974). Fertilizer requirements of vegetable crops. *Proceedings of the Fertilizer Society* No. 145, pp. 1-30.

Greenwood, D. J., Wood, J. T., Cleaver, T. J. & Hunt, J. (1971). A theory for fertilizer response. *Journal of Agricultural Science, Cambridge* **77**, 511-523.

Haworth, F. & Cleaver, T. J. (1964). Growth and mineral composition of vegetable seedlings. *Journal of Horticultural Science* **39**, 34-41.

Jungk, A & Wehrmann, J. (1978). Determination of nitrogen fertilizer requirements by plant and soil analysis. *Plant Nutrition. 8th International Colloquium on Plant Analysis and Fertilizer Problems. New Zealand,* Vol. 1, pp. 209-223. Eds. A. R. Ferguson, R. L. Bieleski & I. B. Ferguson. Auckland, New Zealand: Department of Scientific and Industrial Research.

Kohnke, H. & Bertrand, A. R. (1956). Fertilizing the subsoil for better water utilization. *Soil Science Society of America Proceedings* **20** , 581-586.

Loneragan, J. F. & Snowball, K. (1969). Rate of calcium absorption by plant roots and its relation to growth. *Australian Journal of Agricultural Research* **20**, 479-490.

McEwen, J. & Johnston, A. E. (1979). The effects of subsoiling and deep incorporation of P and K fertilizers on the yield and nutrient uptake of barley, potatoes, wheat and sugar beet grown in rotation. *Journal of Agricultural Science, Cambridge* **92**, 695-702.

Marschner, H. (1974). Calcium nutrition of higher plants. *Netherlands Journal of Agricultural Science* **22**, 275-282.

Palzkill, D. A., Tibbitts, T. W. & Williams, P. H. (1976). Enhancement of calcium transport to inner leaves of cabbage for prevention of tipburn. *Journal of the American Society for Horticultural Science* **101**, 645-648.

Robbins, W. R. (1937). Relation of nutrient salt concentration to growth of the tomato and to the incidence of blossom-end rot of the fruit. *Plant Physiology, Lancaster* **12**, 21-50.

Rowse, H. R., Stone, D. A. & Goodman, D. (1978). The Wye Double Digger. *Report of the National Vegetable Research Station for 1977*, pp. 45-46.

Russell, E. W. (1957). The effects of very deep ploughing and of subsoiling on crop yields. *Journal of Agricultural Science, Cambridge* **48**, 129-144.

Scaife, M. A. & Clarkson, D. T. (1978). Calcium-related disorders in plants – a possible explanation for the effect of weather. *Plant and Soil* **50**, 723-725.

Schulte-Karring, H. (1975). A new soil amelioration technique. *Bulletin No. 4. National Committee of the German Federal Republic on Irrigation and Drainage (ICID).*

Shear, C. B. (1975). Calcium-related disorders of fruits and vegetables. *Hortscience* **10**, 361-365.

Simon, E. W. (1978). The symptoms of calcium deficiency in plants. *New Phytologist* **80**, 1-15.

Spoor, G. (1975). Shatter matters. *Power Farming* **54**, 14-15.

Stone, D. A. (1978). Subsoiling and deep incorporation of nutrients. *Report of the National Vegetable Research Station for 1977,* pp. 43-45.

Swain, R. W., Thomas, R. & Little, R. C. (1969). Subsoiling. *Agriculture* **76**, 291-294.

Thibodeau, P. O. & Minotti, P. L. (1969). The influence of calcium on the development of lettuce tipburn. *Journal of the American Society for Horticultural Science* **94**, 372-376.

Wallace, A., Frolich, E. & Lunt, O. R. (1966). Calcium requirements of higher plants. *Nature, London* **209**, 634.

Warboys, I. B., Wilkes, J. M., Gooderham, P. T. & Wilkins, S. M. (1976). The development of a double digging machine. *The 7th Conference of the International Soil Tillage Research Organisation, Sweden* **46**, 1-46.

Wiersum, L. K. (1966). Calcium content of fruit and storage tissues in relation to the mode of water supply. *Acta Botanica Neerlandica* **15**, 406-418.

Wood, J. T., Greenwood, D. J. & Cleaver, T. J. (1972). Interactions between the beneficial effects of nitrogen phosphate and potassium on plant growth. *Journal of Agricultural Science, Cambridge* **78**, 389-391.

Analysing and predicting the response of the glasshouse crop to environmental manipulation

B. Acock
USDA-SEA-AR Crop Simulation Research Unit, PO Box 5367, Mississippi State, MS 39762 USA

Introduction

Plant physiologists rarely find that they can claim responsibility for innovations leading to increased crop yields. Usually, the farmers, agronomists, or horticulturists try out a new idea and, if it results in a yield improvement, the physiologists later discover the reason for the response. The one notable exception that comes to mind is the discovery of the photoperiodic control of flowering by Garner and Allard (1920). The physiologists of the early 1900s even missed the opportunity of claiming responsibility for introducing CO_2 enrichment because the CO_2 they used was mixed with other phytotoxic products of combustion (Wittwer and Robb, 1964). However, as with the eventual successful application of CO_2 enrichment to crops, there are many instances where the physiologists' insights into plant functions have inspired innovations in other disciplines.

Perhaps the physiologists' role will always be to support or service other disciplines with information about plant function. However, the opportunity now exists for physiologists to make their contribution obvious and explicit because there is a need, and even a demand, from other disciplines for the knowledge acquired by careful analysis of plant behaviour to be resynthesized into dynamic simulation models of crop growth.

The need for crop models

The need for crop models has arisen because agricultural scientists have made as much progress as is possible by considering only the main effects of environmental and nutritional factors. Future advances must come from a consideration of interactions. In other words, we have now reached the limit of information that can be gained by manipulating one factor and failing to control others or even record their levels in an experiment. Of course, multifactorial experiments with nutritional factors are cheap and have been done for many years, but facilities for controlling aerial factors are expensive and are rarely, if ever, provided in sufficient numbers for more than one aerial factor to be examined at a time (e.g. Calvert, 1972). The one viable alternative in these days of cheap computing power is to try to simulate, from what we know already, the crop's response to new combinations of factors.

The demand for such dynamic simulation models is more evident in the USA than it is in the UK. It comes from plant breeders who want to know what specific physiological characteristics to incorporate in future generations of crop plants so that yields can be improved methodically. The demand comes from agricultural engineers who want to simulate the effect of various tillage practices on crop growth so that they can concentrate on implement design rather than field testing. Entomologists want crop models to use with their own dynamic simulation models of pest and predator populations, since it is essential to understand the subtle interplay of crop, pest and predator if we are ever to make effective use of biological control and integrated pest management. Educators want models as teaching aids. Government wants models to continuously update predictions of crop yields in order to know what the surpluses or deficits are likely to be in its own and other countries. The demand from other agencies is so great that the US Department of Agriculture is increasing its support for crop modelling at a time when its overall research budget is decreasing (Strommen, 1979).

Missing the opportunity

Unfortunately, many plant physiologists do not have a favourable reaction to crop modelling. There are several possible reasons for this.

Modelling has been presented as a new activity which is superior to experimentation. The truth, of course, is that experimentalists usually have a conceptual model of the system they are studying. Simulation models are merely conceptual models that have the relationship between the variables defined explicitly and precisely in one or more mathematical equations. In the sense that they are immediately testable and refutable, simulation models are more scientific than conceptual models, but they are not fundamentally different. Crop models depend on the availability of good experimental data for their development and testing, and modelling reveals gaps in our knowledge that can only be filled by experimentation.

Extravagant claims have been made for crop models. Although there are many models extant (Duncan, 1975; Fick, Williams and Loomis, 1973; McKinion, Baker, Hesketh and Jones, 1975; Curry, Baker and Streeter, 1975; Holt, Bula, Miles, Scheiber and Peart, 1975, and Baker, Lambert, Phene and McKinion, 1976), these have all been developed and tested on limited data bases and all have known problems. Despite this, some of the models are already giving useful predictions (Jenkins, Baker, McKinion and Parrott, 1975; Gutierrez, Falcon, Loew, Leipzig and Vandenbosh, 1975, and Landivar, Baker and Jenkins, 1979), and we can expect their utility to increase in step with their precision.

Plant physiologists have hitherto worked on the analysis of plant response to environment and have emphasized its complexity. Models ignore much of the complexity known to exist, and to some physiologists their simplicity is offensive. However, plants are able to compensate for many environmental changes, and it may be sufficient to model the long-term response rather than the compensating mechanism. Only time and experience will eventually tell us what level of detail is necessary for a particular purpose in a model.

Because of their antipathy towards modelling, plant physiologists are in danger of losing their birthright. The demand for models is being met mainly by mathematicians, physicists, and engineers. They are qualified by their numeracy and familiarity with computers, but not by their knowledge of plant behaviour. Instead of criticizing their efforts, we physiologists should be working with them or improving our own computing ability so that we can build better models. (For further discussion see Loomis, Rabbinge and Ng, 1979).

The special relevence of modelling to glasshouse crops

The preceding remarks apply to plant physiologists generally, but they are especially pertinent to those who work with glasshouse crops. In glasshouses we have partial control of the environment, and any investment in modelling should give us a quick return in commercial application of the results. This potential was first seen by Dr Warren Wilson in 1966 when he went to head the Plant Physiology Department at the Glasshouse Crops Research Institute and initiated a crop modelling programme. Although he and I have since left GCRI, most of the team and the equipment he collected is still concerned with crop modelling. In what follows, I describe some of the work of that team under headings corresponding with the three major environmental factors that may be varied in glasshouses.

Crop response to environmental factors

Temperature

The rate of leaf net photosynthesis, P_n, versus leaf temperature has been measured on many species (e.g. El-Sharkawy, Loomis and Williams, 1968, and Hofstra and Hesketh, 1969). In every species there is an optimum P_n somewhere between 25 and 40°C and lower values of P_n at more extreme temperatures. It has been assumed by Gates (1965) and subsequently by Idso (1968) that the amplitude of the curve varies with light-flux density, but that the shape is unaltered. This suggests that canopy net photosynthesis will have a temperature response similar to that of single leaves. The data of El-Sharkawy *et al.* (1968) for *Amaranthus edulis* support Gates' assumption except at the lowest light-flux density of the range tested, so the assumption may be valid for C_4 species. However, Ludwig and Withers (1978, and personal communication) have shown that for tomato, a C_3 plant, the temperature optimum for P_n decreases as light-flux density decreases. Also, the curves for different light flux densities come together at low temperatures but are widely separated at 40°C. In order to examine the relationship between leaf and crop responses for C_3 plants, this behaviour can be simulated as follows:

The rectangular hyperbola can be used to calculate the rate of leaf gross photosynthesis in the absence of any temperature limitation, P_g' (for example see Acock, Thornley and Warren Wilson, 1971):

$$P_g' = \alpha I \tau C / \alpha I + \tau C,$$

where α = leaf photochemical efficiency, I = light-flux density incident on the leaf, τ = CO_2 conductance, and C = ambient CO_2 concentration. Next it may be assumed that the maximum possible rate of leaf gross photosynthesis limited only by temperature, P_g'', is given by

$$P_g'' = fT,$$

where $f = 0{\cdot}07$ mg CO_2 m^{-2} s^{-1} $°C^{-1}$, and T = temperature in °C. The form of the relationship and the value of the constant f have been chosen to give the almost linear increase in net photosynthesis up to a value of 2·4 mg CO_2 m^{-2} s^{-1} at 35°C that was reported by Ludwig and Withers (1978) for leaves in 400 W m^{-2} PAR and 1200 vpm CO_2. Then the actural rate of leaf gross photosynthesis, P_g, equals the lesser of the two values P_g' and P_g''. Finally, Charles-Edwards' (1978) model for the ratio of the rates of photorespiration, R_1, and gross photosynthesis may be used to calculate net photosynthesis, P_n,

$$R_1/P_g = (O/C)\,(0{\cdot}98 \times 10^{-4}\ \exp(0{\cdot}045T))$$

where O = ambient O_2 concentration;
therefore,

$$P_n = P_g - R_l = P_g\,(1 - (O/C)\,(0{\cdot}98 \times 10^{-4}\ \exp\,(0{\cdot}045T)))$$

Simulations of the response of leaf net photosynthesis to temperature are shown in Fig. 1. They are for leaves on plants raised in summer and the parameter values used where $\alpha = 12{\cdot}1 \times 10^{-3}$ mg CO_2 J^{-1} and $\tau = 2{\cdot}54 \times 10^{-3}$ m s^{-1}. The model accounts for 98 per cent of the variation in the available experimental data (Ludwig and Withers, 1978, and personal communicaton). It overestimates P_n at low temperatures but the error is rarely larger than 0·1 mg CO_2 m^{-2} s^{-1}. The inhibition of leaf photosynthesis at temperatures below 15°C is of no commercial significance because the growth and development of tomato plants is so slow at these temperatures. In fact, leaf net photosynthesis responds

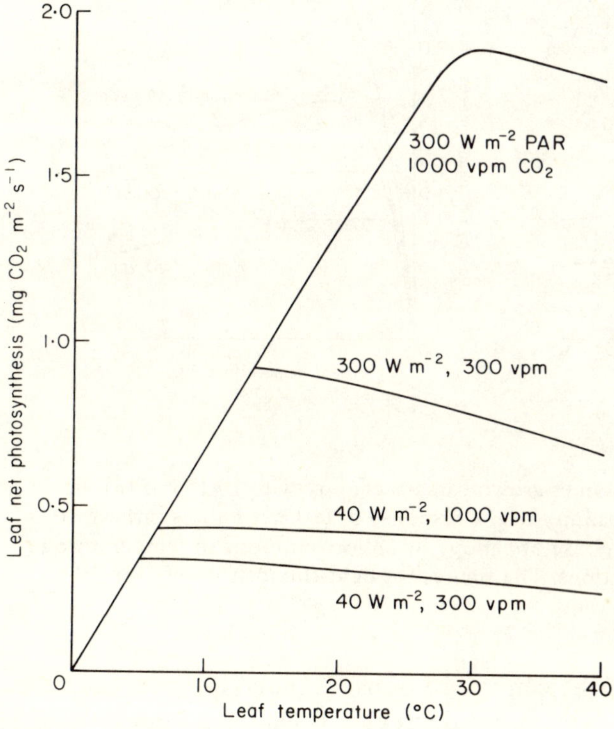

Fig. 1 A stimulation of the response of leaf net photosynthesis to temperature in various environmental conditions for leaves on tomato plants grown in summer. Based on data of Ludwig and Withers (1978, and unpublished).

appreciably to temperatures above 15°C only when the CO_2 concentration and light-flux density are both high.

When a crop canopy experiences similar conditions of high CO_2 and light, very few of the leaves are fully exposed. Most leaves in the canopy are operating at lower light-flux densities. In Fig. 2(a) leaf net photosynthesis is shown as a function of temperature for leaves at various depths in a canopy. Variation in τ with depth in the canopy has been calculated using the relationship given in Acock *et al.* (1978b). In Fig. 2(b) the contributions of the leaf layers have been summed to show the response of canopy net photosynthesis to leaf temperature. Even

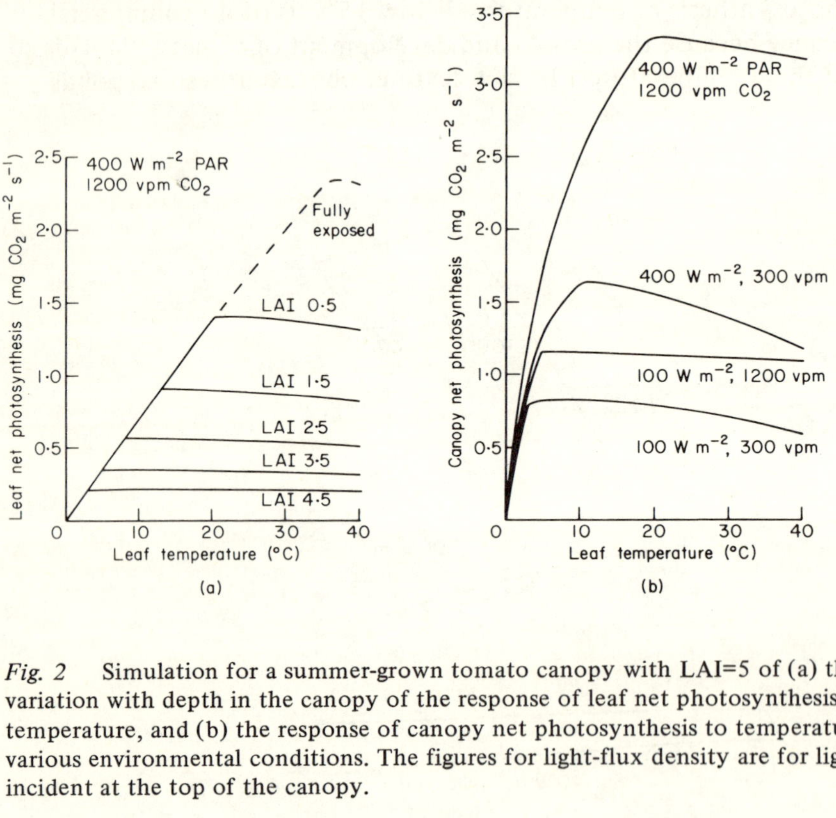

Fig. 2 Simulation for a summer-grown tomato canopy with LAI=5 of (a) the variation with depth in the canopy of the response of leaf net photosynthesis to temperature, and (b) the response of canopy net photosynthesis to temperature in various environmental conditions. The figures for light-flux density are for light incident at the top of the canopy.

with high CO_2 and light the canopy response is much less pronounced than the single leaf response, and it decreases as leaf area index increases.

Given the small size of the response of canopy photosynthesis to temperature and the errors in measuring canopy photosynthesis, it is not surprising that a response is rarely observed. The data in Fig. 3

(Hand and Acock, previously unpublished) for a thin canopy at an elevated concentration show no differences in canopy net photosynthesis at 16·5 and 27°C over the range of light-flux densities commonly encountered in the glasshouse. Baker (1965) has found a similar lack of response to temperature in cotton and these observations are supported by Warren Wilson's report (1966) that unit leaf rate (net assimilation rate) varies little with temperature.

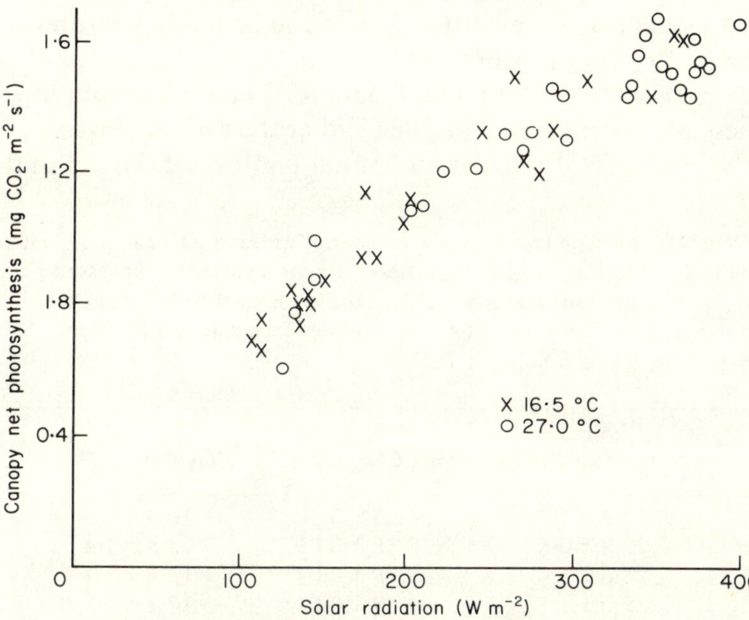

Fig. 3 The response of canopy net photosynthesis to solar radiation at two temperatures for a summer-grown tomato canopy with LAI=2·04, measured in 1200 vpm CO_2 (Hand and Acock, unpublished). In the conditions of this experiment, PAR=0·635R – 2·37 x $10^4 R^2$, where R = solar radiation (Acock *et al.*, 1978b).

Since plant growth is obviously temperature sensitive, even in low light and 300 vpm CO_2 when there is little response of leaf net photosynthesis, some other explanation of the sensitivity is needed. This explanation is not to be found in the partitioning of dry matter. The proportion of dry matter partitioned to leaves has often been observed to be nearly constant over a wide range of environmental conditions (e.g. Hughes and Cockshull, 1971; Hurd and Thornley, 1974, and Acock, Charles-Edwards and Sawyer, 1979). However, specific leaf area and hence leaf-area ratio are found to increase with increasing temperature in many spaced plants and at least one closed canopy. See results and references cited by Acock, Charles-Edwards and Sawyer (1979). Thus, plants at high temperatures, given a fixed amount of dry matter for leaves, make 'thinner' leaves of larger area than do plants at low temperatures. These 'thin' leaves are just as efficient per unit leaf area as the 'thicker', low-temperature leaves. Therefore, spaced plants will grow faster in high temperatures because they display more leaf area, intercept more light, and fix more CO_2 per unit leaf dry weight than do plants in low temperatures. However, after the canopy has closed, the extra leaf area confers little benefit and high temperatures no longer enhance dry weight gain.

These relationships are illustrated with data for Chrysanthemum in Table 1. In these data there is some evidence that the thicker leaves formed at 10°C had a slightly higher rate of net photosynthesis per unit

Table 1. Incremental specific leaf area, S_A (= change in leaf area/change in leaf dry weight), and rates of leaf gross photosynthesis at light saturation expressed per unit leaf area, P_{max} and per unit leaf mass, $P_{max} \times S_A$ for Chrysanthemum canopies grown in various temperatures (Acock, Charles-Edwards and Sawyer, 1979). Values given are means ± S.E. (18 D.F.)

Temperature (°C)	S_A ($m^2\ kg^{-1}$)	P_{max} ($mg\ CO_2\ m^{-2}\ s^{-1}$)	$P_{max} \times S_A$ ($mg\ CO_2\ kg^{-1}\ s^{-1}$)
10	15·9 ± 1·5	0·65 ± 0·10	10·3 ± 1·6
15	23·8 ± 0·6	0·58 ± 0·09	13·8 ± 1·4
20	26·6 ± 1·4	0·49 ± 0·04	13·6 ± 1·1
25	29·0 ± 1·1	0·59 ± 0·05	17·1 ± 1·5
30	28·7 ± 0·9	0·61 ± 0·07	17·5 ± 2·0

leaf area and the canopy achieved a higher final dry weight as a result. It is unlikely that many other crops would show such complete adaptation to temperature.

Even though dry-weight gain of a closed canopy is so little affected by temperature, the rate of plant development is greatly affected, and it may still be desirable to maintain high temperatures after the canopy has closed.

Carbon dioxide concentration

Increasing the CO_2 concentration in which leaves are growing increases their net photosynthetic rate by increasing the CO_2 gradient into the leaves and by suppressing photorespiration (Ludwig, 1972). The latter is important because nearly all glasshouse crops are C_3 species. A high CO_2 concentration also decreases respiration during the subsequent night and this results in a further increase in net CO_2 fixed during a 24-hour period (Ludwig, Charles-Edwards and Withers, 1975). The response of closed crop canopies to CO_2 enrichment is similar to that of single leaves (Hand, 1973; Acock, Hand, Thornley and Warren Wilson, 1976, and Acock *et al.* 1978b).

Just as for temperature, dry-matter partitioning to leaves is little affected by CO_2 concentration (Ford and Thorne, 1967, and Hurd, 1968). However, when plants are grown continuously in high CO_2 concentrations, the growth rate gradually falls because the specific leaf area decreases. Thus, plants in high CO_2 make thicker leaves with a smaller area which intercept less light. Eventually, this reduction in leaf area more than compensates for the increased rate of net photosynthesis per unit leaf area caused by the CO_2 enrichment. Apparently the thicker leaves formed in high CO_2 have the same net photosynthetic rate as their thinner, low-CO_2 brethren when compared at any given CO_2 concentration, i.e. there is no adaptation other than in thickness (Ford and Thorne, 1967). After about 4 weeks, the relative growth rate of spaced plants in high CO_2 falls below that of plants in normal ambient CO_2 (Ford and Thorne, 1967, and Hurd, 1968). However, the initial boost to growth given by CO_2 enrichment is such that no reversal of absolute growth rates has ever been recorded, as far as I am aware.

Although the effect of CO_2 enrichment on spaced plants appears to be transitory, the plants might suffer if the extra CO_2 was withdrawn after several weeks. Anyway, the question is academic because in a cropping situation the canopy eventually closes. Thereafter, all the available light is being intercepted and the differences in specific leaf area are of little consequence. Canopy net photosynthetic rate continues to be higher in high CO_2 than in low CO_2 to the end of the crop's life, resulting in increases of about 37 per cent in the yield of tomatoes grown at 1000 vpm CO_2 (Calvert and Slack, 1975).

Because leaf and canopy photosynthetic rates are so much higher in atmospheres enriched with CO_2, starch is sometimes observed to collect in the leaves. When this happens, dry-matter accumulation rates normally fall, and it is increasingly being suggested that photosynthesis is inhibited by the high starch concentration which, in turn, is caused by the plant being sink limited, i.e. unable to use carbohydrate as fast as it can produce it. See, for example, Warren Wilson (1966), Thorne and Koller (1974), Hofstra and Hesketh (1975), and Nafziger and Koller (1976). This phenomenon is illustrated in Fig. 4 with data for melon plants (Pasternak and Acock, previously unpublished).

In Fig. 4 the specific leaf starch + sugar content at dawn is plotted against the unit leaf rate on a dry weight basis, E_W. The dawn value for starch + sugar is used because that is the balance remaining at the end of one accounting period and the start of the next. However, much starch accumulates during a day; a source-limited plant will use it before the start of the next day. From Fig. 4 it is apparent that melon plants normally maintain a balance of about 3 per cent starch + sugar in their leaves. The lowest value of E_W was for plants in 20°C, 300 vpm. Increasing the temperature to 40°C increased E_W by 34 per cent and increasing CO_2 concentration to 1500 vpm gave a further 52 per cent increase. However, increasing CO_2 concentration from 300 to 1500 vpm without increasing temperature increased E_W by only 21 per cent. It also increased the starch + sugar content of leaves at dawn to about 8 per cent. At the light level used in this experiment (52 W m^{-2} PAR) even single-leaf net photosynthesis would not show such a marked interaction between temperature and CO_2 concentration (see Fig. 1). So it appears that at 20°C the growth rate (cell expansion rate) of melon is too slow to make use of all the CO_2 fixed by the plant in 1500 vpm CO_2. Hence, starch accumulates and inhibits photosynthesis. The

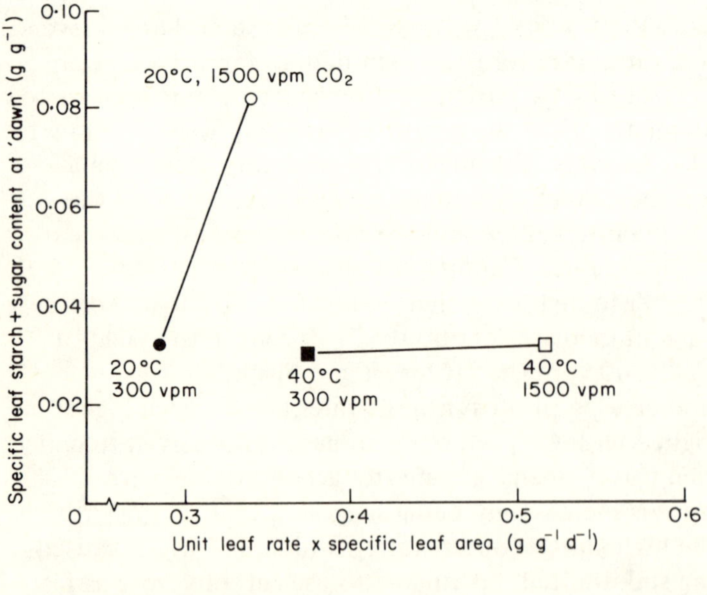

Fig. 4 The relationship between specific leaf starch + sugar content at 'dawn' and unit leaf rate expressed on a dry-weight basis for melon plants grown in various environmental conditions with 52 W m^{-2} PAR for a 14-hour day (Pasternak and Acock, unpublished).

phenomenon has now been described in tomato, soybean, sugarcane, rape, rye, melon and several other species, so it is widespread.

Humidity

Humidity is rarely controlled in glasshouses *per se*, but it varies with ventilation rate and evapotranspiration rate. It is easier and cheaper to maintain high temperatures and CO_2 concentrations if the ventilation rate is low and this results in high humidities. The use of thermal screens (devices to reduce energy loss at night) also leads to high humidities, and we need to know what effects these have on plant growth.

Experiments to examine the response of leaf net photosynthesis to humidity have given conflicting results. Rawson, Begg and Woodward (1977) in presenting their own data for several species and reviewing the literature suggested that leaves only respond directly to humidity—by changing stomatal resistance, and hence, net photosynthetic rate—when there is some degree of water stress in the leaf. Single leaves attached to well-watered plants do not respond, whereas single leaves on moderately water-stressed plants do. However, they found that exposing whole barley plants to different humidities produced a response. They assumed this was because placing whole plants in the measurement conditions rather than single leaves reduced leaf water potential, even though the plants were well watered. This does not mean that the response was mediated by leaf water potential but that leaf water potential controlled the extent to which the stomata responded directly to humidity. It appears that at high leaf water potentials (i.e. near zero) stomata are unresponsive to humidity. When the leaf is under a moderate water stress the stomata are responsive, but at leaf water potentials below about −1 MPa (−10 bar) stomatal opening may be limited by leaf water potential; so the stomata cannot respond to humidity (Barrs, 1972; Lawlor and Milford, 1975; Hall and Hoffman, 1976).

In addition to the barley plants mentioned above, a response of net photosynthesis to humidity has been demonstrated for spaced plants of sugar beet (Milford and Lawlor, 1975) and for a tomato canopy (Acock, Charles-Edwards and Hand, 1976). In the latter experiment, the response was much smaller at a CO_2 concentration of 1200 vpm than it was at 400 vpm, indicating another interaction to consider in future work. However, despite the evidence that even well-watered plants respond to humidity, plants do not always grow better or faster in higher humidity. The conflicting results are well illustrated by the results and references cited by Ford and Thorne (1974). Apparently there is a short-term beneficial effect and a long-term detrimental effect of exposure to high humidity (Acock, 1974). This can be seen in Ford and Thorne's (1974) data for wheat (their Table 2) where the plants in high humidity had a higher dry weight than those in low humidity at 4 weeks, but at 8 weeks the order was reversed.

Recent experiments with spaced tomato plants (Acock, Hurd and Pasternak, unpublished) provide a possible explanation of this detrimental effect of high humidity. The plants were grown in 20°C, 350 vpm CO_2 and 80 W m^{-2} PAR for 16-hour days. They were subjected to water vapour pressure deficits of 1·0 kPa (57 per cent RH) and 0·1 kPa (96 per cent RH) for four weeks from cotyledon expansion. At first the plants in 96 per cent RH grew faster than those in 57 per cent RH, but by the end of the fourth week the dry weights of the tops were not significantly different (Table 2). This was mainly caused by a change in specific leaf area. The specific leaf area of new leaves formed during each week of the experiment is shown in Table 3. It decreased

Table 2. Shoot dry weight and leaf area for tomato plants after growing for 1-4 weeks in two humidities (Acock, Hurd and Pasternak, unpublished). Values given are means ± S.E. (11 D.F.).

	Shoot dry weight (g)		Leaf area (cm^2)	
Week	57% RH	96% RH	57% RH	96% RH
1	0·20 ± 0·01	0·23 ± 0·01	52 ± 2	66 ± 2
2	1·80 ± 0·09	2·54 ± 0·08	379 ± 15	599 ± 17
3	6·54 ± 0·22	8·99 ± 0·28	1302 ± 50	1482 ± 28
4	17·54 ± 0·51	17·78 ± 0·31	2611 ± 118	1651 ± 37

Table 3. Incremental specific leaf area, S_A, and unit leaf rate expressed on a dry weight basis, E_W, for tomato plants grown for 4 weeks in two humidities (Acock, Hurd and Pasternak, unpublished). Values given are means ± S.E. (11 D.F.).

	S_A ($m^2\ kg^{-1}$)		E_W ($g\ kg^{-1}\ day^{-1}$)	
Week	57% RH	96% RH	57% RH	96% RH
1	26·3 ± 0·4	30·4 ± 0·2	437 ± 13	462 ± 11
2	22·6 ± 0·4	25·8 ± 0·6	340 ± 9	385 ± 7
3	22·9 ± 0·5	16·0 ± 0·5	211 ± 7	208 ± 7
4	15·5 ± 0·8	3·3 ± 0·8	167 ± 5	119 ± 7

with time in 57 per cent RH, but in 96 per cent RH it was initially higher, then decreased sharply. As a result of these changes, leaf area per plant was initially higher in 96 per cent RH than in 57 per cent RH, but during the fourth week there was virtually no change in leaf area in 96 per cent RH (Table 2). Also, there were changes in unit leaf rate expressed on a dry weight basis which were similar to the changes in specific leaf area (Table 3). The expected differences between treatments in unit leaf rate on a leaf-area basis were not found even during the first week. The most plausible explanation for this is the one offered by Ford and Thorne (1974), that the initial surge in leaf expansion in high RH so spreads out the photosynthetic apparatus that photosynthesis per unit leaf area is the same in both treatments even though photosythesis per unit of photosynthetic apparatus is enhanced in high humidity. If this explanation is correct, the effect of humidity on photosynthesis per unit leaf area is transitory, but the effects on cell expansion (specific leaf area) and possibly on cell division in the apex (Pareek, Sivanayagam and Heydecker, 1969) take longer to appear. The rate at which they appear is probably a function of the humidity and the growth rates of the plants.

We still have a lot to learn about plant response to humidity, but a pattern is slowly emerging from the confusion and this should stimulate further research. The problems will not go away just because it is difficult both to measure and control humidity reliably.

Conclusions

Canopy net photosynthetic rate is not very responsive to temperature but the rate is increased by both high CO_2 and high humidity. Partitioning of dry matter to the leaves is little affected by any environmental variable, but specific leaf area, S_A, is responsive to all those examined here. S_A increases with increase in temperature, but decreases with increase in CO_2 concentration. The initial response of S_A to high humidity is an increase – probably resulting in a reduction of photosynthesis per unit leaf area – but the long-term response is a decrease. The changes in S_A caused by temperature and CO_2 and the long-term changes caused by humidity do not cause changes in the net rate of photosynthesis (or unit leaf rate) per unit leaf *area.* This is in contrast to the changes in S_A caused by different light regimes, in which case the net rate of photosynthesis per unit leaf volume or weight remains constant (Charles-Edwards and Ludwig, 1975, and Acock, Charles-Edwards and Sawyer, 1979). These changes in S_A occur in both spaced plants and closed canopies and result in different rates of leaf-area growth. Leaf-area growth rate is an important determinant of rate of growth in dry weight in spaced plants, but is much less important in closed canopies. Because of this, it is appropriate to compare assimilation rates on a leaf-area basis when looking for effects in closed canopies and to use a leaf weight (or volume) basis when looking for effects on young spaced plants. This latter almost amounts to examining the relative growth rates because dry-matter partitioning to the leaves is so nearly constant.

In experiments where the plants are transferred to the test environment after being raised elsewhere, the old leaves present on the plants obscure the changes in S_A. If incremental or new leaf S_A is calculated from the change in leaf area divided by the change in leaf dry weight, the rapidity of the response of S_A to environmental factors becomes apparent.

In addition to the changes in S_A, exposure to very low temperatures inhibits photosynthesis and even at moderately low temperatures photosynthesis may be limited by starch accumulating over a period of time. Also, high humidities may cause changes in cell division in the plant apices. Clearly, crop responses to environmental factors are not additive; the interactions are important. Furthermore, they are dynamic. Fitting arbitrary curves to crop data will teach us nothing about why the crop behaved as it did. Dynamic, interactive behaviour requires us to fit dynamic models to our data and eventually these models will be useful for predictive purposes in any environment.

The analytical work done so far on glasshouse crops has only produced models capable of predicting dry weight gain. So far there is no evidence that any new combinations of environmental conditions will dramatically increase crop growth rates, and there is therefore little prospect of further increasing crop yields by environmental manipulation alone. Instead, it seems more likely that the first use of glasshouse crop models will be to help the growers decide how best to use scarce resources, especially fuel, so that they maximize the crop yield per unit of resource used. At present, and in the forseeable future, it is pointless to talk about increasing glasshouse crop yields without considering the cost of doing so. However, this does not mean that crop modelling is unlikely to result in yield increases. Models will help plant breeders select characteristics for future cultivars and help entomologists understand pest movements and life cycles. Indeed, we can expect a synergistic effect from the models, providing a much-needed means of communication between disciplines so that each can better see where they might contribute to an overall improvement in the glasshouse structure-environment-genotype-pest and disease-management complex.

Acknowledgements. This account of glasshouse crop modelling owes much to insights gained from my former colleagues at the Glasshouse Crops Research Institute and to work done with them. Our joint publications frequently listed authors alphabetically, which gave me undeserved prominence.I am especially grateful to those who have permitted me to use data that have not yet been published in full. One of those is Dr Pasternak of Ben-Gurion University of the Negev who was a visiting worker in 1977-1978.

References

Acock, B. (1974). The design and use of growth chambers for investigating the effects of environmental factors on plant growth. *Acta Horticulturae* **39**, 15-38.

Acock, B., Charles-Edwards, D. A. & Hand, D. W. (1976). An analysis of some effects of humidity on photosynthesis by a tomato canopy under winter light conditions and a range of carbon dioxide concentrations. *Journal of Experimental Botany* **27**, 933-941.

Acock, B., Charles-Edwards, D. A. & Sawyer, S. (1979). Growth response of a *Chrysanthemum* crop to the environment. III. Effects of radiation and temperature on dry matter partitioning and photosynthesis. *Annals of Botany* **44**, 289-300.

Acock, B., Thornley, J. H. M. & Warren Wilson, J. (1971). Photosynthesis and energy conversion. In *Potential Crop Production,* pp. 43-75. Eds. P. F. Wareing & J. P. Cooper. London: Heinemann.

Acock, B., Hand, D. W. Thornley, J. H. M. & Warren Wilson, J. (1976). Photosynthesis in stands of green peppers. An application of empirical and mechanistic models to controlled-environment data. *Annals of Botany* **40**, 1293-1307.

Acock, B., Charles-Edwards, D. A., Fitter, D. J. Hand, D. W. & Warren Wilson, J. (1978b). Modelling canopy net photosynthesis by isolated blocks and rows of chrysanthemum plants. *Annals of Applied Biology* **90**, 255-263.

Acock, B., Charles-Edwards, D. A., Fitter, D. J., Hand, D. W., Ludwig, L. J., Warren Wilson, J. & Withers, A. C. (1978a). The contribution of leaves from different levels within a tomato crop to canopy net photosynthesis: An experimental examination of two canopy models. *Journal of Experimental Botany* **29**, 815-827.

Baker, D. N. (1965). Effects of certain environmental factors on net assimilation in cotton. *Crop Science* **5**, 53-56.

Baker, D. N., Lambert, J. R., Phene, C. J. & McKinion, J. M. (1976). GOSSYM: A simulator of cotton crop dynamics. In *Computers Applied to the Management of Large-scale Agricultural Enterprises: Proceedings of U.S.-U.S.S.R.*

Seminar, Moscow, pp. 100-133. Ed. S. Shulman. National Bureau of Economic Research.

Barrs, H. D. (1972). Effects of combinations of atmospheric and soil water stress on plants. *Annual Report of C.S.I.R.O. Division of Irrigation Research 1971-1972,* pp. 5-6.

Brouwer, R. L. & de Wit, C. T. (1969). A simulation model of plant growth with special attention to root growth and its consequences. In *Root Growth: Proceedings of the 15th Easter School in Agricultural Science, University of Nottingham,* pp. 224-244. Ed. W. J. Whittington.

Calvert, A. (1972). Effects of day and night temperatures and carbon dioxide enrichment on yield of glasshouse tomatoes. *Journal of Horticultural Science* **47**, 231-247.

Calvert, A. & Slack, G. (1975). Effects of carbon dioxide enrichment on growth, development and yield of glasshouse tomatoes. I. Responses to controlled concentrations. *Journal of Horticultural Science* **50**, 61-71.

Charles-Edwards, D. A. (1978). An analysis of the photosynthesis and productivity of vegetative crops in the United Kingdom. *Annals of Botany* **42**, 717-731.

Charles-Edwards, D. A. & Ludwig, L. J. (1975). The basis of expression of leaf photosynthetic activities. In *Environmental and Biological Control of Photosynthesis*, pp. 37-44. Ed. R. Marcelle. The Hague: W. Junk.

Curry, R. B., Baker, C. H. & Streeter, J. G. (1975). An overview of SOYMOD, simulator of soybean growth and development. In *Proceedings of the 1975 Summer Computer Simulation Conference*, pp. 954-960. Ed. J. P. Locher III. La Jolla: Simulation Councils Inc.

Duncan, W. G. (1975). SIMAIZ: A model for simulating corn growth and yield. In *The Application of System Methods to Crop Production,* pp. 32-48. Eds. D. N. Baker, R. G. Creech & F. G. Maxwell. Departments of Agronomy and Entomology, Mississippi State University.

El-Sharkawy, M. A., Loomis, R. S. & Williams, W. A. (1968). Photosynthetic and respiratory exchanges of carbon dioxide by leaves of the grain amaranth. *Journal of Applied Ecology* **5**, 243-251.

Fick, C. W., Williams, W. A. & Loomis, R. S. (1973). Computer simulation of dry matter distribution during sugar beet growth. *Crop Science* **13**, 413-417.

Ford, M. A. & Thorne, G. N. (1967). Effect of CO_2 concentration on growth of sugar-beet, barley, kale and maize. *Annals of Botany* **31**, 629-644.

Ford, M. A. & Thorne, G. N. (1974). Effects of atmospheric humidity on plant growth. *Annals of Botany* **38**, 441-452.

Garner, W. W. & Allard, H. A. (1920). Effect of the relative length of day and night and other factors of the environment on growth and reproduction in plants. *Journal of Agricultural Research* **18**, 553-606.

Gates, D. M. (1965). Energy, plants and ecology. *Ecology* **46**, 1-13.

Gutierrez, A. P., Falcon, L. A., Loew, W., Leipzig, P. A. & Vandenbosh, R. (1975). An analysis of cotton production in California: A model for Acala cotton and the effects of defoliators on its yields. *Environmental Entomology* **4**, 125-136.

Hall, A. E. & Hoffman, G. J. (1976). Leaf conductance response to humidity and water transport in plants. *Agronomy Journal* **68**, 876-881.

Hand, D. W. (1973). A null balance method for measuring crop photosynthesis in an airtight controlled-environment cabinet. *Agricultural Meteorology* **12**, 259-270.

Hofstra, G. & Hesketh, J. (1969). Effects of temperature on the gas exchange of leaves in the light and dark. *Planta* **85**, 228-237.

Hofstra, G. & Hesketh, J. D. (1975). The effects of temperature and CO_2 enrichment on photosythesis in soybean. In *The Environmental and Biological Control of Photosynthesis*, pp. 61-70. Ed. R. Marcelle. The Hague: W. Junk.

Holt, D. A., Bula, R. J., Miles, G. E., Scheiber, M. M. & Peart, R. M. (1975). Environmental physiology, modeling and simulation of alfalfa growth. I. Conceptual development of SIMED. *Research Bulletin 907, Agricultural Experiment Station, Purdue University.*

Hughes, A. P. & Cockshull, K. E. (1971). The effects of light intensity and carbon dioxide concentration on the growth of *Chrysanthemum morifolium* cv. *Bright Golden Anne. Annals of Botany* **35**, 899-914.

Hurd, R. G. (1968). Effects of CO_2 enrichment on the growth of young tomato plants in low light. *Annals of Botany* **32**, 531-542.

Hurd, R. G. & Thornley, J. H. M. (1974). An analysis of the growth of young tomato plants in water culture at different light integrals and CO_2 concentrations. I. Physiological aspects. *Annals of Botany* **38**, 375-388.

Idso, S. B. (1968). A holocoenotic analysis of environment-plant relationships. *Technical Bulletin 264, Agricultural Experiment Station, University of Minnesota.*

Jenkins, J. N., Baker, D. N., McKinion, J. M. & Parrott, W. L. (1975). Systems analysis and the evaluation of morphogenetic characters in cotton. In *The Application of System Methods to Crop Production,* pp. 49-90. Eds. D. N. Baker, R. G. Creech & F. D. Maxwell. Departments of Agronomy and Entomology, Mississippi State University.

Landivar, J. A., Baker, D. N. & Jenkins, J. N. (1979). The application of the cotton simulation model GOSSYM in genetic feasibility studies. *Bulletin of Mississippi Agriculture and Forestry Experiment Station* (in press).

Lawlor, D. W. & Milford, G. F. J. (1975). The control of water and carbon dioxide flux in water-stressed sugar beet. *Journal of Experimental Botany* **26**, 657-665.

Loomis, R. S., Rabbinge, R. & Ng, E. (1979). Explanatory models in crop physiology. *Annual Review of Plant Physiology* **30**, 339-367.

Ludwig, L. J. (1972). The relationship between photosynthesis and respiration. In *Crop Processes in Controlled Environments*, pp. 305-326. Eds. A. R. Rees, K. E. Cockshull, D. W. Hand & R. G. Hurd. London: Academic Press.

Ludwig, L. J. & Withers, A. C. (1978). Effect of temperature on single leaf photosynthesis and respiration in tomato. *Report of the Glasshouse Crops Research Institute 1977,* pp. 51-52.

Ludwig, L. J., Charles-Edwards, D. A. & Withers, A. C. (1975). Tomato leaf photosynthesis and respiration in various light and CO_2 environments. In *The environmental and biological control of photosynthesis,* pp. 29-36. Ed. R. Marcelle. The Hague: W. Junk.

McKinion, J. M., Baker, D. N., Hesketh, J. D. & Jones, J. W. (1975). SIMCOT II: A simulation of cotton growth and yield. In *Computer Simulation of a Cotton Production System: User's Manual,* pp. 27-82. USDA Bulletin ARS-S-52.

Milford, G. F. J. & Lawlor, D. W. (1975). Effects of varying air and soil moisture on the water relations and growth of sugar beet. *Annals of Applied Biology* **80**, 93-102.

Nafziger, E. D. & Koller, H. R. (1976). Influence of leaf starch concentration on CO_2 assimilation in soybean. *Plant Physiology* **57**, 560-563.

Pareek, O. P., Sivanayagam, T. & Heydecker, W. (1969). Relative humidity: A major factor in crop plant growth. *Report of the School of Agriculture, University of Nottingham 1968-69*, pp. 92-95.

Rawson, H. M., Begg, J. E. & Woodward, R. G. (1977). The effect of atmospheric humidity on photosynthesis, transpiration and water use efficiency of leaves of several plant species. *Planta* **134**, 5-10.

Strommen, N. D. (1979). Yield model development. *Project Plan for the AgRISTARS Program,* USDA, NASA, USDC, USDI and AID.

Thorne, J. H. & Koller, H. R. (1974). Influence of assimilate demand on photosynthesis, diffusive resistances, translocation and carbohydrate levels of soybean leaves. *Plant Physiology* **54**, 201-207.

Warren Wilson, J. (1966). Effect of temperature on net assimilation rate. *Annals of Botany* **30**, 753-761.

Wittwer, S. H. & Robb, W. (1964). Carbon dioxide enrichment of greenhouse atmospheres for food crop production. *Economic Botany* **18**, 34-56.

Optimizing radiation interception and the direction of assimilates into the crop of fruit trees

J. E. Jackson
Pomology Department, East Malling Research Station, Maidstone, Kent ME19 6BJ

Introduction

Two main challenges face the crop physiologist who seeks to increase the economic yield of English orchards.

Firstly, only a relatively low proportion of the available radiant energy is intercepted by orchard trees over their lifetime. The trees in a five-year-old orchard may intercept only 30 per cent of available light at full leaf, while the corresponding figure for mature orchards seldom exceeds 70 per cent (Jackson, 1975). This obviously puts a low upper limit to dry-matter production (Monteith, 1977) and economic yield.

Secondly, there is the need to optimize consistently the proportion of the dry-matter production which is directed into the fruits. Year-to-year variations in fruit yield which are not positively associated with variations in vegetative growth pose one of the most serious problems of the English fruit industry.

Modifying canopy structure to optimize energy interception

The unique problems posed by tree, cane and bush-fruit orhcards and plantations are a consequence of the size and complexity of the individual plants. Fruit trees are planted at spacings appropriate to their mature size, which, although usually maintained well below its potential maximum by pruning, is not naturally attained until many years after planting. Spacing trials have shown fruit yield to be a linear function of orchard light interception, up to a value of at least 60 per cent (Jackson, 1978), but if the spacing is close enough for efficient light interception soon after planting the orchard may subsequently become too dense at maturity (Verheij, 1972). This poses a problem because fruit-bud production, fruit retention, growth and colour development are all reduced by shade to a greater extent than is vegetative growth as shown

Table 1. Effects of shading in 1970 and 1971 on growth and cropping in 1971 of Cox's Orange Pippin/Malling 26

	Light level* in 1970 and 1971			
	100	37	25	11
Shoot growth 1971 (m)	38·6	40·8	35·7	11·9
Girth increment 1971 (cm^2) †	61·4	43·1	42·8	24·6
Fruit bud clusters/tree 1971 ‡	159	96	69	26
Fruits harvested/tree 1971 ‡	151	74	51	6
Yield/tree 1971 (kg) ‡	17·6	8·8	4·7	0·6
% of fruits > ¼ red §	47	30	16	13

* percentage daylight inside shades in relation to that above canopy
Data from: † Jackson and Palmer 1977a
‡ Jackson and Palmer 1977b
§ Jackson, Palmer, Perring and Sharples 1977

by shoot length and by increment in girth (Table 1 and Jackson and Palmer, 1977b). The latter is closely related to the increment in dry weight of the vegetative parts of an apple tree (Moore, 1978).

By placing shades over whole trees to reduce the irradiance inside to 25 per cent of that above the canopy, fruit yield was reduced to about a quarter of that from the controls. Shading also adversely affected fruit colour, so that the yield of fruits with more than a quarter of their surface coloured red was only about 10 per cent of that of the controls. The small reduction in shoot growth and girth increment to 70 per cent of that of the controls was to be expected. Avery (1975a) concluded that 80 per cent of full photosynthesis could be obtained at between 10 and 40 per cent of full sunlight, and the data of Sirois and Cooper (1964), quoted by Lakso and Seeley (1978), indicated that the rate of photosynthesis of apple-trees is reduced to only 70 per cent of its maximum bright-light value when irradiance is 25 per cent of full sunlight. The density of canopy which might be ideal for dry-matter production is thus well in excess of that which can produce good quality fruits: the ideal is clearly a canopy which is shallow or open enough to produce good-quality fruits throughout, without excess depth needing expensive pruning, spraying etc., but arranged so that nearly all of the light is intercepted and does not provide energy to grow grass in alleyways. Such a canopy should reach its maximum size soon after planting and be easy to maintain at maturity. The traditional large-tree orchard does not begin to approach such a specification. Its light interception in the early years is low and at maturity only the upper and outer part of the trees, which are expensive to prune, spray and pick because of their height, produce good quality fruits (Heinicke,

1966; Jackson, 1970; Jacnya, 1978). The necessary complete re-design of orchard structure involves three central problems in crop and plant physiology.

The theory of light interception by discontinuous canopies

While the traditional hemispherical tree is being replaced in many parts of the world by 'hedgerow' orchards, which consist of narrow vertically arranged canopies separated by tractor alleyways, a number of revolutionary concepts in orchard design are being tested in observation plots or even on a larger scale. These range from shallow horizontal canopies like the East Malling bed systems (Preston, 1978) and the Long Ashton meadow orchard (Hudson, 1970; Luckwill, 1978; Erez, 1978), through orchards with T-shaped (Dunn, 1974) or V-shaped (Chalmers and van den Ende, 1976) hedgerows designed to combine shallow-canopy characteristics with ease of mechanical harvesting, to multiple-row versions of conventional hedgerows (Wertheim, 1978) with a lower ratio of tractor alleyways to rows of trees. Comparative field evaluation of all possible canopy shapes and arrangements would be inordinately slow and expensive: a modelling approach to provide even a preliminary sieve is needed. Classical light-interception models where the transmission of light by the canopy is experimentally related to LAI are clearly inappropriate because of the non-random distribution of leaves (Monsi, Uchijima and Oikawa, 1973). The way in which the foliage is clumped, in terms of tree height, tree thickness and between-tree spacing, determines the pattern of cast shadows. Jackson and Palmer (1972) showed how the dimensions and arrangement of continuous hedgerows would affect light interception if they were opaque, and so effectively established limits to the light interception by hedgerow orchards of any given geometry. Charles-Edwards and Thorpe (1976) and Palmer (1977) developed models to calculate the transmission of direct-beam radiation through hedgerows taking into account their leaf-area densities and, in the latter case, the distribution of fruits and branches. These computer models all require that the overall canopy geometry be defined in relatively simple mathematical terms and the transmission models need considerable computational resources. More recently Jackson and Palmer (1979) described a simple general equation for light interception by any discontinuous canopy.

$$I/I_0 = T_F + (1 - T_F)e^{-KL'}, \qquad (1)$$

where I/I_0 is the average fraction of the incident light reaching the orchard floor, T_F is the transmission due to the overall form of the canopy (i.e. the fraction of available light which would reach the ground if the trees were solid), K the measured within-tree light-extinction coefficient and L' is LAI/$(1-T_F)$. For simple shapes T_F can be calculated, for example it is one minus the fractional interception as calculated by Jackson and Palmer (1972). For more complex shapes it could be calculated from measurements of interception made on non-transmitting scale models. This is at present being attempted using painted polystyrene models which are placed in appropriate arrangements (to simulate a complete orchard) on a plastic diffuser over an array of integrating photocells. The measurements are made outdoors under representative conditions. Once T_F for any particular orchard geometry has been determined in this way it can be substituted in eq.(1), together with appropriate values for LAI and K, to calculate I/I_0 for orchards of the chosen tree size, shape, arrangement and leaf density.

Jackson and Palmer (1979) found this equation to give a good estimate of light interception by hedgerow orchards if K was assumed to be 0·6, which Jackson (1978) had shown to be an average value for apple. For other tree crops, or, indeed for apple orchards of different types, it might be better to determine K directly by using actual orchard measurements of I/I_0 and LAI in conjunction with a physical scale model of the orchard and eq. (1). In this case all the terms in eq. (1) would necessarily be consistent with each other. Further development of this system is in progress and it is hoped to be able to use it not only to estimate potential light interception, hence yield, of a very wide range of orchard systems at varying ages (it being quicker to cut out polystyrene shapes than grow orchards) but also to estimate the contribution of different tree zones by a 'stratified clip' approach using the scale models.

The control of tree vigour

Three main approaches are being used, separately and in conjuction, to contain trees to the desired canopy dimensions and densities. The first is genetical. The use of the dwarfing Malling 9 rootstock is crucial to the success of most modern hedgerow orchards and that of the even more dwarfing Malling 27 to the experimental 'bed orchard' mentioned above. The physiological mechanisms of these rootstock effects still await elucidation – which could lead the way to new chemical control measures. The second is mechanical – pruning by hand or machine in either summer or winter. Severe pruning which releases many buds from apical dominance can lead to excessive, shade-inducing regrowth as shown especially by Cain (1971). Late summer pruning may minimize this and also the incorporation of 1 per cent α-napthaleneacetic acid (NAA) in the wound-dressing paint used after winter pruning greatly reduces the number and vigour of the shoots which emerge (Jackson, Quinlan and Preston, 1978). This latter practice is now widely used in commercial fruit growing. The third is the direct physiological approach using growth retardants such as daminozide. This has proved effective in checking shoot-extension growth (Luckwill, 1970) but, because of side effects such as a reduction in fruit size, its practical application is restricted to certain varieties and specific uses rather than general growth control. The search for alternative growth regulators continues.

The control of light exposure in the fruit bearing zone

It is obvious from Table 1 that the ideal fruit tree would have its crop on the outside without shade from the current season's shoots. Such a tree could be expected to have a much more favourable partitioning of assimilates with a greater proportion going to fruits because of favourable effects on fruit-bud formation, fruit set and fruit growth. It might also have greater total production of dry matter. This is because, as Avery (1975b) has shown, high fruit demand leads to increased photosynthesis and Thorpe (1974) has suggested that direct exposure of fruits to sunlight, by increasing their temperatures, makes them into more demanding and efficient 'sinks'. Some approaches to vigour control, e.g. late summer pruning which increases the exposure of fruits to light, might therefore increase productivity and, in fact, the highest yields are obtained with such systems (McKenzie and Rae, 1978). Of even greater promise are approaches using the chemical regulation of shoot growth and fruit distribution. Quinlan and Preston (1978) reported that by the use of chemical branching agents (e.g. long-chain fatty acids like Off-Shoot O) they could check or kill the shoot tips of main branches and induce the development of many other, much shorter shoots which formed terminal fruit buds and greatly increased yield in the tip-bearing variety Bramley's Seedling. For non-tip-bearing varieties very efficient growth retardants which would prevent shoot growth above the main spur-bearing canopy could be similarly advantageous.

Variation in cropping from year to year

Once the canopy has been well designed to intercept sunlight the next problem is to convert the assimilates into economic yield. Much is already known about the effect on partitioning of pruning and rootstock (Barlow, 1970) and of scion variety (Moore, 1978). A number of approaches used in canopy management also improve the ratio of cropping to growth – for example, the 'chemical pruning' treatments reported by Quinlan and Preston (1978) approximately trebled total yield in the third and fourth years after planting while reducing vegetative growth as measured by stem girth and tree height. Less well understood are the year-to-year variations in cropping which are poorly related to those in vegetative growth (Rogers and Booth, 1964; Moore, 1978). This problem of variation in economic yield, which is apparently not due to variation in dry-matter production as a whole, is perhaps the dominant problem in English fruit growing.

The yields of Cox over the years from 1949 have shown a progressive upwards trend, due to advances in technology, but also massive variations around this trend which, statistically, are largely accounted for by variations in the following weather parameters:

1. February, March and April temperatures – high temperatures being adverse;
2. temperatures over the pollination period – high temperatures being beneficial;
3. temperatures in June – low temperatures being adverse.

These factors, together with the time-trend, account for almost 80 per cent of the variations in national yield of Cox in England (Jackson and Hamer, 1980, in press), the actural yields and those predicted from the appropriate multiple-regression equation are shown in Fig. 1.

The physiological mechanism of the first effect is likely to involve the poor quality and low fertility of apple flowers developed under high-temperature conditions. Abbott (1971) showed such effects in controlled temperature studies on Lord Lambourne apple but much further work remains to be done.

The effect of blossom-time temperatures on yield was most probably due to the effect of temperature on pollen-tube growth. Calculation of time taken to complete pollen-tube growth in each season and its use in the multiple-regression equation gave a better fit than did the use of average temperatures over the period immediately following full bloom. This, in turn, gave a better fit than did temperatures averaged over periods immediately before or appreciably after bloom. This reinforces the view that failure in the pollination process is a frequent cause of low yield (Williams and Wilson, 1970).

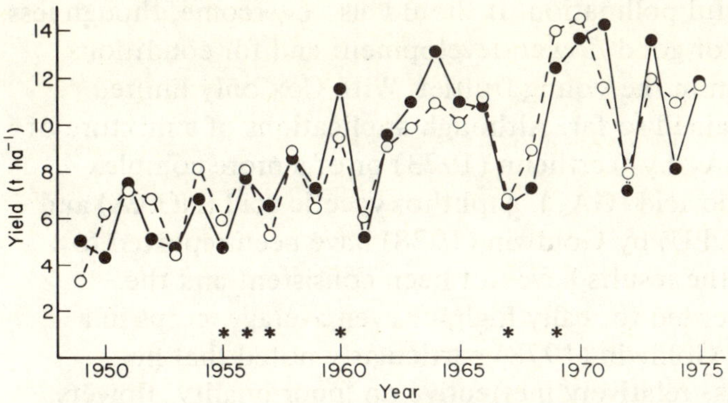

Fig. 1 The relationship between UK average annual yields (t ha^{-1}) of Cox's Orange Pippin and those estimated from the equation

$$Y = 4{\cdot}91 + 0{\cdot}254t - 1{\cdot}127T_{max}\text{ FMA} - 0{\cdot}622G_{Pt} + 0{\cdot}732T_{max}\text{ J},$$

where t = years from 1948, T_{max} FMA = mean of daily temperature maxima in February, March and April, G_{Pt} = calculated days to complete pollen-tube growth and T_{max} J = mean of daily temperature maxima in June. Actual yield = ●——●, Estimated yield = ○----○
* = year of potential frost damage.

The adverse effect of low June temperatures is probably associated with both effects on cell division and on June drop of young fruits (Jackson and Hamer, 1980, in press).

Two physiologically based approaches to these problems are being pursued. The first is to use evaporative cooling of the buds. Water sprinkling can reduce bud temperatures from dry-bulb to approximately wet-bulb values, which could turn a 'physiologically warm' early spring into a 'physiologically cool' one. Cooling the buds slows down their development so that under English conditions a delay of 17 days in the date of full bloom would be possible (Hamer, 1980, in press). Since temperatures increase linearly throughout the normal months of Cox blossoming (April/May) such a delay should lead to more favourable temperature conditions for pollination and, possibly, higher temperatures in the early cell-division stage of fruit growth.

In the second approach, instead of trying to optimize conditions for natural fruit-bud development, fruit set and fruit growth we are attempting direct hormonal control of the processes of fruit set and growth. If one assumes that transport of assimilates to the fruit (hence its growth) is initially controlled by hormones produced by the seeds (Luckwill, 1970) and later, possibly, by flesh as well as seed-produced hormones, the direct application of growth regulators should overcome the need for successful pollination. It should also overcome, though less certainly, the need for good flower development and for conditions suited to cell division in the young fruitlet. With Cox only limited success has been attained so far. Although applications of a mixture of gibberellins A_4 and A_7 by Wertheim (1973) or of a more complex mixture of gibberellic acid (GA_3), naphthoxyacetic acid (NOXA) and N,N-diphenylurea (DPU) by Goldwin (1978) have been reported to increase Cox yields the results have not been consistent and the treatments have never led to really high, or even average, crops in a low-yielding season. Goldwin (1978) particularly noted that his hormone mixture was relatively ineffective on 'poor-quality' flowers, which suggests that it would not overcome adverse effects of high temperatures in February, March and April. It may, however, simply be that additional hormones or different timings are needed on Cox since results on other fruit crops, including different apple varieties, have been very promising.

GA_3 is now used commercially to set crops of Conference pear especially after frost has killed a high proportion of the blossoms. The major problem is often the avoidance of overset, with so many fruits being retained that they are small and fruit-bud production is inhibited; GA_3 spraying is capable of increasing yield fivefold (Wickenden, 1979).

Goldwin and Webster (1978) showed that a mixture of GA_3 and NOXA applied at the time of 50 per cent petal fall and again later could greatly increase the yield of Early Rivers cherries whether or not DPU was added to the mixture. Yields ranged up to 23 times as high as those of the controls but the highest yields were accompanied by severe suppression of fruit-bud formation. Modlibowska (1976-79) used lower-concentration mixtures of GA_3 and the auxin 2, 4, 5-trichlorophenoxy-propionic acid (2, 4, 5-TP) on Van cherries, treating the same halves of the trees in each of the years 1975-78. Yields from the treated halves were 260, 395, 817 and 932 per cent of those of the control halves in the successive years, emphasizing the potential benefits of controlled direction of assimilates to the fruits of cherry trees.

In the case of Victoria plum application of the hormone sprays at the petal-fall stage proved deleterious to fruit set but treatment 4 weeks later led to retention and continued growth of many fruits that otherwise would have shed. Webster and Goldwin (1978) found that a mixture of 150 ppm GA_3 with 300 ppm DPU and 50 ppm NOXA applied at this time consistently increased yields. Yields were nine times that of the controls in 1975 and four times the control yields in 1976. In 1977 and 1978 a simpler mixture of 150 ppm GA_3 and 10 ppm 2, 4,

5-TP increased yields from 24 to 61 kg/tree and from 14 to 111 kg/tree respectively. In the latter case a double application was used and the effect was excessive in that it led to branch breakage under the weight of crop.

Among apples the most promising results have been obtained on Bramley's Seedling and on another triploid variety, Crispin. Modlibowska (1972) induced frost-damaged flowers of Bramley's Seedling to set using a mixture of gibberellins A_4 and A_7. Her subsequent work has shown that a cheaper mixture of GA_3 and 2, 4, 5-TP, applied at petal-fall and again 2 weeks later, is also effective, having given an average yield response of about 30 per cent over several years. Modlibowska and Wickenden (1977), however, noted that sprayed fruits developed more 'bitter pit' (a calcium-deficiency disorder) and advised against attempting their long-term storage. This result is in keeping with Bangerth's (1976) finding that growth-regulator induced parthenocarpic fruits of many kinds frequently show Ca deficiency symptoms and low Ca content (while subsequent work at East Malling has shown even seeded fruits of trees sprayed with GA_3 + 2, 4, 5-TP to have a lower concentration of Ca than controls).

This leads on to the more general question of whether exogenous growth regulators can substitute for the normal, apparently seed-dependent, mechanisms of attracting ions as well as photosynthates into fruits. If not, the former may well have to be applied in separate sprays or in post-harvest dips, especially of Ca salts, to ensure the membrane stability which is essential for long-term fruit storage. There can be little doubt, however, that for many of our fruit crops we now have the technical capability to control partitioning of assimilates by increasing the proportion which goes to form fruits, and further research and development effort should be well repaid.

References

Abbott, D. L. (1971). Physiology of fruit set in apple. Effect of Spring temperature. *Report of Long Ashton Research Station for 1970,* pp. 30-31.

Avery, D. J. (1975a) Effects of climatic factors on the photosynthetic efficiency of apple leaves. In *Climate and the Orchard*, pp. 25-31. Ed. H. C. Pereira. Farnham Royal: Commonwealth Agricultural Bureaux.

Avery, D. J. (1975b). Effects of fruits on photosynthetic efficiency. In *Climate and the Orchard,* pp. 110-112. Ed. H. C. Pereira. Farnham Royal: Commonwealth Agricultural Bureaux.

Bangerth, F. (1976). A role for auxin and auxin transport inhibitors on the Ca content of artificially induced parthenocarpic fruits. *Physiologia Plantarum* **37**, 191-194.

Barlow, H. W. B. (1970). Some aspects of morphogenesis in fruit trees. In *Physiology of Tree Crops*, pp. 25-43. Eds. L. C. Luckwill & C. V. Cutting. London: Academic Press.

Cain, J. C. (1971). Effects of mechanical pruning of apple hedgerows with a slotting saw on light penetration and fruiting. *Journal of the American Society for Horticultural Science* **96**, 664-667.

Chalmers, D. J. & van den Ende, B. (1976). The Tatura trellis: a new design in high yielding orchards. *Journal of Agriculture, Victoria* **73**, 473-475.

Charles-Edwards, D. A. & Thorpe, M. R. (1976). Interception of diffuse and direct beam radiation by a hedgerow apple orchard. *Annals of Botany* **40**, 603-613.

Dunn, J. S. (1974). Mechanical harvesting of apples and raspberries; an agronomic approach. *Span* **17**, 112-113.

Erez, A. (1978). Adaptation of the peach to the meadow orchard system. *Acta Horticulturae* **65**, 245-250.

Goldwin, G. K. (1978). Improved fruit setting with plant growth hormones. *Acta Horticulturae* **80**, 115-121.

Goldwin, G. K. & Webster, A. D. (1978). Improving the fruit set of sweet cherry, *Prunus avium* L., cv. Early Rivers with plant hormones. *Journal of Horticultural Science* **53**, 283-290.

Hamer, P. J. C. (1980). A model to evaluate evaporative cooling of apple buds as a frost protection technique. *Journal of Horticultural Science* **55**, 157-163.

Heinicke, D. R. (1966). Characteristics of McIntosh and Red Delicious apples as influenced by exposure to sunlight during the growing season. *Proceedings of the American Society for Horticultural Science* **89**, 10-13.

Hudson, J. P. (1970). Maximum intensity orchards, growing fruit without trees. *Proceedings of the 18th International Horticultural Congress, Tel Aviv.* **1**, 56.

Jackson, J. E. (1970). Aspects of light climate within apple orchards. *Journal of Applied Ecology* **7**, 207-216.

Jackson, J. E. (1975). Patterns of distribution of foliage and light. In *Climate and the Orchard*, pp. 31-40. Ed. H. C. Pereira. Farnham Royal: Commonwealth Agricultural Bureaux.

Jackson, J. E. (1978). Utilization of light resources by high density planting systems. *Acta Horticulturae* **65**, 61-70.

Jackson, J. E. & Hamer, P. J. C. (1980). The cause of year to year variations in the average yield of Cox's Orange Pippin apple in England. *Journal of Horticultural Science* **55**, 149-156.

Jackson, J. E. & Palmer, J. W. (1972). Interception of light by model hedgerow orchards in relation to latitude, time of year and hedgerow configuration and orientation. *Journal of Applied Ecology* **9**, 341-357.

Jackson, J. E. & Palmer, J. W. (1977a). Effects of shade on the growth and cropping of apple trees. I. Experimental details and effects on vegetative growth. *Journal of Horticultural Science* **52**, 245-252.

Jackson, J. E. & Palmer, J. W. (1977b). Effects of shade on the growth and cropping of apple trees. II. Effects on components of yield. *Journal of Horticultural Science* **52**, 253-266.

Jackson, J. E. & Palmer, J. W. (1979). A simple model of light transmission and interception by discontinuous canopies. *Annals of Botany* **44**, 381-383.

Jackson, J. E. Quinlan, J. D. & Preston, A. P. (1978). Chemical pruning and pruning aids. *Acta Horticulturae* **65**, 199-208.

Jackson, J. E., Palmer, J. W., Perring, M. A. & Sharples, R. O. (1977). Effects of shade on the growth and cropping of apple trees. III. Effects on fruit growth, chemical composition and quality at harvest and after storage. *Journal of Horticultural Science* **52**, 267-282.

Jacnya, T. (1978). Influence of light penetration on the quality of McIntosh apple fruits. *Fruit Science Reports, Skierniewice, Poland* **5**, 7-17.

Lakso, A. N. & Seeley, E. J. (1978). Environmentally induced responses of apple tree photosynthesis. *HortScience* **13**, 646-650.

Luckwill, L. C. (1970). The control of growth and fruitfulness of apple trees. In *Physiology of Tree Crops,* pp. 237-254. Eds. L. C. Luckwill & C. V. Cutting. London: Academic Press.

Luckwill, L. C. (1978). Meadow orchards and fruit walls. *Acta Horticulturae* **65**, 237-243.

McKenzie, D. W. & Rae, A. N. (1978). Economics of high density apple production in New Zealand. *Acta Horticulturae* **65**, 41-46.

Modlibowska, I. (1972). The effect of gibberellins and cytokinins on fruit development of Bramley's Seedling apple. *Journal of Horticultural Science* **47**. 337-340.

Modlibowska, I. (1976-1979). In *Reports of the East Malling Research Station for 1975,* pp. 42-43, *1976,* pp. 47-49, *1977,* pp. 46-47, *1978,* p. 59.

Modlibowska, I. & Wickenden, M. F. (1977). Timely hormone spray could help frosted Bramley flowers survive. *The Grower,* 7 April, 805-806.

Monsi, M., Uchijima, Z & Oikawa, T. (1973). Structure of foliage canopies and photosynthesis. *Annual Reviews of Ecology and Systematics* **4**, 301-327.

Monteith, J. L. (1977). Climate and the efficiency of crop production in Britain. *Philosophical Transaction of the Royal Society, London, Series B* **281**, 277-294.

Moore, C. S. (1978). Biometrical relationships in apple trees. *Journal of Horticultural Science* **53**, 45-51.

Palmer, J. W. (1977). Diurnal light interception and a computer model of light interception by hedgerow apple orchards. *Journal of Applied Ecology* **14**, 601-614.

Preston, A. P. (1978). A bed system for planting apples on the dwarfing rootstock M.27. *Acta Horticulturae* **65**, 229-235.

Quinlan, J. D. & Preston, A. P. (1978). The use of branching agents to replace hand pruning of young trees of Bramley's Seedling apple. *Journal of Horticultural Science* **53**, 39-43.

Rogers, W. S. & Booth, G. A. (1964). Relationship of crop and shoot growth in apple. *Journal of Horticultural Science* **39**, 61-65.

Sirois, D. L. & Cooper, G. R. (1964). The influence of light intensity, temperature and atmospheric carbon dioxide concentration on the rate of apparent photosynthesis of a mature apple tree. *Maine Agricultural Experiment Station Bulletin* **626**.

Thorpe, M. R. (1974). Radiant heating of apples. *Journal of Applied Ecology* **11**, 755-760.

Verheij, E. J. W. (1972). Competition in apple, as influenced by alar sprays, fruiting, pruning and tree spacing. *Mededeelingen van de Landbouwhoogeschool Wageningen* 72-4.

Webster, A. D. & Goldwin, G. K. (1978). The use of hormone mixtures to increase the set of plum, *Prunus domestica* L., cv Victoria. *Journal of Horticultural Science* **53**, 123-129.

Wertheim, S. J. (1973). Fruit set and June drop in Cox's Orange Pippin apple as affected by pollination and treatment with a mixture of gibberellins A_4 and A_7. *Scientia Horticulturae* **1**, 85-105.

Wertheim, S. J. (1978). Intensive apple orchards with slender spindles. *Acta Horticulturae* **65**, 209-216.

Wickenden, M. F. (1979). Use of fruit-setting hormone sprays. *Report of the East Malling Research Station for 1978*, p. 57.

Williams, R. R. & Wilson, D. (1970). (Editors) *Towards Regulated Cropping.* London: Grower Books.

Limits to apple yields imposed by weather

J. J. Landsberg
Long Ashton Research Station, Long Ashton, Bristol, BS18 9AF

Introduction

Analysis of the effects of weather on the productivity of apple trees is complicated by a number of factors. Firstly, apples are woody perennials and the weather and condition of trees in season *n* affect their performance in season *n* + 1. Secondly, the economic end product is the result of a complex of processes beginning in season *n* − 1 including a phase (pollination) in which insect activity and the availability of compatible pollen are important. The growth and production cycle must, therefore, be considered phase by phase, and the effects of weather on the physiological processes which are most important at each phase must be evaluated. The cycle is depicted diagrammatically in Fig. 1.

Bud development commences shortly after full bloom and proceeds through the summer, but not all buds will produce floral primordia. The

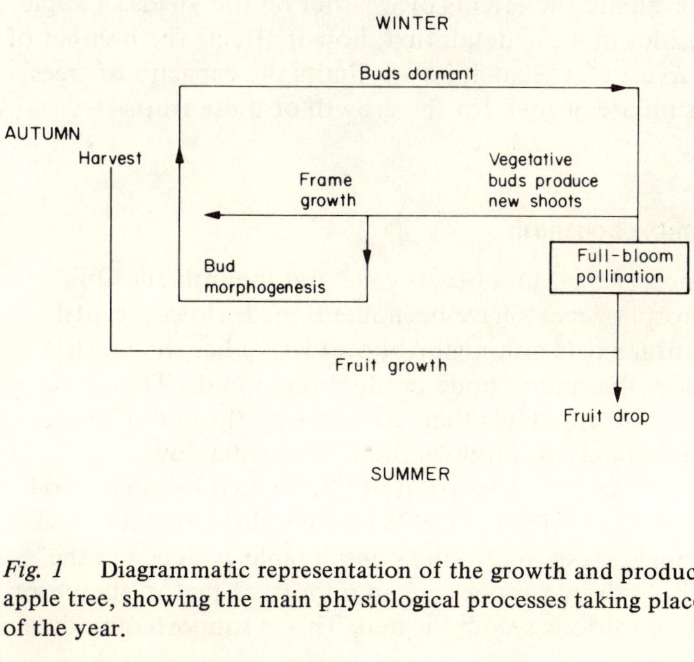

Fig. 1 Diagrammatic representation of the growth and production cycle of an apple tree, showing the main physiological processes taking place at different times of the year.

number that do may be affected by weather conditions and is likely to be reduced if the tree is carrying a large number of fruits.

Buds become dormant in autumn and normally require a period of chilling to break dormancy, after which subsequent growth to full bloom depends on temperature. This phase is discussed in some detail later. At full bloom, which occurs in late April-early May in the northern hemisphere and late October-early November in the southern hemisphere, the flowers are pollinated by insects. Clearly, the number of fruitlets produced depends on the number of flowers present and the effectiveness of pollination. There is then invariably some fruitlet loss (fruit drop), which is proportionately heavier if the number of fruitlets set is large. Fruit drop continues for about a month after pollination; there may be some subsequent losses but the number of fruits which a tree will carry to harvest is usually determined within about 6 weeks of pollination.

Since the fruit yield (Y) of an apple tree is given by fruit number (FN) $\times$ average fruit weight (FW), and since fruit weights vary less than number, FN is likely to be the major determinant of yield in any particular region. It has been shown that much of the variation in yield of mature Cox's Orange Pippin and young Golden Delicious trees, over several years, is accounted for by variation in fruit number (Landsberg, 1977). Although FW may range from about 50 to 250 g, it is unlikely to vary by more than a factor of about 3 in a given region. It is clear, therefore, that to evaluate the effects of weather on the yields of apple trees we must consider in more detail first, how it affects the number of fruits carried to harvest and second, how it limits the capacity of trees to produce the assimilate needed for the growth of these fruits.

Weather and bud morphogenesis

Both the importance of fruit number to yield and the influence of weather on bud morphogenesis have been mentioned. However, little definitive information exists although Abbott (1977) has shown that cool weather reduces the rate of node production in buds. Therefore, since buds which have formed less than 20 nodes by the end of the growing season are unlikely to produce floral primordia, low temperatures may reduce the proportion of buds which become floral (fruit buds). Landsberg and Thorpe (1975) reviewed the factors affecting bud morphogenesis and found considerable evidence in the literature that carbohydrate supplies are an important factor, therefore, leaf area and light conditions are implicated. This is supported by data and conclusions of Jackson and Palmer (1977), who showed that shading in year n resulted in fewer fruit buds in year $n + 1$. High radiation levels, therefore, are likely to lead to a higher proportion of fruit buds. Luckwill (1970) mentions that summer drought often results in prolific flowering of maiden trees, suggesting that this is associated with the early cessation of terminal growth, but Landsberg and Jones (1980) conclude from the literature that bud morphogenesis is more likely to proceed to the point where floral primordia are produced if trees are not subject to significant water deficits during the summer.

Despite the importance of environmental factors on bud morphogenesis the dominant influence on the process appears to be the number of fruits carried on a tree. This is well documented and is clearly illustrated by data from irrigated trees in Israel, where certainly light and temperature were not limiting (Table 1).

Table 1. Fruit numbers per tree in Israeli experimental orchards over 4 years. The data show strong bienniality, reflecting the effects of fruit number on flower number in the following year (Data by courtesy of I. Levin, R. Assaf, B. Bravdo).

Cultivar	Trees ha^{-1}	Year			
		1974	1975	1976	1977
Golden Delicious	83	100	311	304	1452
Golden Delicious	1100	197	1075	362	1442
Jonathan	1100	100	1177	82	828
Calville de St Sauveur	1100	313	925	447	1070

Dormancy and progression to full bloom; pollination

Dormancy in apple buds is a physiological state which requires a period of cold to break it; it is not a cold-induced condition, although apple buds usually enter dormancy at about the time when the weather becomes cold in late autumn-early winter. The advent of unseasonably early cold conditions may arrest development, leaving a number of buds in an incomplete transition (vegetative to floral) phase; such buds may become floral if conditions are suitable in the following spring, but they may remain vegetative. Early cold conditions may also affect the strength of blossom by reducing the period during which the buds grow in the autumn, and a cold autumn may affect ovule fertility. The first concept is qualitative and the effects in terms of pollination success or subsequent fruit retention have not been defined. There are no data on the effects of early autumn cold spells on ovule fertility.

The chilling required to break dormancy in apples can be quantitatively defined because the rate of growth of buds after the advent of favourable temperatures depends on the amount of chilling which the bud has experienced (Landsberg, 1974, 1979). If chilling is inadequate for complete breaking of dormancy the rate of bud growth in spring will be reduced and the period for which there are flowers at the full-bloom stage will be extended. In areas where the winters are not cold enough to break dormancy completely it is necessary, in commercial practice, to use oil sprays (e.g. in parts of Israel and South Africa).

There are many regression analyses of date of full bloom in relation to temperature summations. Landsberg (1974) has produced a model of the effects of temperature on the progression through dormancy and bud growth in spring, to full bloom, which invokes a mechanism for the process. The equation of the model is:

$$BG = \frac{A}{1 + b \exp(-k(I)P)} \tag{1}$$

where BG is an index of development and A is the asymptotic value, the units of BG and A are arbitrary, P is a temperature summation from a base of 5°C, i.e. $P = [\Sigma(\overline{T} - 5)]\ \Delta t$, $\overline{T}$ being mean daily temperature and t time in days. The value of the rate of coefficient (k) is controlled by a dormancy index I.

$$I = \int_0^i 1/\overline{T}(t)\ \mathrm{d}t. \tag{2}$$

If $\overline{T} < 5$°C it is taken as equal to 5°C, k is given by

$$k(I) = k_{max}\ (1 - \exp(-I/6{\cdot}7)). \tag{3}$$

Calculations are based on daily mean temperatures and start from the time buds enter dormancy, taken to be the first frost in autumn, or some other suitable (subjectively selected) starting point. A and b in eq. (1) are set to 100 and the median date of full bloom is reached when $BG = 95$, t days after commencement of dormancy.

The rate coefficient tends to k_{max} as I increases, reaching $0{\cdot}95k_{max}$ when $I = 20$, which implies a requirement for about 100 days with $\overline{T} \leqslant 5$°C before dormancy is completely broken. A reaches 95 when the term (b.exp($-k(I)\,P$)), in the denominator of eq. (1), reduces to about 0·05; with $k = k_{max} \simeq 0{\cdot}02$ this occurs when $P \simeq 380$. P must be greater than this if $k < k_{max}$. This model has been tested and found to be remarkably accurate (Landsberg, 1979).

Equations (1)–(3) can be used to explore the likely dates of full bloom in any region for which suitable temperature data are available,

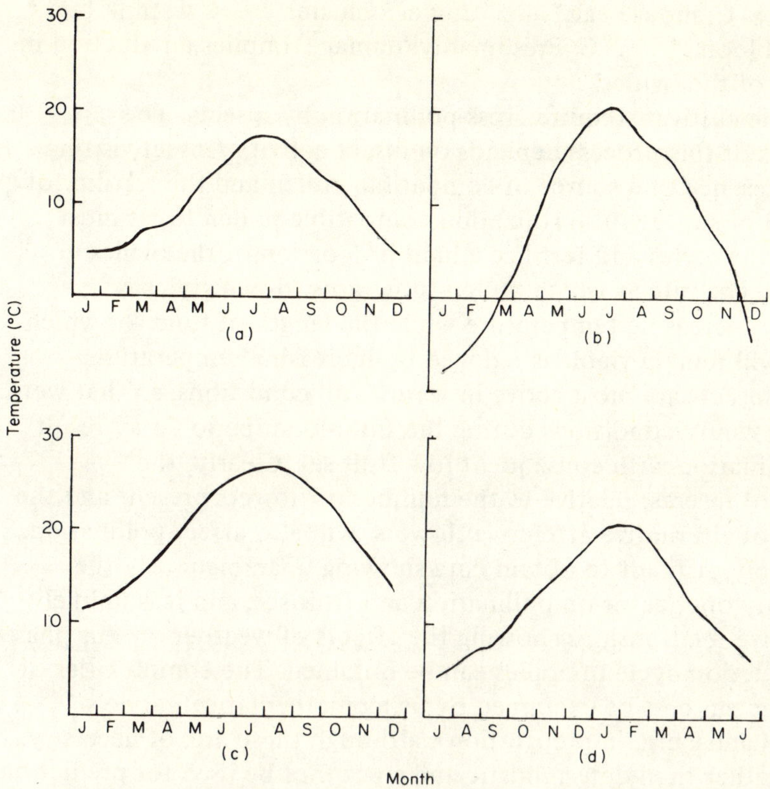

Fig. 2 Average monthly temperatures (°C) in (a) south-west England (b) Ontario, Canada, (c) central Israel, (d) New South Wales, Australia.

and hence to assess the risk of frost damage to flowers. Figure 2 shows the range of average monthly temperatures for four regions where apples are grown. Hogg (1970) states that on average the last spring frost occurs in the first 10 days of March in southern European apple-growing areas, and even in years when it is late it does not occur after mid-April. Further north there is a tendency for the last frost to occur, on average, before 20 April, although this is not true of the main English apple-growing areas. Similar dates should be available for all growing regions, but they may vary quite considerably within a region, depending on topography. Hogg makes the point that frost data are of greater value when expressed as probabilities of occurrence on or after given dates during the spring – they should be used in this way with the bud-development model.

The effects of frost on apple buds and blossoms varies with the stage of development reached. Teskey (1972) suggests the following guide to temperatures which may damage or destroy apple flower buds: buds breaking,* − 8°C; mouse ear,* − 5°C; green cluster, *− 4°C; pink bud,* − 2°C; full bloom,* − 2°C. Presumably 'damage' implies a reduction in the fertility of the ovules.

Most apple cultivars require cross-pollination by insects. The effectiveness of this process depends on insect activity (insect visiting rate), the presence of a source of compatible pollen and the fertility of the ovules. For successful fertilization compatible pollen tubes must grow down the styles and fertilize a high proportion of the ovules in each flower. The rate at which pollen tubes grow down styles is increased by increasing temperature while the length of time for which the ovules will remain viable is reduced by increasing temperature. Pollinating insects are most active in warm, still conditions, so that wet, cold and/or windy conditions during the full-bloom period may result in poor pollination with consequent low fruit set. Clearly, the population of insects, relative to the number of flowers present and the availability of alternative attractive flowers, will also affect pollination. It is extremely difficult to obtain data showing unambiguously the effects of any one factor on pollination and fruit set, and it is unlikely that definitive relationships showing the effects of weather on this phase of the production cycle of apples can be obtained. The complexities of the situation can best be examined by using mathematical models (Brain and Landsberg, in preparation), although these are, of necessity, stochastic rather than deterministic and so cannot be used for prediction.

Fruitlet retention depends not only on the number of fruits present, relative to carbohydrate supply, but also on the number of seeds per fruit (Quinlan and Preston, 1968; Luckwill, Weaver and Macmillan, 1969). It is therefore strongly dependent on the effectiveness of pollination: if a considerable proportion of flowers are poorly pollinated, so that there are only a few seeds in each fruitlet, losses will be high. Landsberg and Brain (1978) have successfully modelled fruit drop in terms of seed number per fruit.

Water requirements

Apples are mesophytic plants not adapted to extreme water stress. In many of the temperate production areas of the world water is not a

* The progress of bud growth in apples in spring is charted, in horticultural parlance, by a series of stages: swelling, breaking, bud burst, mouse ear, green cluster, pink bud and petal fall. At full bloom each bud produces a cluster of about 6 flowers. Fruitlets are readily identifiable 2-3 weeks after petal fall. These stages are useful markers but cannot be said to have any real physiological significance.

problem and it is often difficult to demonstrate significant differences between irrigated and non-irrigated trees, or between different irrigation treatments. In low rainfall areas irrigation is standard practice.

Fruitlet retention is definitely affected by water stress. Powell (1974) used soil covers to induce stress in 9-year-old Cox trees in England and found that fruit set, as a percentage of flower number, ranged from 8·5 per cent on trees which received no water between the end of March and June, through 15·7 per cent in trees watered by rain, to 24 per cent in well irrigated trees. The reduction in fruit drop was caused by an increase in the number of fruitlets retained per flower cluster, with a considerable reduction in the number of flower clusters which contained no fruitlets. Since water stress is a function of atmospheric conditions (evaporative demand) and soil moisture, with evaporative demand being the more important component in many situations, fruit drop is likely to be greater in areas where spring is dry and hot than in areas where spring is mild and moist.

The general effects of water stress in apple trees are similar to those in other plants: leaf growth is reduced (Chapman, 1973), trunk girth is reduced (Assaf, Bravdo and Levin, 1974; Assaf, Levin and Bravdo, 1975; Goodes, Higgs and Hyrycz, 1978) and fruit diameters are reduced (Assaf *et al.* 1974, 1975). There have been many irrigation experiments on apple trees but there are virtually no good estimates of the extent to which water stress, induced by lack of rainfall, constitutes a limit to productivity. Estimates could be made by calculating rainfall probabilities and the rates of water use of orchards, although such calculations are complicated by considerations such as the type of inter-row vegetation (if any) and unknown factors such as losses due to interception by trees and losses by evaporation from bare soil after wetting. In general it is probably safe to say that, if the soil profile is full in the spring and if rainfall is not less than water use by the trees until about mid-season, then water stress will not significantly limit the productivity of apple trees. The reasons for this lie in stomatal control of water loss and in the large soil volume exploited by apple roots.

The stomata of apple leaves have been shown to be sensitive to air humidity (West and Gaff, 1976; Warrit, Landsberg and Thorpe, 1980); stomatal conductance decreases with increasing leaf-to-air vapour-pressure deficit. This has the consequence that the rate of transpiration by apple trees reaches a limit and may in fact be lower at high radiant-energy levels and in dry conditions than under conditions of lower evaporative demand (Thorpe, Warrit and Landsberg, 1980; Landsberg and Butler, 1980). These authors have analysed this phenomenon in detail and it is illustrated in Fig. 3, where mean daily transpiration rates of apple leaves in four regions with very different climates are presented (cf. temperature data in Fig. 2, radiation received during the growing season, Table 3). This control of transpiration rates by stomatal

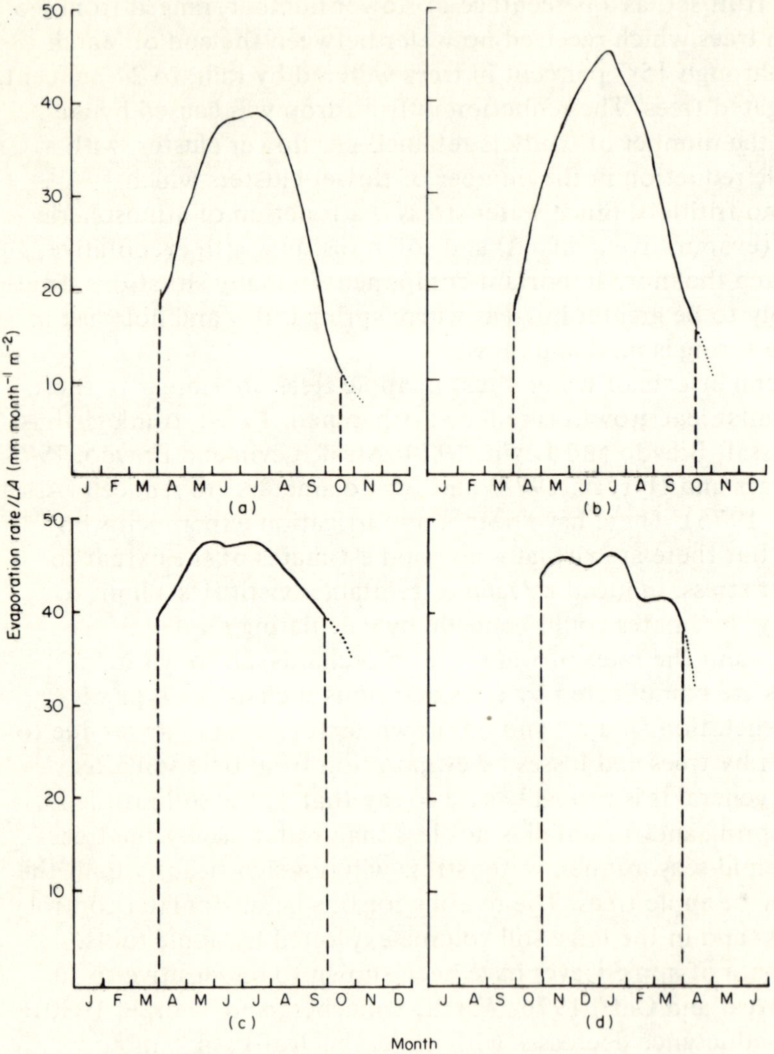

Fig. 3 Calculated average monthly transpiration rates (E, mm month^{-1}) per unit leaf area of apple trees (m^2), in (a) south-west England, (b) Ontario, Canada, (c) central Israel, (d) New South Wales, Australia. The calculations were made using energy-balance equations and a model (Thorpe *et al.*, 1980) describing the tendency of stomata to close in dry conditions. The vertical lines are drawn at the (approximate) dates of leaf appearance and fruit harvest.

response to humidity implies that tissue water potentials in apple trees in markedly different regions will be similar, but it also means that photosynthesis may be reduced in dry regions.

Carbohydrate production and partitioning

The previous discussion of the effect of weather on fruit numbers leads to a consideration of the carbon balance of apple trees, which must obey the law of conservation of mass so that, given fruit numbers, the problem reduces to one of determining the mass of carbon compounds which a tree can produce, and the fate of those compounds. Nutrition is not dicussed since we may assume that limits to yield imposed by nutrition can be eliminated, while as far as the carbohydrate balance is concerned the contribution of inorganic substances to the total mass is small and, for present purposes, can be taken as constant.

Of the assimilates produced by photosynthesis some are utilized for respiration and the remainder are allocated to the growth of leaves, tree frame (shoots, branches, trunk), roots and fruits. In apple trees a variable proportion may be stored as high-molecular-weight carbohydrates which can be metabolized for later use. This proportion is thought to influence the growth of buds in the following spring, and subsequently of fruits, but there is no quantitative evidence for this. For present purposes we will consider only net carbohydrate production and the way it is used for growth.

If ΔW is the change (increase) in dry weight of a tree over a time period Δt, then the weight change of any component, j, of that tree is $p_j \, \Delta W$, where $p_j \leqslant 1$. Dry-weight gain and partitioning by an apple tree are defined completely, therefore, by

$$\frac{\Delta W}{\Delta t} = \frac{\Delta W}{\Delta t}\,(p_{\mathrm{L}} + p_{\mathrm{S}} + p_{\mathrm{R}} + p_{\mathrm{F}}), \tag{4}$$

where the subscripts L, S, R and F denote leaves, frame, roots and fruits, and $p_{\mathrm{L}} + p_{\mathrm{S}} + p_{\mathrm{R}} + p_{\mathrm{F}} = 1$. The implications of eq. (4) are obvious: any condition which leads to an increase in $\Delta W/\Delta t$ without affecting the values of p_j will cause an increase in growth of all components in the same proportions, and anything which causes a change in any p_j value will alter the growth pattern of the tree. The aim of the horticulturist, therefore, is to achieve the maximum possible value of p_{F} commensurate with the highest possible value of ΔW.

Dry-weight gain is a result of assimilate production by photosynthesis, which depends on the interception of short-wave radiant energy by leaves. Thorpe *et al.* (1978) have shown that it is possible to estimate assimilate production of an apple tree by calculating the interception of light by the tree and combining this with information on the photosynthetic characteristics of leaves. Heim, Landsberg, Watson and Brain (1979) showed that differences in dry-matter production between trees in two different regions were proportional to leaf-area duration and total short-wave radiant energy. They also showed that the proportion of dry-weight increment allocated to fruits was a function of FN/LA the ratio of number of fruits per tree to unit leaf area (see Fig. 4).

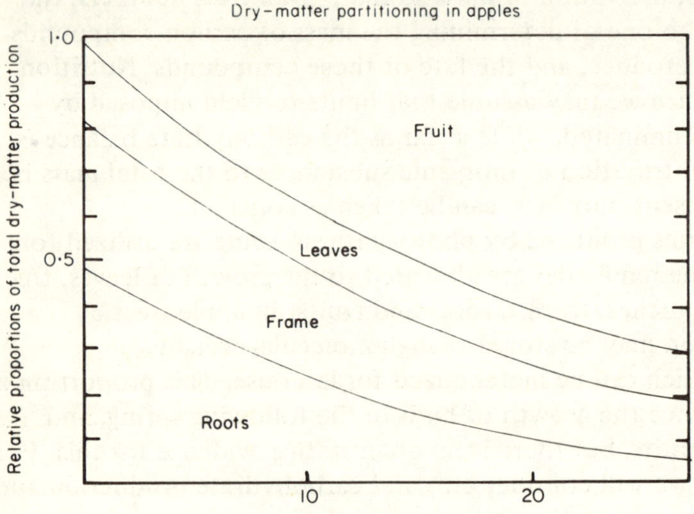

Fig. 4 Diagram showing the influence of fruit number per unit leaf area on the proportion of total dry-weight gain allocated to different components of apple trees. Re-drawn from Heim *et al.* (1979).

The importance of *FN/LA* has been recognized for some time (e.g. Hansen, 1969) and the factors causing variation in fruit size, colour, dry-matter content etc., at any given value of *FN/LA*, have been discussed in detail by Hansen (1977). Despite these variations it is clear from Fig. 4, and the smooth curves of fruit weight against *FN/LA* given by Hansen (1977, Fig. 1), that this parameter provides a useful measure of a fruit-sink strength and hence p_F; in fact the curves in Fig. 4 can be described by equations of the form

$$p_j = a + b \ln (FN/LA). \qquad (5)$$

The relationships discussed above indicate that, given information on fruit numbers, an analysis of apple yields in relation to weather should be based on estimates of radiation interception and p_F. We can assume that $\Delta W \propto$ intercepted radiation and $p_F \propto FN/LA$, respectively.

Fruit yields in relation to radiant energy

In this section, data on fruit yields per tree, corresponding fruit numbers, tree size and weather conditions during several growing seasons, kindly provided by scientists at a number of centres round the world, have been analysed using the principles outlined above. In each case the data came from several experimental orchards of varying types (tree size, population, cultivar).

It was stated earlier that fruit number per tree is likely to be the major determinant of yield. Therefore, data on yield per tree from Holland, Denmark, Australia, South Africa and the UK were included in a regression of yield on fruit number, without regard to tree size, cultivar or any other parameter. The result was:

$$Y = 0{\cdot}15(\pm 0{\cdot}013)FN - 0{\cdot}13(\pm 5{\cdot}25), \qquad r^2 = 0{\cdot}73 \qquad (6)$$

(r^2 = the proportion of the variation in Y accounted for by the regression) The intercept can be taken as zero.

Since fruit number accounts for more than 70 per cent of the variation in yield, it follows that the remainder must be accounted for by fruit weight. Following the reasoning outlined earlier we may assume that the total dry matter produced between full bloom and harvest (ΔW) is proportional to the radiant energy intercepted by the trees during the same period. This can be approximated by multiplying the total short-wave radiant energy incident on a horizontal surface (the radiation integral, $\Sigma Q\ \Delta t$ MJ m^{-2}) by the leaf area of the trees (LA), i.e.

$$\Delta W \propto \Sigma Q\ \Delta t \times LA.$$

The faction of ΔW allocated to fruits (p_F) is proportional to FN/LA, therefore

$$\Delta W_F\ (= Y) = \Delta W p_F,$$

that is,

$$Y \propto \Sigma Q \Delta t\ LA \times FN/LA;$$

hence

$$Y \propto \Sigma Q \Delta t\ FN.$$

Using the mean daily radiation values for each month (measured, except in the case of Denmark for which it was calculated from sunshine hours) $\Sigma Q \Delta t$ was calculated for the period between the dates of full bloom (given) and harvest (estimated) for 47 sets of data, corresponding to those used to establish eq. (6). (Data from Israel were not included because only long-term average radiation data were available. No fruit numbers were available from Ontario.)

Regressing Y on $\Sigma Q \Delta t \, FN$ gave

$$Y = 0{\cdot}53(\pm 0{\cdot}018)\Sigma Q \Delta t \, FN \times 10^{-4} - 2{\cdot}5(\pm 2{\cdot}2), \qquad r^2 = 0{\cdot}95. \quad (7)$$

The intercept can be taken as zero.

This is a great improvement compared with the regression on FN alone and, when we consider the range of cultivars, tree sizes etc. included in the regression, the r^2 value is remarkably high. It is of considerable interest that tree yields from Israel, where $\Sigma Q \Delta t$ was 4600 MJ m^{-2}, were considerably higher than elsewhere (see Table 2) and would not have fallen around the line defined by eq. (7); the slope of Y against $\Sigma Q \Delta t$ for the Israeli data would be about 0·32.

Equation (7) indicates that for a given FN, tree yields in 'high-energy' regions will be larger than in regions receiving less radiant energy, implying that the difference must be due to higher average fruit weights (FW). Since $Y = FN \times FW$, the fruit weights likely in particular regions (or seasons) can be calculated from eq. (7); alternatively we may calculate the amount of radiant energy needed to produce fruits of a given (average) weight. For example, if the market requires fruits of about 100 g (equivalent to a diameter of 60-65 mm), then $FN \times FW/0{\cdot}53 \times 10^{-4} \times FN = FW/0{\cdot}53 \times 10^{-4}$ gives the required value of $\Sigma Q \Delta t$, in this case about 1900 MJ m^{-2}. The confidence limits of the regression coefficient (95 per cent limits are from approximately 0·49 to 0·57) indicate that average fruit weights in such an area (or season) are likely to be between 0·092 and 0·108 kg. The effects of radiation and temperature on fruit weight are examined in more detail later.

To evaluate other implications of eq. (7) we require some information on tree sizes and fruit loads. Using data obtained from Australia (J. Campbell, personal communication), from published sources and collected at Long Ashton Research Station, the relationship between leaf area tree^{-1} (LA, m^2) and trunk girth (G, cm) or diameter (D, cm) was determined. This is

$$LA = 0{\cdot}0023\,(\pm 0{\cdot}0014)G^{2{\cdot}8(\pm 0{\cdot}19)} \qquad (r^2 = 0{\cdot}96) \qquad (8)$$
$$= 0{\cdot}057\,D^{2{\cdot}8}\,.$$

Although there is considerable between-tree variation in LA, which is affected by growing conditions, pruning practice etc., this is a useful relationship. Optimum FN may be defined as the maximum number of fruits which a tree can carry without reducing average fruit weight below commercially acceptable levels. (This is not necessarily the optimum in terms of bud morphogenesis). Information on optimum *FN* tree^{-1} was obtained by selecting the highest yields tree^{-1} from among the data available (two or three values were selected from each region) and assuming that when these yields were attained *FN* was near the optimum for the trees in question (in no case was it too large since average *FW* in all cases selected was greater than 0·1 kg). In each case *LA* was calculated using eq. (8), hence *FN*/*LA* could be obtained. When $\log_e$ (*FN*/*LA*) was plotted against *D* for the trees from which the yields came, the points fell close to a straight line (Fig. 5), which is defined by

$$FN/LA(\text{opt.}) = 1{\cdot}0 \exp(4{\cdot}0 - 0{\cdot}133D). \qquad (9)$$

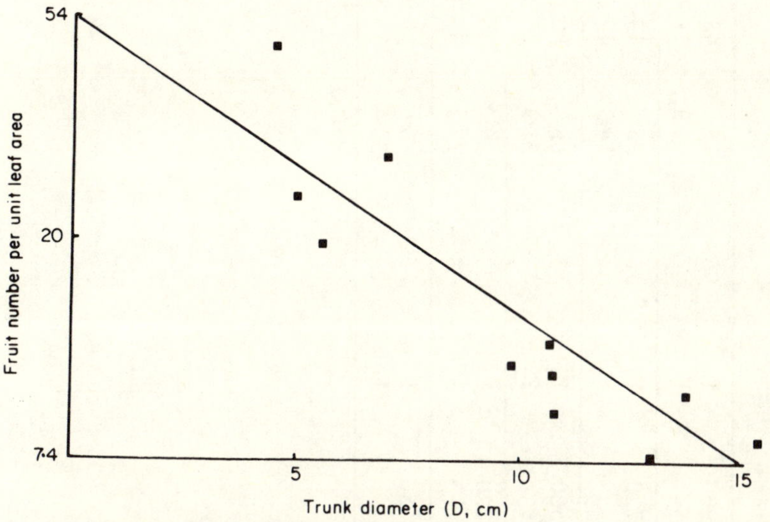

Fig. 5 Estimates of the optimum fruit number/unit leaf area *FN*/*LA* plotted on a logarithmic scale against trunk diameter (cm). The line is drawn by eye.

Because of the assumption and approximations on which eq. (9) is based it was not fitted statistically; it should only be regarded as a useful approximation and significance levels and standard errors may suggest a spurious level of precision. However, confidence in the relationship is increased by the fact that, for trees with $D \simeq 2$ cm, the size on which Fig. 4 is based (see Heim *et al.* 1979), the predicted *FN*/*LA* (optimum) is 41, which is close to that indicated by Fig. 5. It is also close to the optimum value indicated by Hansen (1977, Fig. 3) from his work on young trees. Equation (9) indicates, as intuition would suggest, that the number of fruits supported and grown to useful size by unit leaf area on large trees is considerably lower than on small trees. For example, for $D = 5$ cm *FN*/*LA* (optimum) = 28 m^{-2}, while for $D = 12$ cm the value is 11. This is to be expected because of the increase in total respiration

Table 2. Optimum apple yields for different tree populations in different energy environments. Yields are presented in kg tree^{-1} (Y_1) and t ha^{-1} (Y_2) and $\Sigma Q \Delta t$ is the total solar radiation during the growing season (MJ m^{-2}).

Trees ha^{-1}	Likely trunk diameter at maturity (cm)	Maximum LA tree^{-1} (m^2)	Optimum FN tree^{-1}	Yields $\Sigma Q \Delta t$ =	2000	2500	3000	3500	4000	4500
83	15	120	900	Y_1	95	119	143	167	191	215
				Y_2	7·9	9·9	11·9	13·9	15·8	17·8
150	12	65	720	Y_1	76	95	114	134	153	172
				Y_2	11·4	14·3	17·2	20·0	22·9	25·8
330	9	30	500	Y_1	53	66	80	93	106	119
				Y_2	17·5	21·9	26·2	30·6	35·0	39·4
1000	6	9	220	Y_1	23	29	35	41	47	53
				Y_2	23·3	29·2	35·0	40·8	46·6	52·5
4600	3	1·5	50	Y_1	5·3	6·6	8·0	9·3	11	12
				Y_2	24·4	30·4	36·8	42·8	48·8	54·7

with increase in the mass of wood present. We should note that the definition of 'optimum fruit number' as used here is applied to a single season and it does not imply an optimum in terms of effects on yields in following seasons.

The data analysed here related to orchards with populations ranging from about 80 to 4600 trees ha^{-1} and indicated that the corresponding trunk diameters of mature trees in such orchards range from 15 to 1·5 cm. To calculate optimum fruit yields for such a range of orchard types in regions with different radiation integrals a set of 5 trunk sizes was selected. Leaf areas were calculated using eq. (8) and the appropriate tree populations were selected by making use of another useful approximation which has become apparent from the study of available information, viz. that the leaf-area index for apple orchards as a whole tends to be about unity. Optimum *FN*/*LA* was then calculated using eq. (9) and yields per tree were calculated using eq. (7). The results are presented in Table 2, where a threefold difference in yield potential between high density and large bush orchards in any particular energy environment is indicated.

Equation (7) implies that differences in yield between trees with the same *FN*, in different regions or seasons, are attributable to differences in radiant-energy income. However, since it is reasonable to assume that fruit-growth rate is a function of temperature, and hence that final fruit size is related to the integral of temperature in terms of time, this was calculated as the summation of daily mean temperatures × days (day-degrees, $\Sigma \overline{T} \, \Delta t$), and fruit weights were regressed on $\Sigma \overline{T} \, \Delta t$.

The regression was highly significant and gave the equation

$$FW = -44(\pm 30) + 0{\cdot}07(\pm 0{\cdot}012)\, \Sigma T \, \Delta t, \qquad r^2 = 0{\cdot}48.$$

This implies that $\Sigma T \, \Delta t$ must exceed about 2000 day-degrees for average fruit weights to exceed 100 g, which is equivalent to an average daily temperature of about 16·5°C for a 120 day growing season. However, the regression only accounts for about 50 per cent of the variation in *FW*. Inclusion of *FN*/*LA* in the regression did not significantly improve it, although this may have been due to the inaccuracies inherent in the estimates of leaf area. Better estimates of *FW* will therefore be obtained from eq. (7). Radiation and temperature are of course correlated, although the correlation is not as good as is sometimes thought; Fig. 6 presents mean daily temperatures (°C) for each month of the growing season for all the regions for which data were available, plotted against the corresponding mean daily short-wave energy income values ($MJ\ m^{-2}$). There is considerable scatter in the relationship. However, since fruit temperatures are strongly dependent on the radiant-energy load to which the fruits are subject (Thorpe, 1974) the regression of fruit weight on radiant energy includes temperature effects.

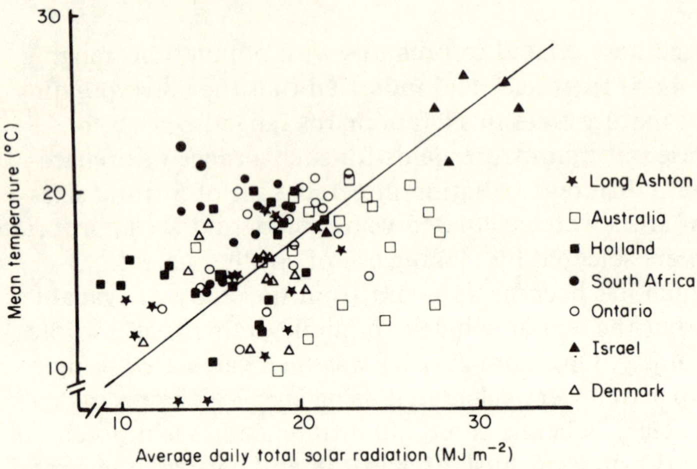

Fig. 6 Average daily temperatures for each month of the growing season plotted against average daily total solar radiation for the corresponding months. Data are for several years from Australia (□), Denmark (○), Israel (☆), Long Ashton (★), Netherlands (●), Ontario (△) and South Africa (■). Note that the correlation between temperature and radiation is not good.

Discussion

From the information presented in this paper, we can conclude that apple yields may be limited by weather acting through its effect on bud morphogenesis via the carbon balance, interacting with fruit load, and the effects of temperature on rates of node production. Apple trees are not particularly susceptible to extreme cold in winter (witness their success in Ontario) but inadequate winter chilling may influence yield by causing delayed and irregular flowering and consequent problems at pollination time. Furthermore the rate at which flowers grow in spring, and the time of full bloom, are strongly dependent on spring temperatures. Frost in spring is a major limiting factor and more statistical studies on frost occurrence in relation to flower bud development must be valuable in assessing the probability of a successful crop in any given region. Because irrigation is widely used in areas normally subject to severe moisture deficits, water does not generally limit apple yields. In temperate regions responses to irrigation vary with season and the returns on capital investment may be small.

For any given *FN* the major factor limiting apple yields is radiant energy. The analysis presented here has ignored factors such as tree shape and its influence on radiation interception (see Jackson, p. 151), but the regression of tree yield on fruit number multiplied by the energy integral, obtained using data from all over the world, provides a very useful tool for assessing the influence of radiant energy on apple yields. The apparently low efficiency with which trees in Israel, a high-energy environment, converted radiant energy to fruit, suggests that the photosynthetic mechanisms of the leaves of those trees were light-saturated for much of the time. It is also likely that photosynthesis is restricted by stomatal closure. We may tentatively suggest that a radiation integral of about 4000 MJ m^{-2} over the growing season is optimum for apples. Equation (7), and the data in Table 2, suggest that apple production is unlikely to be economic in regions where the radiation integral is less than about 2000 MJ m^2.

Table 3. Summary of climatic data during the growing season for different apple-growing regions.

	Total solar radiation (MJ m^{-2})	Average temperature (°C)	Accumulated temperature (°C × days)
Holland	2200	16·7	2077
New South Wales, Australia	3500	17·7	2675
Cape Province, South Africa	2600	16·7	2780
Southern Ontario, Canada	2500	17·3	2440
Israel	4600	24·8	3700
Denmark	2000	14·9	1821
South-west UK	2400	14·9	2100

Table 3 is a summary of the climatic data during the growing season for the apple-growing regions from which the data analysed in this paper have come. These data probably encompass the conditions under which apples can be produced successfully.

The equations presented here (eqs. (7)-(9)) provide a means of assessing potential apple productivity in any region or season, and eq. (7) provides a means of estimating actual yields, which may well be worth testing against yields from individual orchards. In this case the equation should be used in association with information on the variation in *FN* $tree^{-1}$ within orchards, adapting the method of 'every-tree approximation' used in forestry. The result would be a series of yield classes, each containing a known proportion of the population, and orchard mean yield would be derived from a weighted average of these classes. Regression analyses of the effects of the previous crop, and autumn and spring weather, on *FN*, incorporating trunk girth to account for the effects of tree size, are being done at the moment. It is hoped that this work will provide an equation which can be used, in the early summer, to predict *FN*. This will greatly enhance the predictive and analytical value of the equations presented here.

Acknowledgements. I am grateful to the scientists listed below for their willing help in sending me the data on apple yields and weather which provided the basis for yield/weather analysis in this paper:
Dr J. Tromp, Proefstation voor de Fruitteelt, Wilhelminadorp, Netherlands. Data contributed by Dr Tromp and Dr S. J. Wertheim.
Dr Paul Hansen, State Research Centre for Horticulture, Blanstedgaard, Denmark.
Dr Israel Levin, Institute of Soils and Water, The Volcani Centre, Bet Dagan, Israel. Data contributed by Dr Levin, Dr R. Assaf and Dr B. Bravdo. Climatic data from Professor G. Stanhill.
Professor J. T. Proctor, The University of Guelph, Ontario Agricultural College, Department of Horticultural Science. Data from Horticultural Research Institute of Ontario contributed by Professor Proctor and Mr. W. F. Pierce.
Mr O. Bergh, Fruit and Fruit Technology Research Institute, Stellenbosch, Cape Province, Republic of South Africa.
Ms Jill Campbell, Organge Agricultural Research Centre, NSW, Australia.
I am indepted to Mr W. J. Redmond for a translation of Hansen's (1977) paper.

References

Abbott, D. L. (1977). Fruit bud formation in Cox's Orange Pippin. *Report of Long Ashton Research Station for 1976,* pp. 167-176.

Assaf, R., Bravdo, R. & Levin, E. (1974). Effects of irrigation according to water deficit in two different soil layers, on the yield and growth of apple trees. *Journal of Horticultural Science* **49**, 53-64.

Assaf, R., Levin, I. & Bravdo, B. (1975). Effect of irrigation regimes on trunk and fruit growth rates, quality and yield of apple trees. *Journal of Horticultural Science* **50**, 481-493.

Chapman, K. R. (1973). Effect of four water regimes on performance of glasshouse-grown nursery apple trees. *Queensland Journal of Agricultural and Animal Sciences* **30**, 125-135.

Goode, J. E., Higgs, K. H. & Hyrycz, K. J. (1978). Nitrogen and water effects on the nutrition, growth, crop yield and fruit quality of orchard-grown Cox's Orange Pippin apple trees. *Journal of Horticultural Science* **53**, 295-306.

Hansen, P. (1969). ^{14}C-studies on apple trees. IV. Photosynthate consumption in fruits in relation to the leaf-fruit ratio and the leaf-fruit position. *Physiologia Plantarum* **22**, 186-198.

Hansen, P. (1977). Blad-frugt relationer og frugtudvikling hos aebletraer. *Tidsskrift for Planteavl.* **81**, 123-132.

Heim, G., Landsberg, J. J. Watson, R. L., & Brain, P. (1979). Eco-physiology of apple trees : dry matter production and partitioning by young Golden Delicious trees in France and England. *Journal of Applied Ecology* **16**, 179-194.

Hogg, W. H. (1970). Climatic factors in apple growing. *Experimental Horticulture* **21**, 67-92.

Jackson, J. E. & Palmer, J. W. (1977). Effects of shade on the growth and cropping of apple trees. II. Effects on components of yield. *Journal of Horticultural Science* **52**, 253-266.

Landsberg, J. J. (1974). Apple fruit bud development and growth: analysis and an empirical model. *Annals of Botany* **38**, 1013-1023.

Landsberg, J. J. (1977). Studies on the effect of weather on the growth and production cycle of apple trees. *Journal of the Royal Horticultural Society of England* **138**, 116-133.

Landsberg, J. J. (1979). From bud to bursting blossom: weather and the apple crop. *Weather* **34**, 394-407.

Landsberg, J. J. & Brain, P. (1978). Model of apple tree growth: fruit drop. *Report of Long Ashton Research Station for 1977,* pp. 119-120.

Landsberg, J. J. & Butler, D. R. (1980). Stomatal response to humidity: implications for transpiration. *Plant, Cell and Environment* **3**, 29-33.

Landsberg, J. J. & Jones, H. G. (1980). Apple orchards. In *Water Deficits and Plant Growth,* Vol. VI. Ed. T. T. Kozlowski. London: Academic Press. (in press).

Landsberg, J. J. & Thorpe, M. R. (1975). The mechanisms of apple bud morphogenesis: analysis and a model. *Annals of Botany* **39**, 689-699.

Luckwill, L. C. (1970). The control of growth and fruitfulness of apple trees. In *Physiology of Tree Crops,* pp. 237-254. Eds. L. C. Luckwill & C. V. Cutting. London: Academic Press.

Luckwill, L. C., Weaver, P. & Mcmillan, J. (1969). Gibberellins and other growth hormones in apple seeds. *Journal of Horticultural Science* **44**, 413-424.

Powell, D. B. B. (1974). Some effects of water stress in late spring on apple trees. *Journal of Horticultural Science* **49**, 257-272.

Quinlan, J. D. & Preston, A. P. (1968). Effects of thinning blossom and fruitlets on growth and cropping of Sunset Apple. *Journal of Horticultural Science* **43**, 373-381.

Teskey, J. E. (1972). Critical temperatures for fruit blossoms. *Ontario Ministry of Agriculture and Food: Factsheet.*

Thorpe, M. R. (1974). Radiant heating of apples. *Journal of Applied Ecology* **11**, 755-760.

Thorpe, M. R., Warrit, B. & Landsberg, J. J. (1980). Responses of apple leaf stomata : a model for single leaves and a whole tree. *Plant, Cell and Environment* **3**, 23-27.

Thorpe, M. R., Saugier, B., Auger, S., Berger, A. & Methy, M. (1978). Photosynthesis and transpiration of an isolated tree: model and validation. *Plant, Cell and Environment* **1**, 269-277.

Warrit, B., Landsberg, J. J. & Thorpe, M. R. (1980). Responses of apple leaf stomata to environmental factors. *Plant, Cell and Environment* **3**, 13-22.

West, D. W. & Gaff, D. F. (1976). The effect of leaf water potential, leaf temperature and light intensity on leaf diffusion resistance and the transpiration of leaves of *Malus sylvestris. Physiologia Plantarum* **38**, 98-104.

Section 3 **Cultural and Technological Aspects**

Effects of manipulating root and air temperatures on tomato growth and the efficient use of energy

J. Moorby
Glasshouse Crops Research Institute, Worthing Road, Rustington, Littlehampton, Sussex, BN16 3PU.
Present address: Agricultural Research Council, 160 Great Portland Street, London W1N 6DT.

Introduction

Growers are concerned with economic returns and these do not often equate with total biological yields. In much of horticulture, and especially tomato production, the major aim is to maximize yields in the early part of the season when biological yields are low but prices are high.

Until a few years ago the average costs of producing an early tomato crop were divided approximately equally between labour, heating and the remaining other costs such as spraying and marketing. It was in this situation that the blueprint followed by most growers was developed (Calvert, 1972, Ministry of Agriculture, Fisheries and Food, 1976). Neglecting small variations in temperature imposed to prevent abortion of the first truss of fruit, the day and night temperatures recommended in the blueprint are 20 and 13-16 °C respectively. The lower night temperatures tend to be used in the more northerly growing areas where solar radiation is less and planting takes place later.

Since the development of the blueprint, the price of oil for heating has at least quadrupled and therefore the whole basis for assessing the most sensible growing conditions has been changed. In addition to these economic considerations, there is the more altruistic question of whether it is justifiable to use between 25 and 33 per cent of the total quantity of oil used in agriculture to heat glasshouses (Winspear, 1978b). I will ignore this last question and confine myself to means of saving energy whilst maintaining or, if possible, increasing yield.

Energy requirements

In order to set limits to the problem it is necessary to provide some indication of the energy costs of heating a glasshouse. This would seem to be a reasonable requirement, but it is difficult to satisfy. A single boiler usually supplies a range of houses, often run at different temperatures. Further, it would be foolhardy to extrapolate from a single set of data because the energy required is very dependent on the exposure and orientation of the house and the presence or absence of windbreaks. Because of these problems some information has been extracted from theoretical tables produced at the National Institute of Agricultural Engineering (Bailey, 1977). The tables were derived using a value of 6·2 W m^{-2} K^{-1} for the thermal transmittance, an average annual transmission of direct and diffuse radiation of 62 per cent and measurements of solar radiation made at Kew. Allowance is made for

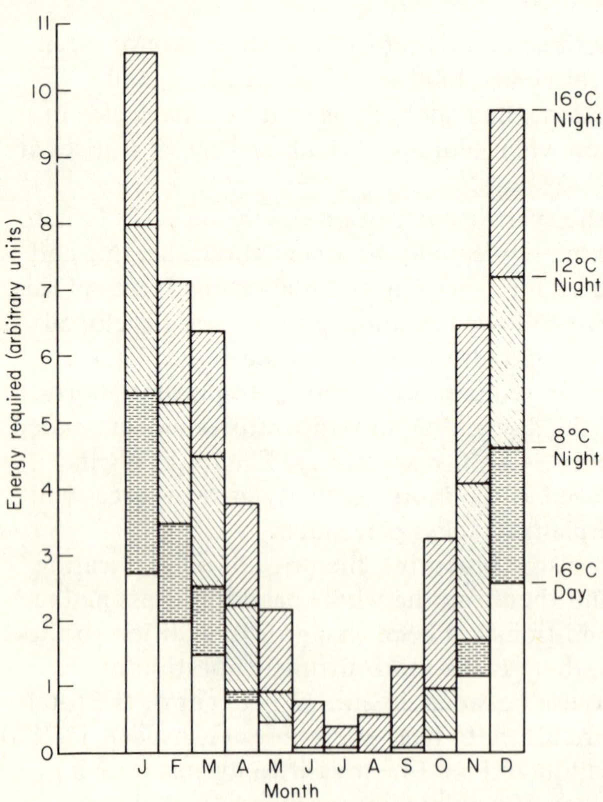

Fig. 1 The energy required to maintain a day temperature of 16°C (□) and a night temperature of 8°C (▤); the extra energy required to maintain night temperatures of 12°C (▧) and 16°C (▨). Calculated from Bailey (1977).

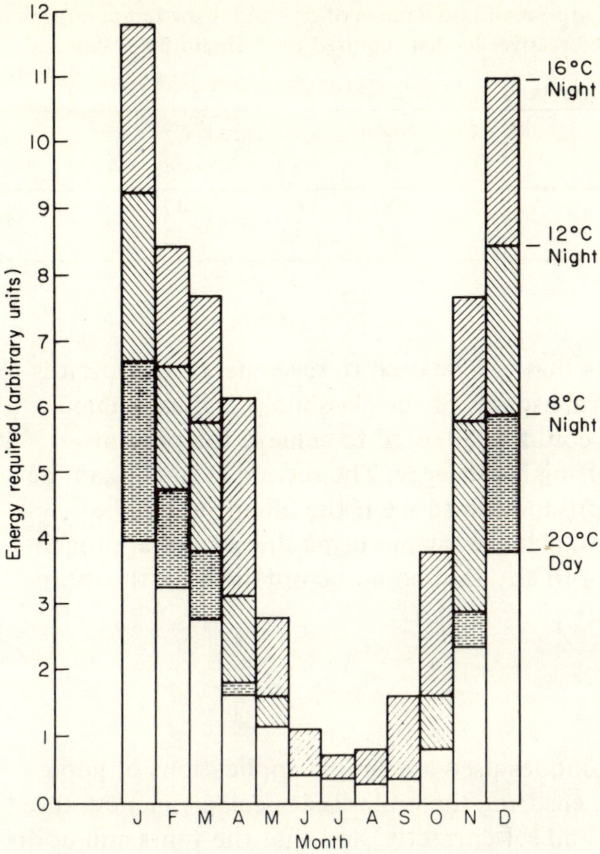

Fig. 2 The energy required to maintain a day temperature of 20°C and night temperatures of 8°C, 12°C and 16°C. Symbols as in Fig. 1. Calculated from Bailey (1977).

the thermal gain from solar radiation; any further increase in temperature demands the use of other forms of heating. Figures 1 and 2 show the energy required throughout the year to maintain two different day temperatures and a range of night temperatures. It can be seen that about twice the amount of energy is required to maintain a night temperature of 16 °C than to maintain the day temperature at 20 °C. Further, a decrease of 4°C in the night temperature saves about twice the amount of energy as a 4°C decrease in day temperature. The different amounts of energy used *per annum* to maintain a range of day and night temperatures are presented in Table 1 relative to the amount required to maintain blueprint conditions.

Table 1 The energy needed to maintain a range of day and night temperatures throughout the year expressed relative to that required to maintain 20°C day and 16°C night.

Day temperature (°C)	Night temperature (°C)		
	16	12	8
20	100	69	47
16	84	54	32

Energy conservation

There are two approaches that can be used to save energy. The first is to look closely at the physical aspects of the glasshouse and associated equipment to see if they could be adapted to achieve the present blueprint conditions by using less energy. The second is to re-examine the biological factors in production to see if the blueprint can be changed. If it is possible to achieve savings using the second approach they would be in addition to any that could be obtained by the more physical techniques.

Physical techniques

Most of the physical techniques used are simple applications of good husbandry. For example, ensuring the boiler is at peak efficiency, the thermostats are accurate and set correctly, and that the vents and doors close properly; about 10 per cent of the average heat losses are from leakage. Wind can also increase heating costs significantly, the losses from a glasshouse being approximately doubled as windspeed increases from zero to 6·7 m s^{-1} (15 mph) (Sheard, 1978). These losses can be reduced easily by planting hedges or installing plastic windbreaks.

An engineering solution to the problem, but one with definite biological implications, is the use of thermal screens (Dawson and Winspear, 1976). It was noticed that the use of blackout screens in chrysanthemum production reduced the amount of energy required. A thermal screen is essentially similar, i.e. a sheet of polythene or similar plastic material which can be drawn over the plants at night. The effect of a screen can be seen in Fig. 3. The screen reduces the volume of air to be heated and also intercepts the long-wave radiation emitted by the plants and soil. These effects seem to reduce the temperature difference between the plant and surrounding air (*cf*. Bailey, 1978). In addition, the provision of an extra surface to intercept the upward flux of long-wave radiation allows the air above it, and especially the roof of

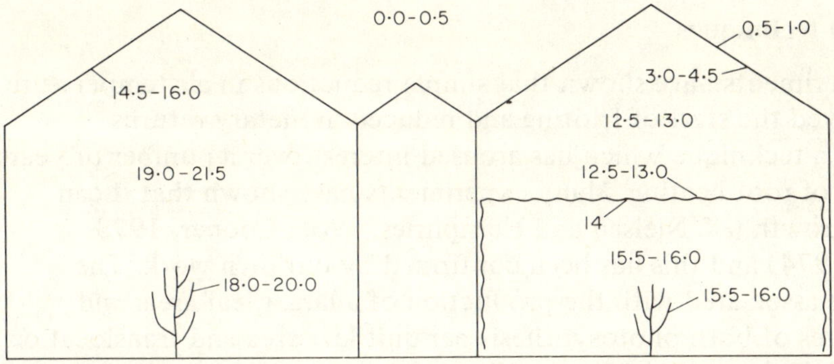

Fig. 3 The mean temperatures (°C) measured in a glasshouse in Kent and in an adjacent glasshouse fitted with a thermal screen on a night when the mean external air temperature was 0·0-0·5°C. The screen was polythene. (N.B. The thermostat setting on the unscreened house was higher than in the screened house.) Data supplied by Dr S. W. Burrage.

the house, to cool. In a screened house the temperature gradient between the house and the external air is, therefore, less than in the unscreened house, so reducing the rate of conduction of heat from the former (Winspear, 1978a). Heat loss is minimized if the screen is fitted with the minimum of gaps between it and the side screens, these latter usually being permanent. It is, however, important to ensure that the installation does not cause unnecessary shading of the crop. In one house it has been estimated that the retracted screen and insulated walls caused a significant decrease in yield. It might be possible to avoid this by the use of some form of double glazing, but the economics are uncertain at the present time. Another problem which can arise on some occasions is that of high humidities under the screen.

A way of saving costs without changing anything would be to obtain the heat from something other than fossil fuel. Many alternatives have been tried, but few have been found to work in practice. Three which are used are geothermal heat in Iceland, waste heat and CO_2 from a whiskey distillery in Scotland and waste heat from power stations in England and France. This latter source is widely available. The large 'base-load' power stations of the Central Electricity Generating Board in the UK which operate virtually continuously, convert about a third of the energy used to electricity. Each power station has to dispose of about the same amount of energy per annum as is used by the whole of the UK glasshouse industry. There are three problems attached to the use of this energy. It is in the form of warm water at a temperature lower than is ideal for use in glasshouse heating, it has to be disposed of throughout the year and not only during the cooler weather, and power stations are not usually sited in areas suitable for glasshouse production.

Biological techniques

Most experiments have shown that simple reductions in air temperature have delayed the start of fruiting and reduced monetary returns. However, a technique which has aroused interest over a number of years is the use of root heating. Many experiments have shown that it can increase growth (*cf.* Nielsen and Humphries, 1966; Cooper, 1973; Nielsen, 1974) and this has been confirmed by our own work. The increase is associated with the production of a larger leaf area, and greater rates of both photosynthesis per unit leaf area and translocation (Miss V. S. Fish, personal communication). In addition, the form and structure of the roots are changed, roots from heated solutions being larger and more extensively branched and having fewer layers of cortical cells and more highly differentiated xylem. These changes are accompanied by greater rates of transpiration and ion uptake (Moorby and Graves, 1980).

Attempts to use root heating in traditional commercial systems with plants grown in soil or peat bags have sometimes given equivocal results (e.g. Calvert, 1956; Boxall, 1970) although other workers have been more successful in consistently improving production (Jones, Sandwell and Talent, 1978). The most promising use for root heating has come from the introduction of recirculating nutrient solutions for crop production; particularly the nutrient film technique (NFT) (Spensley, Winsor and Cooper, 1978; Morgan and O'Haire, 1978). With this technique, plants are grown in plastic gulleys through which the solution flows at a depth of only a few millimetres. The composition of the solutions can be controlled easily. In large-scale systems this is done automatically, sensors being used to measure the conductivity and pH of the solution and to operate pumps which add concentrated nutrient solutions and acid. Regular analyses ensure that the amounts of nutrients supplied are within prescribed limits and that there is not an excessive increase in the concentrations of ions such as sulphate. The attraction of this technique is that the volume of the rooting zone is strictly defined and is very small relative to that of soil-grown plants. In addition, the solution can be heated easily and used to control the temperature of the roots.

When the work on solution heating at the Glasshouse Crops Research Institute was extended to an NFT system it was decided to examine the effects of root heating at day/night air temperatures of 20/9°C. (N.B. This was the thermostat setting in the glasshouse. The temperature did not fall below this, but there was no cooling.) The evidence available at that time suggested that under these conditions growth without root heating would be unacceptably slow and therefore no unheated treatments were included in the experiment. Fruit production under these conditions is shown in Fig. 4. Figure 4 also shows the production of fruit from an identical adjacent glasshouse maintained at 20/18°C with plants grown in an unheated NFT system. The higher night air temperature obviously encouraged earlier fruiting, but the rate of production from both houses quickly became the same. Observations on unheated guard rows showed that these plants also produced a good crop although picking was delayed relative to plants with heated roots.

These results, together with those from earlier crops, showed that for root heating the temperature response had a broad optimum over the range 23-28°C and it was possible to achieve reasonable fruit yields with a considerable saving of fuel. Using the data in Table 1, and assuming a fuel cost of £25 000 ha^{-1} to maintain blueprint conditions, it can be estimated that a reduction of the night temperature from 16 to 9°C could save £10 750 ha^{-1}. Against this has to be set the cost of solution heating. In the experiments the solution was heated using electricity to

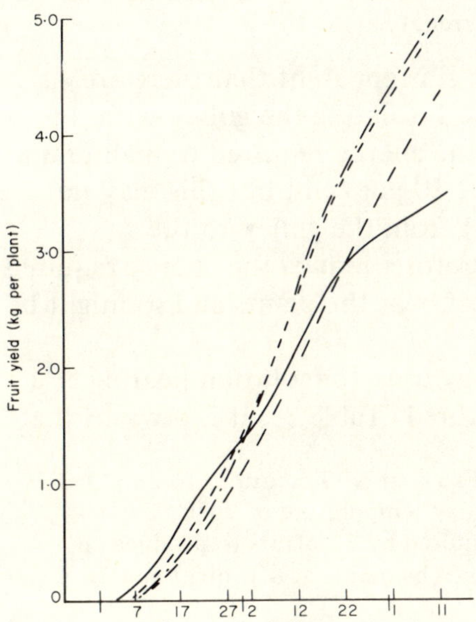

Fig. 4 The yield of fruit produced by a crop of tomatoes grown in a heated NFT system at root temperatures of 20°C (– –), 23°C (– · –) and 32°C (– – –) and day/night air temperatures of 20/9°C respectively. (N.B. The yields obtained at root temperatures of 26°C and 29°C were not significantly different from those at 23°C and 32°C and have been omitted for clarity.) Also included is the yield of a crop planted in a similar, but unheated, NFT system in an adjacent house with day/night temperatures of 20/18°C respectively (——). The data from the unheated NFT was supplied by Mr D. M. Massey.

facilitate monitoring energy consumption. However, assuming the use of gas, and a factor for boiler efficiency, the cost of solution heating can be calculated (Table 2). The errors associated with extrapolating

Table 2. The energy required to heat a nutrient film system throughout the year to the temperatures stated. The day/night air temperature settings were 20/9°C. The length of the gulleys was 15 m and six were supplied from each tank. The costs are calculated assuming the use of gas at 17 p/therm (1 therm = 0·1055 GJ) and a boiler efficiency of 80 per cent. The costs per hectare are only estimates because of the errors inherent in extrapolating experimental results.

Solution temperature (°C)	Energy required GJ	Cost £	£ ha^{-1}
20	4·25	8·6	2961
23	8·11	16·3	5649
26	13·08	26·3	9109
29	19·53	39·3	13606
32	27·94	56·3	19470
26I*	10·21	20·6	7116

(*Insulated with 2·5 cm of expanded polystyrene)

experimental data are well known, but it is apparent that there are possibilities for substantial fuel savings. Insulating the gulley with 2·5 cm of expanded polystyrene reduced the energy required to maintain a solution temperature of 26°C by about 20 per cent, but this may not be necessarily useful. The heat was lost from the gulley to the glasshouse atmosphere and would, therefore, reduce the energy required for air heating. Further, it would be lost near the plants and so might be a very effective form of heating.

In subsequent experiments the energy used for solution heating at a range of night temperatures was monitored (Table 3). It is obvious that

Table 3. Relative amounts of energy used to heat NFT solutions to 26, 23 and 20°C at a range of night temperatures and a day temperature of 20°C. The measurements were made over the period required by 32 plants to produce an 8-truss crop when supplied from a single tank. The plants at 8°C night and 26°C solution temperatures used 7·78 GJ.

Solution temperature (°C)	Night Temperature (°C) 16	12	8
26	71	94	100
23	36	55	64
20	14	24	32

the rate of heat loss from the solution to the atmosphere, and hence the amount of energy required, was less at the higher night temperatures. However, it must be remembered that considerable amounts of heat are required to maintain the night temperature.

Table 4 shows the effects of the treatments on fruit yield, but the results were equivocal because of a problem which became apparent about 4 weeks after transferring the seedlings to the gulleys. Some of the plants at a night temperature of 8°C became chlorotic and showed signs of stunting. These symptoms were seen first at a root temperature of 20°C, later but less severely at 23°C, and slight symptoms occured for a short time at 26°C. The most affected plants eventually developed necroses and swollen and cracked stems. Investigations found that the 'corrosion-resistant' immersion heaters were sheathed in monel which is 60-70 per cent nickel and the symptoms were those of nickel toxicity. The interactions between night and solution temperatures seemed to correlate with the rate of growth. Nickel was present in the plants

Table 4. The weight of fruit produced, percentage in Class I, and cash returns (using ADAS market figures) from an 8-truss tomato crop grown at a range of solution and night temperatures. The air temperature during the day was 20°C throughout.

Solution temperature (°C)		Night temperatures (°C)		
		8	12	16
20	Yield (kg)	3·40	6·10	5·77
	Class I (%)	34·2	65·1	70·1
	Return (£)	1·97	4·38	4·11
23	Yield (kg)	6·72	5·77	6·24
	Class I (%)	65·8	55·9	52·6
	Return (£)	4·63	3·93	4·27
26	Yield (kg)	6·53	5·62	5·95
	Class I (%)	68·0	51·5	47·3
	Return (£)	4·67	3·69	3·97
Unheated	Yield (kg)	7·21	6·37	5·90
	Class I (%)	77·7	73·1	75·6
	Return (£)	5·34	4·62	4·26
	Standard errors	Yield (kg)	± 0·19	
		Class I (%)	± 3·24	
		Return (£)	± 0·16	

grown at 12 and 16°C, but it was only at 8°C, where growth was much less, that the concentration of nickel was sufficient to cause toxicity symptoms. We have since removed the heaters and induced similar symptoms by adding nickel to the solution. The same heaters had been used in the experiment which gave the data in Fig. 4 and Table 2 and so might have had an effect even though there were no visual symptoms. Nickel is, however, unlikely to have increased the yield and it is believed that the conclusions drawn from this experiment are valid.

The results from the most recent experiment (Table 4) are more difficult to interpret because nickel could be found in all the plants from heated solutions. There was an obvious effect of the nickel from the nutrient solution at 20°C in a night air temperature of 8°C and this was also evident in the quality of the fruit and cash returns. In other experiments, where there was no possibility of an effect of nickel, root heating has always increased growth and yield. It is surprising, therefore, that in this experiment the best fruit and greatest returns came from plants grown at 8°C without root heating (Table 4). It must be remembered, however, that these were produced from a limited number of trusses and that fruiting at 16°C night temperature started and finished earlier than at 8°C. If the number of trusses had not been limited the plants at 16°C might have outyielded those at 8°C, but with fruit of lower quality.

Conclusions

The results available at present indicate that crop yield can be maintained or increased by root heating. If this is done at low night temperatures the fuel requirements can be reduced by 25-30 per cent. When these savings are coupled with those to be made by more mechanical means, the total potential savings are substantial, although there might be technical and financial difficulties in achieving them. For example, it can be difficult to install thermal screens in existing structures and the financial benefits may then be more apparent than real. It is necessary, therefore, to substantiate these findings and quantify them more fully, so that as the existing price structure changes management practices can be adapted accordingly. If these suggestions can be realized the immediate prospects for the glasshouse industry do not look entirely bleak. The long term prospects are different and will depend on the relative changes in the prices of fuel, labour and fruit and whether successful techniques are developed for the use of heat wasted by other industries.

Acknowledgements. I would like to acknowledge the help of the following friends who have been involved in various aspects of the work reported: Dr A. J. Cooper, Miss C. Cave, Miss V. S. Fish, Dr C. J. Graves, Dr N. Lepp, Mrs. C. Richardson and Mr P. C. Slough. I would especially like to thank Dr S. W. Burrage and Mr D. M. Massey who supplied data for Figs. 3 and 4 respectively.

References

Bailey, B. J. (1977). The calculation of glasshouse fuel requirements using degree-day data corrected for solar heat gain. *NIAE Department Note* DN/9/798/04013.

Bailey, B. J. (1978). Heat conservation in glasshouse with aluminized thermal screens. *Acta Horticulturae* **76**, 275-278.

Boxall, M. I. (1970). The effects of soil and air temperature on the cropping of tomato plants grown in troughs. *Electricity Council,* ECR/R 255.

Calvert, A. (1956). The influence of soil and air temperature on cropping of glasshouse tomatoes. *Journal of Horticultural Science* **31**, 69-75.

Calvert, A. (1972). Effects of day and night temperatures and carbon dioxide enrichment on yield of glasshouse tomatoes. *Journal of Horticultural Science* **47**, 231-247.

Cooper, A. J. (1973). *Root Temperatures and Plant Growth*. Research Review 4, Commonwealth Bureaux Horticulture Plant Crops, C.A.B. Farnham Royal.

Dawson, J. R. & Winspear, K. W. (1976). The reduction of glasshouse heat losses by internal blinds. *Journal of Agricultural and Engineering Research* **21**, 431-436.

Jones, D. A. G., Sandwell, I. & Talent, C. J. W. (1978). The effect of soil temperature when associated with low air temperatures on the cropping of early tomatoes. *Acta Horticulturae* **76**, 167-171.

Ministry of Agriculture, Fisheries and Food (1976). Production of early and main crop heated tomatoes. *Short term leaflet* 38.

Moorby, J. & Graves, C. J. (1980). The effects of root and air temperatures on the growth of tomatoes. *Acta Horticulturae* **98**, 29-43.

Morgan, J. V. & O'Haire, R. (1978). Heated hydroponic solutions as an energy saving technique. *Acta Horticulturae* **76**, 173-180.

Nielsen, K. F. (1974). Roots and root temperatures. In *The Plant Root and the Environment,* pp. 293-333. Ed. E. W. Carson. Charlottesville: University Press of Virginia.

Nielsen, K. F. & Humphries, E. C. (1966). Effects of root temperature on plant growth. *Soils and Fertilizers* **29**, 1-7.

Sheard, G. F. (1978). *Energy and the future of the glasshouse industry.* Eric Gardner Memorial Lecture to National Farmers Union, Eastbourne 1978.

Spensley, K., Winsor, G. W. & Cooper, A. J. (1978). Nutrient film technique – crop culture in flowing nutrient solution. *Outlook on Agriculture* **9**, 299-305.

Winspear, K. W. (1978a). Vertical temperature gradients and greenhouse energy economy. *Acta Horticulturae* **76**, 97-103.
Winspear, K. W. (1978b). Energy saving strategies and the United Kingdom glasshouse industry. *Acta Horticulturae* **87**, 351-361.

Cultural methods for soft fruits — the physiologist as innovator

P. D. Waister
Scottish Horticultural Research Institute, Invergowrie, Dundee

Introduction

The cultural methods used in the production of soft fruits may be divided into those designed to control weeds, pests, and diseases and those more directly aimed at influencing growth and yield. Only these latter will be considered here. Their effect is mostly on dry matter accumulation and its partitioning, and the plant science that must provide the rationale for the operations is plant physiology. However, examination of current methods suggests that most are derived from experience and empirical investigations rather than scientific understanding. The object of this paper is to consider ways in which the physiologist may exert a greater influence on production methods in the future.

The three most important soft-fruit crops in Britain are strawberries, raspberries and black currants. With the exception of a small acreage of strawberries under glass or plastic protection, all production comes from open-field culture, and is therefore subject to a largely uncontrollable aerial environment. It is only in the intensive propagation period that there are opportunities for very significant manipulation of this environment.

Following planting, the strawberry remains in the ground for up to 5 years, and raspberries and black currants may be retained for 10 years or more.

Approaches to innovation

Before discussing the potential contribution of crop physiology it is worth considering why empirical methods have been the more successful in the past.

Cultural methods are not only interdependent but they also show interaction with the weather and with genotype. This complex interplay of factors is in theory susceptible to mathematical analysis, given the necessary data on the environment and on physiological responses. In practice there is still a long way to go before this approach can compete with field experimentation which integrates the effects of the total environment to which a crop is exposed, including the pest and disease complex. This explains why, for example, such basic matters as plant population and spatial arrangement have been determined empirically, and have remained largely unaffected by progress in physiological understanding.

Field experimentation has been augmented by what amounts to a prolonged evolutionary investigation by growers. There are several thousand producers of soft fruits in Britain and, by accident or design, no two growers will ever subject their crops to exactly the same treatment. It is obvious that if an improvement is effected which leads to a recognizable increase in yield or quality, it will be further tested and incorporated in the system of cultivation. Advances are still being made in this way, as will be seen later when cultural methods for the raspberry crop are examined.

Despite the successes of the empirical method there remain a number of blocks to increased productivity which might still be overcome by cultural techniques. Better control is needed over the growth and development of soft-fruit crops. This could come either from chemical regulation, or from new cultural methods applied to unconventional cultivars. These two possibilities are discussed in the following sections, by reference to individual crops and problems.

Chemical regulation of growth and development

There are as yet no growth regulators for soft fruits on the list of chemicals officially approved for use by growers in Britain (Anon. 1979). However, the potential for use in these crops has been recognized, as the literature shows.

Control of runnering of strawberries

Under normal field cropping conditions strawberries produce an excessive number of stolons and runner plants, and the surplus has to be removed by the grower. The plant itself initiates and terminates the runnering phase in response to day length and temperature, which suggests an opportunity for control by applying growth regulators to the field crop. Both maleic hydrazide (Thompson, 1960) and chlormequat (Guttridge, Anderson, and Stewart, 1966) have been reported to be effective for this purpose, but neither has found a place in commercial practice.

'Running-off' of black currants

A significant cause of reduced yields in the black currant is the condition known as 'running-off'. Varying with season, there is a loss of a proportion of fruiting points on the inflorescences or 'strigs', so that final berry number may be much lower than the number of flower buds originally formed. Some of this loss may be caused directly by frost damage, but a further decrease arises from the shedding of flowers and fruits at or after anthesis. Several reasons have been advanced to explain this 'running-off'. Earlier workers considered it to be primarily a failure of pollination (Wellington, Hatton and Amos, 1921), but others have later concluded that the problem arises from seed abortion (Teaotia and Luckwill, 1956) and have suggested that this results in cessation in the supply of an essential auxin from the seed to the abscission layer. However, attempts to reduce fruitlet shedding by applications of naphthalene acetic acid have produced erratic results (Luckwill, 1956). There seems to have been no real progress on this topic in the past 20 years, so there is still an opportunity here for the physiologist.

Fruit ripening and abscission in black currants

The black currant crop is the first of the three soft fruit crops to have been successfully harvested by machine. Since this is a 'once-over' operation it is important that virtually all fruits should ripen simultaneously. This does not always occur, and attempts have been made to hasten the ripening of the green fruits by application of ethephon, which also promotes easy abscission. Despite many reports of successful experimental use, this material is little used commercially in black-currant growing.

Fruit release in raspberries

Mechanical harvesting in the raspberry crop is a selective process, in which a vibration mechanism is used to shake the ripe fruits from their receptacles while leaving the unripe fruit for a later pass of the machine. There may be as many as six harvests during the 20 day picking season. Machines currently under test in Britain do not adequately discriminate between ripe and unripe fruit, and so cause crop loss. The mechanism of fruit release from the receptacle is not sufficiently understood. The hundred or more cohering drupelets of the raspberry fruit form a shell around the receptacle, each drupelet having its own vascular connection. To allow release of the fruit the vascular strands must first break down and then the clasp of the fruit on the receptacle must loosen. This latter process seems to be the block, and needs to be investigated before appropriate chemical aids can be found. Ethephon has been tried on this crop too, but did not have the desired effect on fruit loosening. It did accelerate ripening, but at the expense of fruit size (Jolliffe, 1974).

Micro-propagation of strawberries

The foregoing are examples of possible field applications of growth regulators, but there are also opportunities in the propagation phase. One of the recent achievements of plant physiology has been the extensive development of tissue-culture methods. When applied to the strawberry crop (Adams, 1972; Boxus, 1974) these techniques were so successful that they have been rapidly exploited by commercial propagators for the multiplication of healthy stocks or of new cultivars.

There are various reasons why reports of successful use of growth regulators in research work have not led to innovation in commercial practice. Deterrents include costs of chemicals and economic problems of clearance and approval for use on minor crops. There is also a degree of unpredictability in the effects of growth regulators, both from season to season and between cultivars, which underline the continuing need for research on mechanisms of action.

Cultural methods for unconventional cultivars

Plant breeding offers an alternative to chemicals as a means of changing crop behaviour but, particularly with perennial crops like the soft fruits, it is a lengthy process. Changes in cultural methods, particularly those associated with mechanical harvesting, make new demands on the breeder. The converse is also true, as is shown by some recently released cultivars which have required alterations of cultural methods to make them acceptable in commerce. Some examples will be examined in detail.

Cane vigour control in raspberries

The new raspberry cultivar, Glen Clova, was released a few years ago in Scotland. It was soon found that its growth was so vigorous as to present a real management problem in commercial practice, and the future of the cultivar was in doubt. By introducing a form of vigour control first used in the USA, the cultivar was not only given an acceptable growth habit but was also induced to yield as much as 40 per cent more fruit (Lawson and Wiseman, 1979). The origins of this practice are interesting. It appears that a grower in British Colombia found himself with a weedy plantation in late spring. He dealt with the problem by applying dinoseb, a desiccant which not only killed the weeds but also destroyed the first flush of new canes. A second flush of canes appeared, and apparently gave adequate growth by the end of the season to provide a crop for the following year. A grower in Oregon, probably recognizing the advantages of having reduced vegetative growth at the time of picking, then adopted this practice of chemically burning out the first new canes. Agronomists in Oregon and Washington examined the implications of control of growth of young cane and showed that, in addition to facilitating picking by hand or machine, there was an appreciable yield benefit (Sheets 1973, Norton 1974). Successful though this approach has been, there is still a need to investigate the mechanism involved in the yield response. Until this is known it will be difficult to decide on the target for plant form in breeding programmes.

This point is illustrated in Fig. 1, which shows the total dry-weight accumulation in the fruiting and primocanes of a plant of cv. Glen Clova, as measured by Wright (1979) in his studies of within-plant competition. The additional (lowest) line indicates the approximate dry-matter accumulation in primocanes of plants subjected to vigour control (Lawson, personal communication).

If the competitive effect of the primocanes is related to their total dry matter, it can be seen that the existence of a critical period for competition cannot be deduced simply by removing young canes at different intervals after emergence, and measuing yield responses from the fruiting cane. A plant breeder wishing to produce a cultivar with enhanced yield potential wants to know whether to select for delayed emergence of young cane, or for emergence at the normal time but with slow growth thereafter. In other words he wants information from the physiologist on the critical period for competition.

In the long run it may prove impossible for the breeder to combine restricted early competition from vegetative canes with rapid growth in the latter part of the season, if that is what is required, in which event this new cultural method may become a permanent feature of raspberry cultivation.

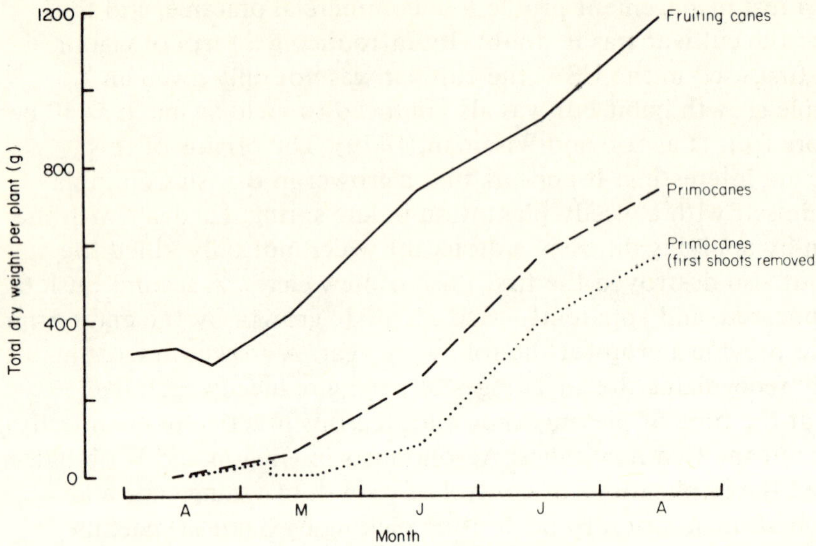

Fig. 1 Dry-weight accumulation per plant in the raspberry cultivar Glen Clova. The upper two curves are from Wright (1979) the third is based on data from Lawson (personal communication).

Response of strawberries to defoliation

The benefits of cane-vigour control in raspberries stem from alteration of the balance between fruiting and vegetative phases. A similar manipulation in the strawberry crop has led to some striking yield responses in certain cultivars. There has long been a practice of leaf removal shortly after harvest, perhaps originally as a hygiene measure and later because there seemed to be some yield benefit. Guttridge (1959) examined the consequences of this treatment and suggested why it is so successful in enhancing yield in some cultivars. Growth and reproduction in this crop are strongly dependent upon temperature and photoperiod. Short days are required for flower initiation in the late summer, a chilling period is needed to promote vegetative and reproductive development the following spring, and the long days and higher temperatures of summer are necessary for runner production. Certain cultivars fail to initiate a full complement of inflorescences unless the fully-expanded leaves are removed immediately after the current season's crop is harvested. Guttridge suggested that this behaviour stems from the production of an inhibitory hormone by the mature leaves, presumably as a response to long days. Removal of these leaves permits flower initiation. A similar effect of defoliation has recently been reported in maize (Crookston and Hicks, 1978). The response in strawberries is strongly cultivar-dependent, as shown in Table 1, derived from the results of Cormack and Brown (1979).

Table 1. Effects of defoliation on yields of strawberry cultivars planted in 1975, and on number of inflorescences produced per crown (in brackets).

	Cultivar		
	Saladin	Jurica	Cambridge Favourite
Yield in 1976 (t ha^{-1}) (not defoliated 1975)	33·5(1·2)	26·5(1·9)	18·7(1·5)
Mean of yields in 1977 and 1978 (t ha^{-1})			
(a) Not defoliated in 1976 and 1977	16·8(0·7)	31·6(1·6)	29·1(1·6)
(b) Defoliated in 1976 and 1977	35·6(1·4)	18·2(1·6)	26·3(1·5)

The positive response of the new cultivar Saladin in the second and third cropping years arose from increased inflorescence initiation, while the negative response of the cultivar Jurica stemmed from a combination of reduction in crown number and in fruit weight per crown, effects which may have resulted from tipping the vegetative to fruiting balance too far. The standard commercial cultivar Cambridge Favourite was little affected by defoliation.

It is not yet known why some cultivars respond so positively to leaf removal and others show no response or a negative one. Nor is there an explanation for the fact that plants which are left intact in the year of planting appear to initiate enough inflorescences for a full crop in the following year (first line of table). Excessive vegetative vigour has been suggested as a contributory factor in positive response, and it is interesting that the largest increases recorded in Scotland have come from cultivars selected at the Scottish Horticultural Research Institute unit at Auchincruive under conditions of heavy infection pressure from the fungus *Phytophthora fragariae.* This is a root pathogen whose effects are thought to be minimized when root regeneration is vigorous (Gooding 1972), and it may be that selecting for tolerance has produced unusually vigorous types. The question for the physiologist is whether there is greater yield potential in genotypes which respond to defoliation or in those that do not.

Growth habit in the strawberry

Another problem to be resolved concerns the relative merits of the runnering and crown-branching forms of strawberries. The cultivar Olympus, released a few years ago in the state of Washington (Barritt and Schwartze, 1973), has outyielded all previous releases in that area. It has a pronounced crown-branching habit and produces relatively few runners, which makes it difficult and expensive to propagate. Experiments on chemical promotion of runnering suggested that this difficulty could be overcome and, for two or three years, commercial propagators used gibberellic acid for this purpose, though with only limited success (Barritt, personal communication). Its use was discontinued when it was found that early planting of runner beds was equally effective, but the runner production still falls well short of that from conventional cultivars. If crown branching forms are proved to have a consistently greater yield potential than those that produce runners, a solution to the propagation problem will have to be found by the physiologist.

Restrictions on the control of plant form

Vigour control in raspberries and defoliation in strawberries both produce marked yield responses, but for different reasons. The effect in strawberries is largely on the flower-initiation process, whereas in raspberries it is on flower and fruit development after inflorescence initiation, and also on the morphological development of the fruit-bearing structure. Cultural practices like these are only possible where the vegetative phase can be distinguished as a separate target for chemical or mechanical treatments. Primocanes can be controlled in raspberries because they arise at ground level and are below the fruiting zone at the time of treatment with the desiccant, and in strawberries the expanded leaves may be removed by mechanical means because they are held above the crowns which will be induced to bear inflorescences for the following year's crop. There appear to have been no comparable developments in the culture of black currants. In this species the vegetative and reproductive structures are closely intermingled and not easy to subject to differential treatments.

Fitting the cultural method to the genotype

The cultivars referred to above were not bred with new cultural practices in mind. It was fortunate that methods were available which permitted their high yield potentials to be realized.

Investigation of the possibilities for more radical change in cultural methods is probably handicapped by the limited range of plant forms available as cultivars. For example, following a series of experiments on the effects of competition between the fruiting and vegetative phases in the raspberry, it appeared that it might be possible to change the annual bearing system into a biennial one, with the two phases separated in time (Waister, Cormack and Sheets, 1977). This system offers a number of advantages, particularly with respect to mechanical harvesting. The full potential of the method has almost certainly not been assessed because the cultivars used have all been bred and selected for their performance under the annual method of culture.

In addition to his now recognized role in helping to specify plant models for the breeder, the physiologist should perhaps be encouraged to extend the specifications to cover a genotype × culture 'package'. It may be that the breeder is discarding valuable material early in the selection process because he lacks information on the cultural methods necessary to evaluate it adequately.

References

Adams, A. N. (1972). An improved medium for strawberry meristem culture. *Journal of Horticultural Science* **47**, 263-264.

Anon. (1979). *Approved Products for Farmers and Growers,* 291 pp. Ministry of Agriculture, Fisheries and Food. London: HMSO.

Barritt, B. H. and Schwartze, C. D. (1973). The Olympus strawberry. *Washington Agricultural Experiment Station Circular* **570**, 6pp.

Boxus, Ph. (1974). The production of strawberry plants by *in vitro* micro-propagation. *Journal of Horticultural Science* **49**, 209-210.

Cormack, M. R. and Brown, J. McD. (1979). Strawberry cultivar trial 1975-78. *Scottish Horticultural Research Institute Occasional Publication* No. 7, 34 pp.

Crookston, R. K. and Hicks, D. R. (1978). Early defoliation affects corn grain yields. *Crop Science* **18**, 485-489.

Gooding, H. J. (1972). Studies on field resistance of strawberry varieties to *Phytophthora fragariae. Euphytica* **21**, 63-70.

Guttridge, C. G. (1959). Evidence for a flower inhibitor and vegetative growth promoter in the strawberry. *Annals of Botany* **23**, 351-360.

Guttridge, C. G., Anderson, H. M. and Stewart, W. S. (1966). The control of strawberry runners in the field with CCC. *Experimental Horticulture* **15**, 92-95.

Jolliffe, P. A. (1974). Effects of ethephon on raspberry fruit ripeness, fruit weight and fruit removal. *Canadian Journal of Plant Science* **55**, 429-437.

Lawson, H. M. and Wiseman, J. S. (1979). Control of cane vigour. *Annual Report of the Scottish Horticultural Research Institute for 1978,* pp. 45-46.

Luckwill, L. C. (1956). Fruit drop in black currants: II. The effect of naphthalene acetic acid. *Annual Report of Long Ashton Research Station for 1955*, pp. 75-77.

Norton, R. A. (1974). Basal cane control of red raspberries – four years' results. *Proceedings of the Lower Mainland Horticultural Improvement Association* for 1974, pp. 16-19.

Sheets, W. A. (1973). Chemical pruning of caneberries with dinoseb at Aurora. *Proceedings of the Western Horticultural Association* **63**, 150-152.

Teaotia, S. S. and Luckwill, L. C. (1956). Fruit drop in black currants. I. Factors affecting 'running-off'. *Annual Report of Long Ashton Research Station for 1955*, pp. 64-74.

Thompson, P. A. (1960). The control of runnering in strawberries with maleic hydrazide. *Journal of Horticultural Science* **35**, 249-259.

Waister, P. D., Cormack, M. R. and Sheets, W. A. (1977). Competition between fruiting and vegetative phases in the red raspberry. *Journal of Horticultural Science* **52**, 75-85.

Wellington, R. Hatton, R. G. and Amos, J. (1921). The 'running-off' of black currants. *Journal of Pomology* **2**, 160-198.

Wright, C. J. (1979). Studies on within-plant competition in the red raspberry (*Rubus idaeus* L.) *Ph.D. thesis,* University of Dundee, pp. 183.

Agronomic research and potato production practice

P. M. Harris
Department of Agriculture and Horticulture, University of Reading

Introduction

In this paper, changes in maincrop potato yields which have taken place since 1955 have been related to changes in production practices. Estimates of the potential yield of the crop are given and suggestions are made as to how the gap between national average yield and the potential yield of the crop may be narrowed and what part may be played by agronomic research.

Discussion is restricted to maincrop potatoes owing to the ready availability of survey data published by the Potato Marketing Board (PMB) (50 Hans Crescent, London SW1X 0NB) for this crop. The period since 1955 has been chosen because annual statistical information is available from this date (*PMB Handbook of Potato Statistics*), while the surveys have been published at intervals since 1958.

I have taken Mitchell's (1976) somewhat daunting definition that agronomy is 'an integration of biological, chemical and physical knowledge into useful (crop) production systems'. Agronomic research is, therefore, characterized by the involvement of a large number of potentially interacting factors which pose considerable practical and intellectual difficulties to the agronomist. Agronomic research to be 'useful' does not necessarily have to result in increases in yield per unit area, but may as usefully lead to reductions in unit costs of production and increased profitability, while improvements in quality may lead to increases in price and profitability without any increase in yield per unit area.

Taking the five-year means of the years 1955-59 and 1973-77, the numbers of registered producers in Great Britain fell by 55 per cent and the area devoted to the crop decreased by 31 per cent. In contrast, national production has remained almost constant. The main positive achievement of the last two decades has been the substantial increase in the productivity of individual producers, made up in part of an increase at 56 per cent in the area of potatoes grown per producer and in part by an increase in yield of 42 per cent (see North, p. 266). It is with yield per unit area that this paper is concerned.

Yield

It has been pointed out by Taylor (personal communication) that the best way of estimating the national yield of potatoes is to use the check weighings on 1600 or more fields of maincrop varieties carried out by the PMB just before harvesting. The total yield of tubers thus obtained is then multiplied by a factor of 0·695; all the statistics of yield published by the PMB are based on data reduced by this factor. The factor has been in use for about 40 years, although there is no known documentary evidence concerning the corrections that were intended to be covered by it. Empirical evidence suggests that the factor corresponds to a yield estimate equivalent to the saleable yield from store in March. Yields from agronomic experiments may or may not take harvesting losses into account, but rarely include the storage losses. In the subsequent discussion yield will imply yield per unit area except where yield data have been taken from PMB statistics.

Changes in yield with time

The change with time in maincrop yield (harvested after 31 July) obtained in England and Wales is shown in Fig. 1; five-year rolling average yields are given. If the very poor yields associated with the drought of 1975 and 1976 are discounted, there has been an almost linear increase in yield since 1955. Even when these years are included in the five-year average yields, the increase is significantly linear (R^2 = 0·85), with yield increasing at a rate of 0·52 t ha^{-1} y^{-1}. These changes must be attributable to improvement in genotype and the environment or to interactions between them; these will be discussed separately.

The influence of genotype

The contribution of changes in variety to yield improvement over the period has been assessed using the technique described by Silvey (1978) for cereals, to which reference should be made for further details. The calculation is based on (a) the relative yields of varieties occupying more than 1 per cent of the maincrop area (Table 1), which have been

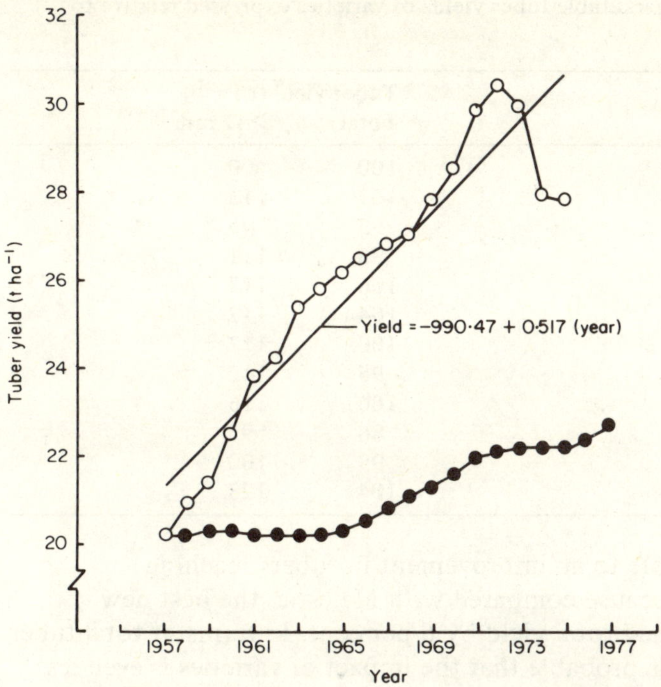

Fig. 1 Five-year rolling average yields of maincrop potatoes in England and Wales (○), and the estimated contribution of varieties to increases in yield (●).

estimated from unpublished variety trials carried out by the National Institute of Agricultural Botany (NIAB) and (b) the proportion of the total maincrop acreage occupied by each variety taken from PMB annual statistics. The analysis is restricted to England and Wales, as data for some of the varieties grown in Scotland were not available. The estimated contribution made by varieties to increases in yield is shown in Fig. 1.

During the period 1957-64, changes in variety were small and made no contribution to the substantial yield increase of 4·1 t ha^{-1} estimated from the linear regression of yield against time. Over the period as a whole, new varieties are estimated to have contributed 2 t ha^{-1} to the overall estimated increase in yield of 9·3 t ha^{-1}, a contribution of 21 per cent. Most of the increase in yield attributable to varieties took place between 1964 and 1971.

Between 1947 and 1975, Silvey (1978) estimated that of the increase in cereal yield that had occurred, 50 per cent was attributable to variety in the case of wheat and 30 per cent in the case of barley, so these crops show greater improvement in yield contributed by breeding than the potato crop. Table 1 shows that most of the increase in yield due to

Table 1. Total and marketable tuber yields of varieties expressed relative to Majestic

Variety	Tuber yield (t ha^{-1}) Total	>41 mm
Majestic	100	100
Désirée	101	113
King Edward	93	95
Maris Piper	95	111
Pentland Crown	106	117
Pentland Dell	104	112
Pentland Hawk	100	113
Pentland Ivory	95	112
Pentland Squire	106	116
Record	86	95
Redskin	99	100
Stormont Enterprise	104	123

varieties is attributable to an improvement in tubers reaching a marketable grade, because compared with Majestic, the best new varieties only increased total yield by 6 per cent. In terms of total tuber dry-matter yield, it is probable that the impact of varieties is even less than for total fresh tuber yield. Pentland Crown and Désirée, which together accounted for 47 per cent of the acreage in 1978, are both rated to be of lower dry matter than Majestic (NIAB, 1976), which accounted for 44 per cent of the acreage in 1955 and less than 1 per cent in 1978. It is rather surprising to conclude that breeding has made such a small contribution to yield increases over the period and that the main reason for any increase is adjustments in tuber size and dry-matter content. Thus genetic changes have led to little if any increase in the yield potential of the crop.

The influence of environment

The contribution of agronomy to yield improvements must be sought in either the manipulation of the crop or in the crop's environment. In this respect the potato crop is unique among British arable crops in that it is propagated vegetatively and the improvement in the crop's environment which contributes to yield increases must include the environment in which the 'seed' tuber is produced and stored, in addition to the environment in which the ware crop is grown.

Space does not permit a full evaluation of the changes in the environment which may have contributed to the changes in tuber yields,

Table 2. Changes in production practices with time in Great Britain

	Date of survey					
	1958	1963	1968	1973	1975/6	1977/8
Fertilizer rate			($kg\ ha^{-1}$)			
N	135	152	170	174	172	183
P_2O_5	145	154	175	182	177	194
K_2O	224	232	246	249	247	262
			($t\ ha^{-1}$)			
Seed rate:						
Certified	2·6	2·9 }	2·7	2·8	2·9	2·9
Once grown	2·1	2·5 }				
			(% crop treated)			
Farmyard manure	55	50	31	36	41	38
Sprouted seed	19	26	37	37	41	38
Seed source:						
Certified	57	62	58	68	70	62
Once grown	41	34	42	29	28	33
Fungicide	47	56	63	84	82	87
Pesticide	—	22	38	45	63	71
Herbicide	—	—	44	67	68	65
Irrigation	—	—	—	3	11	15

but an attempt has been made to place some values on the more important factors. It has been found convenient to divide the period into two: from 1957-1964, when variety had no effect on yield, and from 1964-1975. Changes in practice have been taken from surveys (PMB, 1958, 1963, 1968, 1973, 1975/76 and 1977/78) and are shown in Table 2.

1957-64 Using the survey data to estimate the annual rate of change of certain agronomic practices and using certain experimental data, it is estimated that over this period nitrogen use increased by 25 kg N ha^{-1}, larger seeds were used (59-69 g) and there was an increase in the proportion of the crop sprouted (10 per cent). These changes could account for about 25 per cent, 10 per cent and 12 per cent respectively of the increase yield, or about half of the total increase over the period. Much of the remaining increase must be attributable to an improvement in crop health, through an increase in the use of certified seed and in the proportion of the crop treated with fungicides for the control of blight (*Phytophthora infestans*).

1964-75 As indicated, varieties had all their effect on yield during this period and accounted for about 35 per cent of the improvement. Fertilizer use continued to increase, but at a lower annual rate except for potassium, and the use of sprouted seed increased substantially between 1963 and 1968 before levelling off at about 40 per cent of the total. All of these factors would be expected to contribute to the increase in yield, but the most dramatic changes took place in the use of crop protection chemicals and irrigation. Comparing the 1977/78 survey with that of 1963, herbicide usage has increased from 22 to 71 per cent, blight sprays from 56 to 87 per cent and insecticides from 22 to 71 per cent of the maincrop acreage. The proportion of the maincrop irrigated rose from 3 per cent in 1973 to 15 per cent in 1977 under the stimulus of the two exceptionally dry summers of 1975 and 1976.

Size of enterprise and yield

Taking five-year rolling averages, the average number of hectares per producer increased linearly with time at a rate of 0·124 ha per registered producer per year. As yield also increased linearly with time, it is also highly correlated with the size of potato enterprise, and again based on five-year mean yields, the number of hectares per registered producer accounted for 88 per cent of the variation in yield. The average rate of increase was approximately 4 t for each additional hectare per producer.

This association may be attributed to two main causes. Firstly, with the disappearance of marginal producers, production may have become concentrated in more favourable soil and climatic environments. Secondly, practices associated with increases in yield tend to be adopted more completely by large-scale producers. This possibility is illustrated in Table 3.

Genotypic environment interactions

Interactions between genotype and environment have been little studied and their importance probably underestimated. There is a tendency to neglect the fact that the vegetative material being propagated is affected by the environment in which it is produced and stored and that this may have large or small effects upon subsequent growth and yield depending amongst other things on variety. This is illustrated by the effect of contrasting storage treatments on the subsequent development of leaf area and yield in the varieties Arran Comet and Pentland Crown (Fig. 2).

Table 3. Relationships between the size of maincrop potato enterprises and certain practices

Practice	Size of enterprise						Survey
	(ha)						
	<2	2-4	4-8	8-14	14-30	>30	
	(% area)						
Seed source:							1977/8
Foundation stock	3	5	9	17	16	20	
Uncertified once grown	54	50	39	32	25	22	
	(average number of applications)						
Pesticides	0·7	1·1	1·4	1·3	1·8	2·0	1977/8
Blight sprays	1·6	2·3	2·9	3·2	3·4	4·1	1977/8
	(ha)						
	<2·4	2·4-8·5	8·5-14·2	14·2-30·3	>30·3		
Fertilizer use:	(kg ha^{-1})						1958
N	115	130	135	140	153		
P_2O_5	129	139	144	151	160		
K_2O	189	219	225	239	241		

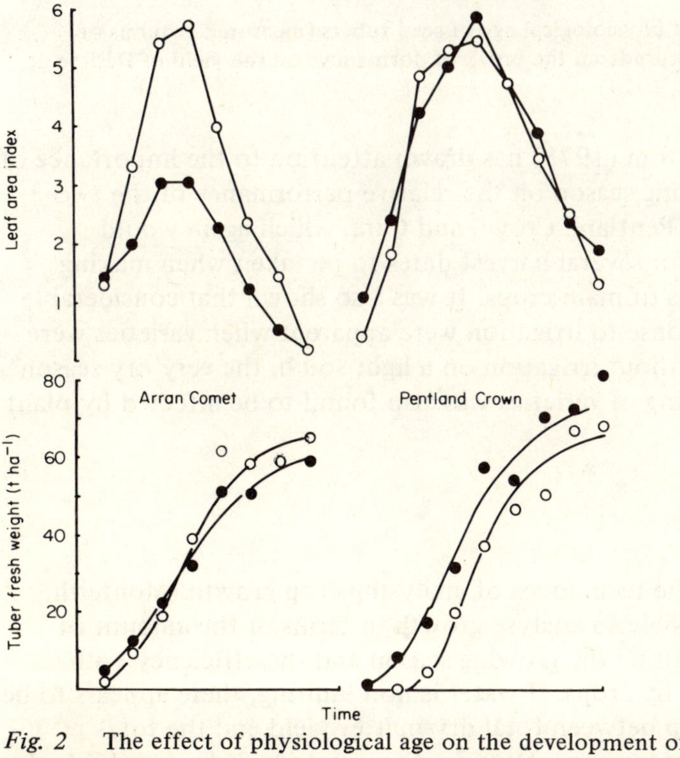

Fig. 2 The effect of physiological age on the development of leaf-area index (LAI) and tuber yield with time in the varieties Arran Comet and Pentland Crown (P. Tostevin, unpublished data). ○, non-sprouted; ● sprouted seed.

In variety testing (Richardson, 1979), it is hoped to eliminate such effects from interfering with varietal comparisons by growing seed at the same site and ensuring that the varieties are at a similar physiological state (as indicated by the degree of sprouting) at the time of planting. This, however, ignores the fact that the particular physiological stage selected may not be optimal for all, or indeed for any, of the varieties under test. The optimum physiological state of the seed, in some varieties at least, will vary with the length of the growing season (Fig. 3).

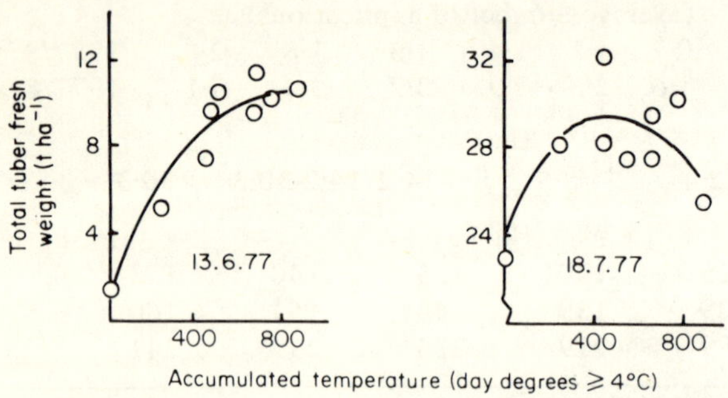

Fig. 3 The effect of physiological age of seed tubers (measured in terms of accumulated temperature from the break of dormancy) on the yield of Désirée at two harvest dates (Ali, 1979).

Recently Ramsbottom (1978) has drawn attention to the importance of the length of growing season on the relative performance of the two maincrop varieties Pentland Crown and Cara, which again would indicate the need for several harvest dates to be taken when making varietal assessments of main crops. It was also shown that considerable differences in response to irrigation were apparent when varieties were grown with and without irrigation on a light soil in the very dry season of 1976. The ranking of varieties was also found to be affected by plant density.

Potential yield

Developments in the techniques of analysing crop growth (Monteith, 1977) make it possible to analyse growth in terms of the amount of light intercepted during the growing season and the efficiency with which light is used by crops. If water is non-limiting, there appears to be a linear relationship between total dry-matter yield and the total quantity of light intercepted. Data for the potato have been published by Scott and Wilcockson(1978) and the relationship between tuber

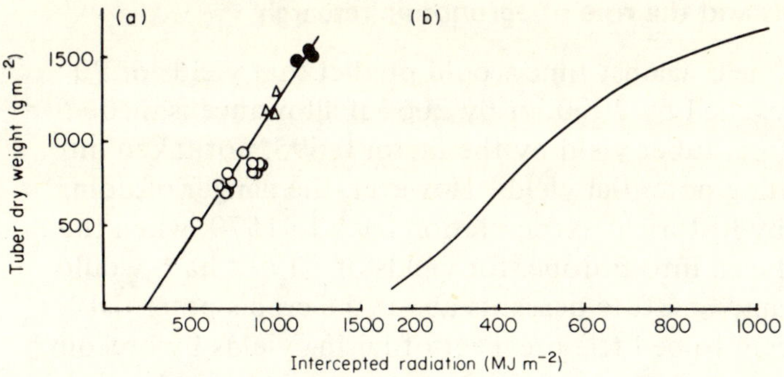

Fig. 4 (a) Tuber dry weight and intercepted radiation data from Sutton Bonnington (○, 1964; △, 1972; ●, 1973). Data from Scott and Wilcockson (1978) (b) Tuber dry weight and intercepted radiation during the course of the growing season for the variety Arran Comet, Sonning 1978 (Tostevin, unpublished data).

dry-matter yield and intercepted radiation is shown in Fig. 4(a). This relationship is linear, but more recent evidence from experiments on the Reading University Farm at Sonning indicate that the relationship might be curvilinear when yield is plotted against light interception over the whole course of growth (Fig. 4(b)).

With a knowledge of the efficiency of conversion of intercepted radiation into tuber dry-matter yield and the radiation intercepted by crops, which will depend on the light environment and the size, disposition and duration of the crop canopy, it is possible to estimate the potential yield for any given environment. For example, Alcock (1967) estimated this to be 92·5 t ha^{-1} for a crop intercepting all the incident radiation during a tuber growth period of 1 June to 10 September at Bangor, N. Wales. More recently Allen (personal communication) has estimated the yield potential of western coastal areas, with high light intensities and a potentially long growing season due to higher spring temperatures, to be of the order of 125-140 t ha^{-1} of fresh tubers. In contrast, where early leaf development is restricted by temperature, as in Scotland, the potential yield may well be less than 80 t ha^{-1}.

Experimentally, yields of these magnitudes have been reached (Scott and Wilcockson, 1978; Evans, Neild and Gunn, 1978) and with a crop mixture consisting of alternating plants of Wilja and Cara, a yield of 100 t ha^{-1} (total tubers) was achieved on small plots at the Reading University Farm, Sonning, in 1978. In maximum yield studies, yields of up to 85 t ha^{-1} have been recorded on 'normally grown' farm crops (Gunn 1978).

Raising yield levels and the role of agronomic research

The regression of yield against time would predict that yields of 90 t ha^{-1} would be reached by 2090, or by 2064 if allowance is made for the reduction in total tuber yield by the factor 0·695, not taken into account in estimating potential yields. However, the danger of doing this is illustrated by historical extrapolation back to 1570, when the potato was introduced into Europe, for yields of −179 t ha^{-1} would have cast doubts among astute peasants about the crop's potential.

There would seem to be little prospect of raising yields by breeding, as existing varieties or mixtures of varieties seem to be capable of giving yields close to the calculated potential. Assuming that no improvement in the efficienty of conversion of light energy into dry matter is likely, the only prospect of achieving a substantial increase in yield potential would be to improve the utilization of light in the early part of the growing season, without a concomitant advancement of leaf senescence at the other end of the growing season, which often occurs (Ivins and Bremner, 1965) but which is not considered to be inevitable (see Fig. 2). To achieve this would demand earlier planting than is currently possible, because of the risk of frost, the more rapid expansion of leaves at low temperatures, and a prostrate leaf canopy early in the crop's development to permit greater interception of light at low leaf area indices. Sources of frost resistance exist in some wild potato species (Hawkes, 1978) and a number have been used in breeding programmes to develop frost resistant varieties for use in the Andes (Estrada, 1978). While incorporating these attributes into commercial varieties for use in the UK might be very difficult (Howard, 1978) it remains a possibility. Much more could be done, however, to exploit the potential of existing varieties by using mixtures of varieties (Schepers and Sibma, 1976) and within single and mixtures of varieties by exploiting the potential of mixtures of seed of differing physiological ages; Hodgson (personal communication), for example, has demonstrated on the Leeds University Farm the possibilities of increasing the yield of Majestic by the latter technique.

The problem of bridging the gap between current average commercial yields and potential yields must rely on the economic manipulation of the various components involved in the production process in order to maintain a crop canopy operating at the maximum efficiency for as long a period as possible. Most of the improvements in yield in recent years have come about through the more widespread adoption and intensification in the use of techniques based on principles which have been well known for many years, such as better mineral nutrition, increased plant density, irrigation and protection against pests and diseases. Raising the general level of yield may be expected to come about through the elimination of the sub-optimal use of resources, and here the widespread adoption of irrigation could result in very rapid increases in yield as tuber growth appears to cease at very low soil-water deficits (Penman, 1971). The better use of resources should be favoured by the gradual increase in the size of enterprise.

In the introduction it was stated that agronomic research is characterized by the involvement of a large number of potentially interacting factors, and there are considerable difficulties in defining the optimum level of each. They may be expected to vary for each farm, which in itself comprises a unique production system with respect to soil, climate, equipment, managerial ability, etc. This dilemma is exemplified by the development of the 'blueprint' for maximizing potato yields (Evans, 1975), which avoids the use of impossibly complex experiments, but sacrifices any possibility of discriminating between factors regarding their importance in determining yields. It has been argued (Spedding, 1975) that 'the essence of applied research is to determine the responses of the important outputs to changes in the main inputs within a relevant context'; and that the main object of applied research in agriculture involves the construction of models of the major agricultural systems and their constituent sub-systems in a process 'which makes it clear what data are needed, why and where and how they will make their contribution to more applied research and thus ultimately to practice'.

The adoption of this approach would require a co-ordination of effort and expertise which so far has not been forthcoming in dealing with agronomic research in this crop. Little real attempt has been made to co-ordinate the effort of individual research workers or groups of workers in such a way as to render their efforts more fruitful to the industry, while still allowing them initiative and intellectual satisfaction in the work they pursue. To allow this to happen would require the creation of a small team given the task of developing a model of maincrop production. The team would have close contacts with the industry and with those engaged in agronomic and other relevant research. This would stimulate an awareness among individual research workers of the relevance of their research to the model. At the same time it would indicate where adjustments to treatments, experimental procedures and recording might be made which would make individual experiments infinitely more useful to an understanding of the causes of variation in yield. This would provide the relevant knowledge on which to base rational advice for individual producers.

If this type of approach is not followed, it is possible that agronomic research could develop in connection with optimizing inputs on specific farms which individually form unique systems as previously indicated. As the tendency is for individual production units to get bigger, there should be opportunities for the employment by farmers, or farming groups, of agronomists competent to conduct research and give advice on specific problems within the limits of the particular systems involved. However, this development would, on its own, give much less benefit to the industry as a whole.

Acknowledgements. I would like to thank D. E. Richardson and M. J. Bailey of the National Institute of Agricultural Botany for supplying data from potato variety trials, and to J. A. H. Taylor (PMB), E. J. Allen (University College of Wales, Aberystwyth) and P. Tostevin (Reading University) for data and helpful discussions.

References

Alcock, M. B. (1967). Maximum crop production. *Arable Farmer* **11**(1), 42-45.

Ali, S. M. J. M. (1979). Effect of temperature during sprouting on growth and yield of the maincrop potato variety Désirée. *M.Sc. thesis*, University College of Wales, Aberystwyth.

Estrada, R. N. (1978). Breeding frost-resistant potatoes for the tropical highlands. In *Plant Cold Hardiness and Freezing Stress*, pp. 333-341. Eds. P. H. Li and A. Sakai. New York: Academic Press.

Evans, S. A. (1975). Maximum potato yield in the UK. *Outlook on Agriculture* **8**, 184-187.

Evans, S. A., Neild, J. R. A. and Gunn, J. S. (1978). Maximum yield of the potato crop – studies by ADAS 1971-1977. The 'blue-print' specification and evaluation of the components. In *Maximising Yields of Crops,* pp. 174-184. Ministry of Agriculture, Fisheries and Food. London: HMSO.

Gunn, J. S. (1978). Maximum yield of the potato crop – studies by ADAS 1971-1977. Field Studies 1975-1977. In *Maximising Yields of Crops,* pp. 185-194. Ministry of Agriculture, Fisheries and Food. London: HMSO.

Hawkes, J. G. (1978). Biosystematics of the potato. In *The Potato Crop,* pp. 15-69. Ed. P. M. Harris. London: Chapman & Hall.

Howard, H. W. (1978). Can future requirements be fulfilled by new varieties? In *Growing Potatoes,* pp. 60-62. NIAB. Fellows Conference No. 2.

Ivins, J. D. & Bremner, P. M. (1965). Growth, development and yield in the potato. *Outlook on Agriculture* **4**, 211-217.

Mitchell, R. L. (1976). *Crop Growth and Culture*. Ames: Iowa State University Press.

Monteith, J. L. (1977). Climate and the efficiency of crop production in Britain. *Philosophical Transactions of the Royal Society, London,* B **281**, 277-294.

NIAB (1976). *Recommended varieties of potatoes.* Farmers leaflet No. 3 Cambridge: National Institute of Agricultural Botany.

Penman, H. L. (1971). Irrigation at Woburn – VII. *Rep. Rothamsted Exp. Stn. for 1970*, part 2, pp. 147-170.

Ramsbottom, J. E. (1978). The variety in relation to husbandry. In *Growing Potatoes,* pp. 22-27, NIAB Fellows Conference Report No. 2.

Richardson, D. E. (1979). Potato variety testing at NIAB *Journal of the National Institute of Agricultural Botany* (in press).

Schepers, A. & Sibma, L. (1976). Yield and dry matter content of early and late potatoes, as affected by monoculture and mixed cultivars. *Potato Research* **19**, 73-90.

Scott, R. K. & Wilcockson, S. J. (1978). Application of physiological and agronomic principles to the development of the potato industry. In *The Potato Crop,* pp. 678-704. Ed. P. M. Harris. London: Chapman & Hall.

Silvey, V. (1978). The contribution of new varieties to increasing cereal yield in England and Wales. *Journal of the National Institute of Agricultural Botany* **14**, 367-384.

Spedding, C. R. W. (1975). *The Biology of Agricultural Systems.* London: Academic Press.

Competition effects within mixed stands

C. J. T. Spitters
Dept. Theoretical Production Ecology, Agricultural University, Bornsesteeg 65, 6708 PD Wageningen, Netherlands.

Introduction

Among crop scientists there is a continued interest in mixed cropping. Most of the experiments reported here concern mixtures which were constituted in a rather arbitrary manner, the yields being compared with the yield of monocultures without further analysis. The experimental design would have been more sophisticated, the analysis more thorough and the focus on the search for overyield mixtures improved if the competition effects had been better quantified. Moreover, fallacies would have been avoided.

The most convenient model to analyse the competition effects within mixed stands is that introduced by de Wit in 1960 (Trenbath, 1978; Spitters, 1979, Chapter 3). With this model some fallacies in the interpretation of mixed cropping experiments will be exposed and some mechanisms for overyielding in mixtures will be discussed.

Design and interpretation of competition experiments

Competition between two species can be studied with a replacement series. A replacement series is the result of generating a range of mixtures by starting with a monoculture of one species and progressively replacing plants of that species with plants of the other species until a monoculture of the latter is produced. All monocultures and mixtures are grown at equal density.

The results of such an experiment are presented in a replacement diagram where for each species the yield per unit area is plotted against the proportion of the total number of seeds sown (Fig. 1). The yield of a species increases linearly with its relative seed frequency in the mixture when there is no competition or when both species are equally competitive (dashed lines from the x axis in Fig. 1(a)). The curves for actual yield in Fig. 1(a) (continuous lines from the x axis) deviate from linearity. Species i shows a convex curve, i.e. i has a higher yield in the mixture than expected if both species were equally competitive. Hence

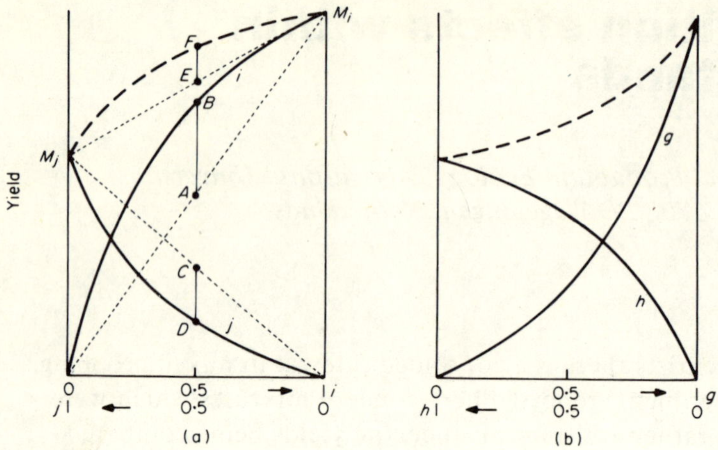

Fig. 1 Replacement diagrams (a) with the species with the higher monoculture yield as the stronger competitor and (b) with the species with the higher monoculture yield as the weaker competitor.

i is a strong competitor. On the other hand, the curve for species *j* is a weak competitor. These effects were quantified by de Wit (1960).

The species compete in the mixture for a limited amount of 'space'. The term 'space' summarizes all growth requisites like light, water and nutrients. Each species expands into its share of the total available space. The output yield (O) of a species in a mixture is considered to be proportional to the space (RY) acquired by that species:

$$O = RY \times M,$$

where M is the monoculture yield. The space occupied by the species is estimated as $RY{=}O/M$ and is called its 'relative yield' (de Wit and van den Bergh, 1965). When the species compete for the same limited growth requisites, the total available space equals unity.

The 'relative yield total' in a mixture of *i* and *j* is then

$$RYT = RY_i + RY_j = \frac{O_i}{M_j} + \frac{O_j}{M_j} = 1.$$

RYT is greater than one when the species compete for only a part of the space, i.e. when they are limited in part by different resources. We will see that this is a necessary but insufficient prerequisite for overyielding mixtures. In some reports the term 'land equivalent ratio' (LER) has been used for RYT, but in the calculation of LER the replacement principle has not always been used.

Pseudo benefit of mixed cropping

Two fallacious approaches are frequently found in the literature: (a) mixtures are compared with monocultures growing in a stand that is half as dense, and (b) the success of mixed cropping is read from the difference between the total yield of the mixture and the mean yield of the monocultures.

Mixtures and monocultures at different spacings

In many of the experiments reported, mixtures have been established by adding the plant populations used in the monocultures. A simple example may illustrate the pitfall behind this 'additive scheme'. Consider two species that do not differ in yield and in density response. In the example of Fig. 2, they both yield 5·6 t ha^{-1} in monoculture at

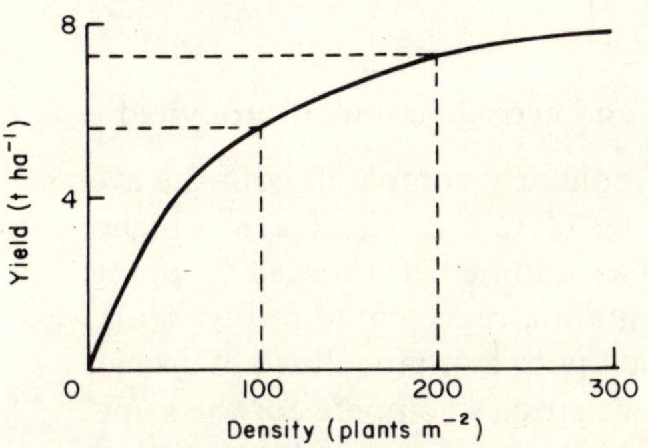

Fig. 2 The effect on yield of doubling the plant population.

100 plants m^{-2}. In an additive scheme, the mixture is grown at 200 plants m^{-2}. The yield of a species in this mixture equals its yield in monoculture when the species are equally competitive. Hence, from Fig. 2, we read 7·3 t ha^{-1} for the total yield of the mixture. The mixture yields 30 per cent more than either of the monocultures. However, the yield benefit is only due to a greater density and may also be obtained with the monocultures growing at the same density as the mixture.

Instead of the additive scheme a replacement scheme should be used, i.e. the mixtures should be established by replacing a certain proportion of one species with the same proportion of the other species, thus keeping the total density constant. From the replacement experiment RYT can be estimated. Only when RYT is greater than one can mixed cropping give a yield benefit, because in part the species occupy different spaces. For the space which the species do not have in common, the replacement experiment degenerates into a density experiment (Fig. 5; Berendse, 1979). This suggests that when RYT is greater than one, the optimal density is higher for the mixture than for either of the monocultures. This is supported by experimental evidence (Willey, 1979). The optimal spacing for the mixture may be determined in a separate experiment with the mixtures at different densities.

The density in the replacement series has, in general, to be in agreement with that used by farmers. However, the species may differ strongly in their optimal density. For example, Phaseolus beans are sown at twice the density of maize. The replacement diagram is then generated by substituting one maize plant with two bean plants (Willey and Osiru, 1972). In the equations, the density is expressed in 'plant units' with one plant unit equivalent to one maize plant and to two bean plants (de Wit, 1960, p. 63).

Difference between mixture yield and average monoculture yield

The yield of the 1 : 1 mixture is frequently compared with the average of the yield of the two monocultures (F vs E in Fig. 1(a)). A higher yield of the mixture is interpreted as argument for mixed cropping. However, when we consider the situation represented in Fig. 1(a), we see that the highest yield is obtained with a monoculture of species i (M_i). It can be shown that when two species compete for the same limited resources, i.e. $RYT=1$, the highest yield is produced with the better monoculture (de Wit, 1960).

On the other hand, when $RYT>1$, the species occupy in part different spaces so that a mixture may exploit the resources more efficiently than either monoculture and the mixture may yield more than the best monoculture. However, a value of RYT greater than one does not guarantee a better yield mixture (van den Bergh, 1968, pp. 6 and 8).

Often the farmer grows more than one crop, e.g. to guard against market risks and to spread labour peaks. We distinguish two situations: $RYT=1$ and $RYT>1$. When $RYT=1$, the same yield of each species may be obtained with monocultures as with a mixture, without changing the total area of land. The area fractions of the monocultures have to be taken equal to the relative yields of the species. In the situation presented in Fig. 3, the yields of the 1:1 mixture are reproduced by sowing 0·75 of the area with the stronger competitor i and 0·25 of the area with the weaker competitor j, both in monoculture. Hence, when

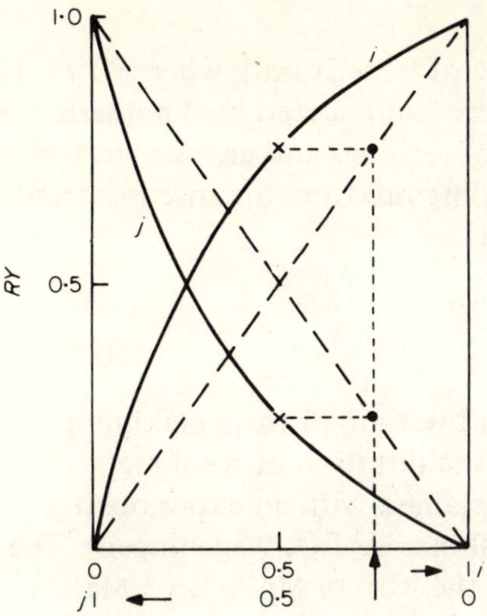

Fig. 3 Replacement diagram with relative yields showing that the yields of a mixture may be reproduced by growing the monocultures on the same area of land when RYT=1.

RYT=1 there is no advantage of growing a mixture instead of the monocultures.

When $RYT>1$, a larger area of land is needed to produce the same yield of each species with monocultures than with a mixture. The value of RYT measures the relative area under monocultures that is required to give the same yield of each species as in the mixture. Hence when $RYT>1$ mixed cropping may be of advantage. However, the industrial farmer usually chooses to grow the crops in monoculture for reasons of mechanization and management.

Comparison of the yield of the mixture with the mean of the yields of the two monocultures may lead to unjustified optimism with respect to the perspectives of mixed cropping. When RYT=1, a 1:1 mixture yields more than the average of the two monocultures when the stronger competitor is the species with the higher yield in monoculture (Fig. 1(a)), whereas the mixture yields less when the reverse is true (Fig. 1(b)). This is because the gain of the strong competitor ($B - A$ in Fig. 1(a)) and the loss of the weak competitor ($C - D$) in the mixture are proportional to their monoculture yields (M_i and M_j) (Spitters, 1979, Chapter 3). As already shown, there is no advantage in mixed cropping when RYT=1.

Real benefit of mixed cropping

As we have seen, mixed cropping may be of benefit only when $RYT>1$. The mechanisms resulting in $RYT>1$ were summarized by Trenbath (1976) and are well-known in mixtures of legumes and grasses. In the following, some other types of over-yielding mixtures are discussed and interpreted in terms of the de Wit model.

Nonsimilar growth periods

Mixed growing of species which differ in the time of their maximum demands on the growth factors extends the duration of resource exploitation. This mechanism will be explained with an experiment with potatoes (unpublished data of L. Sibma, CABO, Wageningen). The variety Tanja was planted on 5 May and the variety Multa on 5 May, 1 July and 1 August. The varieties were grown in monocultures and in 1:1 mixtures. The replacement diagram shows a convex and a concave curve when the varieties were planted at the same time (Fig. 4(a)). *RYT* was about unity indicating that the varieties competed for the same resources. When Multa was planted on 1 August, two density-response curves emerged in the replacement diagram (Fig. 4(c)) showing that neither variety felt the other's presence in the mixture. *RYT* was about 2. It was noted that by the time that Multa was established, Tanja was completely matured.

A similar experiment was done with the varieties Doré and Spartaan in a different year and at a different location. This time *RYT* was about 1 irrespective of the planting time of Spartaan (Figs. 4(d)-(f)). Hence the varieties compete for the same limited resources, even when Spartaan was established after Doré had matured. The replacement diagram for the latter situation (Fig. 4(f)) therefore shows a convex and a concave curve instead of the two density-response curves in the experiment with Tanja and Multa (Fig. 4(c)). Apparently, the early variety Doré consumed almost all of some limiting resource so that there remained little for Spartaan when it was established in the field in August.

In the second experiment where *RYT* was about 1, fertilizer was applied only before the first planting. On the other hand, in the first experiment with a large *RYT*, an additional dressing was given just before planting the late variety. This difference in fertilizer application may explain the discrepancy between the experiments. It illustrates that mixed cropping with species differing in growth-period results in a maximum yield benefit when the limiting resource is delivered

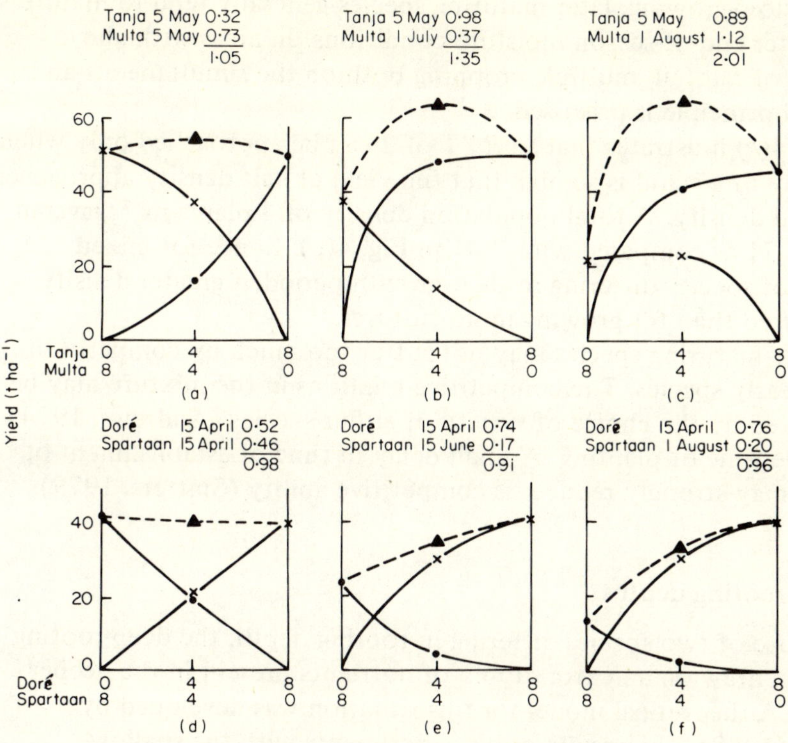

Fig. 4 The effect of intercropping an early-planted potato variety with a later-planted one on tuber yield at a population density of 8 plants m^{-2}. In the headings, the name of each variety is followed by the date of planting and the relative yield (RY).

continuously. This situation is most likely to be met under optimal conditions and with a supply of nutrients spread over the season, either by natural release or split dressing. On the other hand, when the limiting resource is depleted by the early species, i.e. RYT=1, the highest yield is obtained with the better monoculture and the mixture yields may be reproduced with monocultures on the same total area of land.

The previous principles are reflected in the types of mixtures found in areas differing in rainfall (Andrews and Kassam, 1976). In rain-fed agriculture with rainfall below 600 mm simultaneous cropping is practised with crops of similar maturity. Species with a retarded growth cycle are not found in the mixtures because they would have to mature under severe water stress. In contrast, crop mixtures involving different maturities are common in areas of rainfall between 600 and 1000 mm.

Here the slow-growing, later-maturing species generally tend to mature under better end-of-season moisture conditions. In areas with above 1000 mm of rainfall, multiple cropping both on the simultaneous and sequential principle is practised.

Figure 4(c) illustrates that an *RYT* of 2 can be approached only when the density of a stand is so high that the yield at half density approaches that at full density. A total population density of 4 plants m^{-2} gave an *RYT* of 1·71 as compared with 2·01 in Fig. 4(c). Hence for mixed cropping of species differing in their growth period, a greater density must be used than for growing monocultures.

The late-maturing species may not suffer too much by competition with the early species. The competitive relations in the mixture may be manipulated by the choice of variety of either species (Andrews, 1974) and by the time of planting. A small delay in time of establishment of a species may strongly reduce its competitive ability (Spitters, 1979).

Different rooting depths

In a mixture of two species differing in rooting depth, the deep-rooting species can draw on an extra supply of nutrients present in the deeper soil layer. A theoretical model for this situation was developed by Berendse (1979). The results of his experiments with the shallow-rooting grass *Anthoxanthum odoratum* and the deep-rooting herb *Plantago lanceolata* satisfy his model. The species share a common pool of nutrients in the upper layer where both are rooting. Therefore *RYT*=1 and the curves in the replacement diagram for the upper layer are convex and concave (Fig. 5). In the second layer, the replacement experiment degenerates into a density experiment for the deep-rooting species. In the combined replacement diagram (Fig. 5), *RYT* is greater than 1 and the curve of the deep-rooting species is S-shaped. It is assumed that the two soil layers are two strictly separated compartments and that the limiting resource is distributed homogeneously over both soil layers.

A mixture may yield more than the better monoculture only when the shallow rooting species is the more productive one in monoculture.

Multiple resource competition

Braakhekke (1980) elaborated a model for a situation where two species are limited by two different resources. Consider the extreme situation where species *a* is limited in its growth by element 1 and does

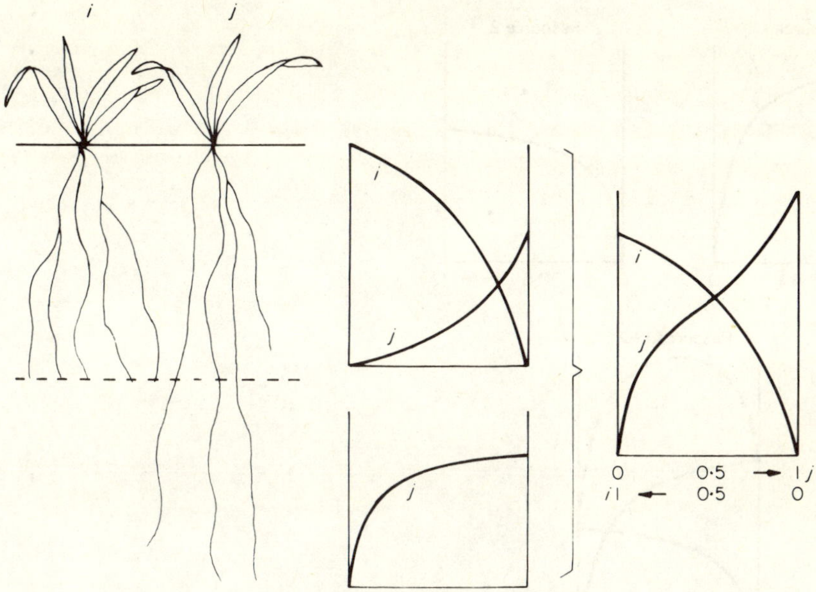

Fig. 5 Competition between a shallow- and a deep-rooting species interpreted as the combined effect of competition for the same resources in the upper soil layer and a density experiment with the deep-rooting species in the deeper soil layer.

not take up element 2 while species *b* is limited by element 2 and does not take up element 1. Neither species senses the others presence in the mixture, which results in two density-response curves in the replacement diagram (Fig. 6). *RYT* approaches the value of 2 at dense sowings which suggests a great advantage in mixed cropping.

The situation where both species *a* and *b* take up both of the elements 1 and 2 is more realistic. Suppose that the amounts supplied of each of the elements are completely taken up in the mixture and in the monocultures. The distribution of the elements among the species in the mixture satisfies the competition equations of de Wit (1960) which gives for each element an uptake replacement diagram with *RYT*=1 (Fig. 7, top). Yield is approximately linear with the uptake of the element that limits the growth ('Law of the minimum' of von Liebig). We obtain the yield replacement diagrams for element 1 limiting the growth of each species by multiplying the uptake curve of the species with its efficiency of resource utilization (Fig. 7, centre). These efficiencies are the slopes of the linear relations between yield and uptake and may be different for each species and for each element. Therefore, the yield replacement diagram when element 2 is the

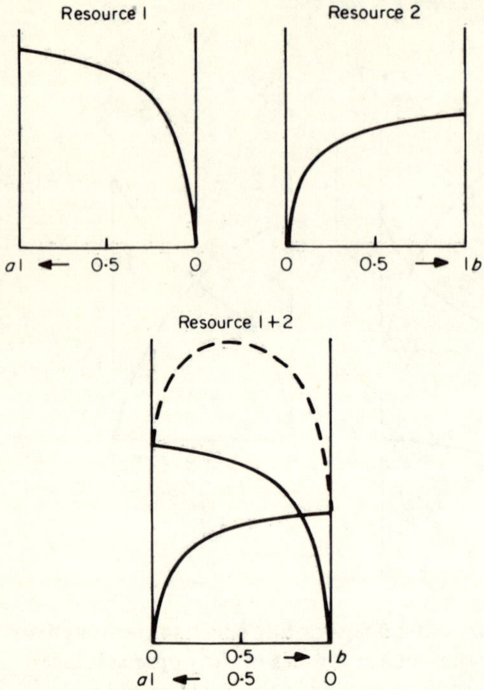

Fig. 6 Competition between species *a* and *b* when *a* is limited in its growth by resource 1 and does not take up resource 2 while *b* is limited by resource 2 and does not take up resource 1. The top diagrams are for the two resources separately whilst the bottom one shows the combined effect.

limiting growth factor may deviate from that for element 1 (Fig. 7, centre). In Fig. 7, the yield curve of species *a* for element 1 is consistently lower than that of *a* for element 2, so that the growth of species *a* is limited by element 1. On the other hand, species *b* is limited by element 2. Therefore, the yield replacement diagram is composed of the *a* curve for element 1 and the *b* curve for element 2 when the species compete for both resources (Fig. 7, bottom). We see that a mixture yields more than the better monoculture and *RYT* exceeds 1. The luxury consumption of non-limiting resources is reduced in the mixture compared with the monocultures so that the nutrients are used more efficiently in the mixture. Hence, mixed cropping may be of advantage when each species is the stronger competitor for the element that limits its own growth.

On the other hand, an *increased* luxury consumption in a mixture is found when each species is the stronger competitor for the element that

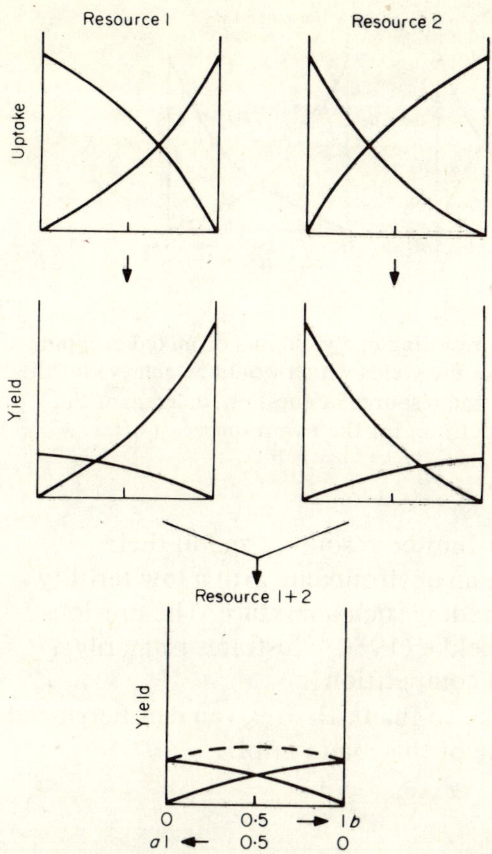

Fig. 7 Multiple-resource competition resulting in a yield benefit of mixed cropping. Diagrams at the top give the uptake of the two resources separately; diagrams in the centre represent the yields which would be achieved if the growth of both species were limited by the resource in question; diagram at the bottom gives the combined effect of competition for the two resources. (Data from Braakhekke, 1980.)

limits the growth of the other species (Fig. 8). Therefore, multiple-resource competition may explain a value of *RYT* smaller than one as well as the occurrence of allelopathic effects.

In conclusion, a mixture may outyield the better monoculture when the two species are limited in their growth by different elements and each species is the stronger competitor for the element that limits its own growth. The use of this mechanism for mixed cropping will be limited because the prerequisites are difficult to meet. Species differing

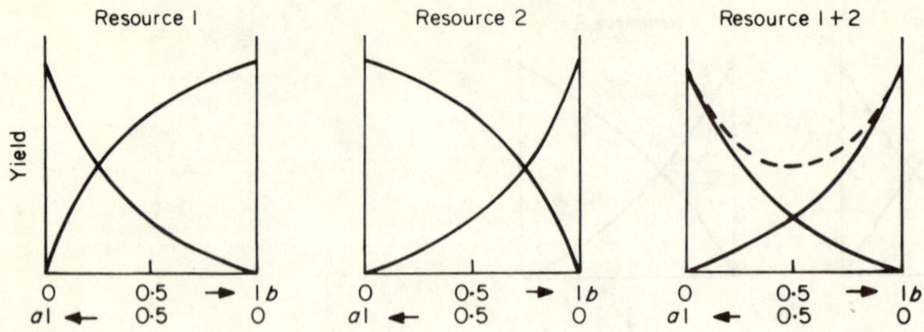

Fig. 8 Multiple-resource competition resulting in a yield loss of mixed cropping. Diagrams at the left and centre represent the yields which would be achieved if the growth of both species were limited by the resource in question; diagram at the right gives the combined effect of competition for the two resources. (After Braakhekke, 1980.).

strongly in competitive ability for limited resources and in their resource utilization efficiencies in an environment with a low fertility offer the greatest probability of finding such a mixture. The previous discussion of the model of Braakhekke (1980) illustrates primarily a way of thinking in the analysis of competition.

Acknowledgements. I would like to thank Dr J. P. van den Bergh and Dr G. S. Innis for a critical reading of this manuscript.

References

Andrews, D. J. (1974). Responses of sorghum varieties to intercropping, *Experimental Agriculture* **10**, 57-63.

Andrews, D. J. & Kassam, A. H. (1976). The importance of multiple cropping in increasing world food supplies, In *Multiple Cropping*, pp. 1-10. Eds. R. I. Papendick, P. A. Sanchez & G. B. Triplett. Madison: American Society of Agronomy, Special Publication 27.

Berendse, F. (1979). Competition between plant populations with different rooting depths. I. Theoretical considerations. *Oecologia* (in press).

Bergh, J. P. van den (1968). An analysis of yields of grasses in mixed and pure stands. *Verslagen landbouwkundig Onderzoek* (Agricultural Research Reports) **714**, 71 pp. Wageningen: Pudoc.

Braakhekke, W. G. (1980). On coexistence: a causal approach to diversity and stability in grassland vegetation. *Verslagen landbouwkundig Onderzoek* (Agricultural Research Reports). Wageningen: Pudoc (in preparation).

Spitters, C. J. T. (1979). Competition and its consequences for selection in barley breeding. *Verslagen landbouwkundig Onderzoek* (Agricultural Research Reports) **893**, 268 pp. Wageningen: Pudoc.

Trenbath, B. R. (1976). Plant interactions in mixed crop communities. In *Multiple Cropping*, pp. 129-169. Eds. R. I. Papendick, P. A. Sanchez & G. B. Triplett. Madison: American Society of Agronomy, Special Publication 27.

Trenbath, B. R. (1978). Models and the interpretation of mixture experiments. In *Plant Relations in Pastures*, pp. 145-162. Ed. J. R. Wilson. Melbourne: CSIRO.

Willey, R. W. (1979). Intercropping– its importance and research needs. *Field Crop Abstracts* **32**, 1-10 and 73-85.

Willey, R. W. & Osiru, D. S. O. (1972). Studies on mixtures of maize and beans (*Phaseolus vulgaris*) with particular reference to plant population. *Journal of Agricultural Science, Cambridge* **79**, 517-529.

Wit, C. T. de (1960). On competition. *Verslagen landbouwkundig Onderzoek* (Agricultural Research Reports) **66**(8), 82 pp. Wageningen: Pudoc.

Wit, C. T. de & Bergh, J. B. van den (1965). Competition between herbage plants. *Netherlands Journal of Agricultural Science* **13**, 212-221.

The importance of mixed stands in tropical agriculture

B. N. Okigbo
International Institute of Tropical Agriculture, Ibadan, Nigeria

Introduction

The tropics are those regions of the earth's surface located between latitudes 23½° north and south of the equator. According to Ruthenberg (1976) based on Troll (1966), the tropics are characterized by the following climates: (1) humid climates with 7 or more humid months (usually more than 1400 mm per annum); (2) semi-humid (wet and dry) climates with 4½-7 humid months of 100-200 mm unimodal or bimodal rainfall patterns, and (3) semi-arid climates with 2-4½ humid months of cool season or warm season rainfall. Temperatures are usually above freezing point but vary according to latitude, season and altitude. In relation to the range of environmental conditions and cultures in the tropics, there is a diversity of farming systems and these are associated with different cropping systems (Table 1). These have been described by Whittlesey (1962), Duckham and Masefield (1971), Laut (1971), Benneh (1972), Greenland (1974), Grigg (1974), and Ruthenberg (1976).

A cropping system defines the pattern of growing crops in terms of crop combinations and sequences in time and space in addition to the practices and technologies with which the crops are produced. Crops may be grown in pure stands or in mixtures. Mixed cropping is the growing of more than one crop plant species simultaneously in contiguous stands on the same piece of land. Mixed cropping may, with subtle differences in meaning, be variously designated as intercropping, interplanting, mixed cultivation, mixed culture, and polyculture. Mixed cropping is only one of the ways in which more than one crop is grown on the same piece of land in the same year (i.e. multiple cropping). This paper is devoted to a discussion of the different types of mixtures, the objectives of mixing, their merits and demerits and the evolution and importance of mixed stands in tropical agricultural systems.

Table 1. Classification of farming systems in relation to occurrence of crops in mixed stands

Farming system complex	Occurrence of mixed stands †
A. Traditional and transitional systems	
1. Extensive	
(a) Nomadic herding	?
(b) Shifting cultivation	**
2. Semi-permanent cultivation	**
3. Permanent intensive rain-fed agriculture	
(a) Compound farming	**
(b) Intensive subsistence agriculture	**
(c) Terrace agriculture	*
(d) Asian wet rice	*
Hydraulic systems	
(a) Floodland cultivation	*
(b) Irrigated agriculture	*
B. 'Modern' farming systems and their local adaptations	
1. Extensive livestock ranching	?
2. Intensive livestock production (poultry, pigs, dairy, beef)	*
3. Mixed farming and regulated ley systems	*
4. Arable rain-fed crop production	*
5. Irrigated cropping systems	*
6. Specialized horticulture	
(a) Plantations	**
(b) Market gardening	*

† ? Rare
* Sometimes
** Widespread or frequent

Kinds of mixtures

The diverse cropping systems in tropical farming systems differ in relation to the intimacy or contiguity of individual stands, relative times of planting, the duration of mixing and the spatial arrangement of components of the mixture. Different types of mixtures encountered in tropical agriculture include (1) *mixed intercropping* – growing of two or more crops simultaneously with no distinct row arrangement; (2) *row intercropping* – growing two or more crops simultaneously with one or more crops planted in rows; (3) *strip intercropping* – growing two or more crops simultaneously in different strips wide enough to permit independent cultivation but narrow enough for the crops to interact agronomically; and (4) *relay intercropping* – growing two or more crops simultaneously, but with one or more of them planted after the first crop has reached reproductive stage of growth but some time before harvesting (Sanchez, 1976; Andrews and Kassam, 1976). Another aspect of intercropping of relevance to some cropping systems of tropical Africa may be termed *patch intercropping*, which involves growing two or more crops simultaneously on small contiguous patches or mounds. Papendick, Sanchez and Triplett (1976) review mixed cropping in the tropics and elsewhere as an aspect of multiple cropping.

The various kinds of intercropping constitute ways of increasing or diversifying production per unit area. Other kinds of mixtures which are used to achieve various objectives include (1) windbreaks, shelterbelts, shade plants and live supports which in horticulture and plantation agriculture are used to modify the environment and enhance crop growth (Table 2); (2) catch crops which are relatively earlier maturing crops interplanted with later maturing or perennial crops (Table 2), and (3) grass and legume mixtures in pastures used to improve nutritive value and productivity of pastures in addition to taking advantage of biological nitrogen fixation in legumes (Table 3).

Reasons for mixing and diversity of mixtures

Farmers in the tropics give various reasons for growing crops in mixtures. Norman (1974) has reported that a majority of the farmers in northern Nigeria gave increased yield as the most important factor. Diversification of production, risk aversion and tradition are also important reasons for growing crops in mixtures. The advantages and disadvantages of intercropping reported by various workers (Webster and Wilson, 1966; Wrigley, 1969; Norman, 1974; Williams and Joseph, 1970; and Okigbo, 1978) are summarized in Table 4.

It is a common practice in West Africa for most important staple food crops and some perennial crops to be produced in widespread but varying intercropping systems (Fig. 1). One of the commonest objectives of mixing is to ensure constant harvests; sometimes crops of the same species or plant types with different periods of maturity may be grown. For example, in various parts of Nigeria photoperiod-sensitive (early) and -insensitive (late) types of okra (*Hibiscus esculentus*) are mixed so as to ensure that the vegetable is available in both the wet and dry season (Fig. 2). Similarly, in the rain-forest areas, maize may be intercropped with rice, and in the savannah areas millet is intercropped with sorghum in such a way that the maize and millet are harvested earlier, while the farmer is waiting for the rice and sorghum, respectively (Fig. 2).

Table 2. List of plantation crops in relation to kinds of intercrops that may be associated with their production in certain locations

		Catch crop	Cover crop	Pasture or forage crops	Shade	Wind break & shelterbelt	Companion crop	Live support
1.	*Ananas comosus* – pineapple	X			X?		X	
2.	*Camellia sinensis* – tea				X			
3.	*Cinchona* – quinine				X			
4.	*Citrus spp.* – oranges, lemons, etc.	X					X	
5.	*Cocos nucifera* – coconut	X	X	X			X	
6. (a)	*Coffea spp.* – coffee	X			X	X		
6. (b)	*Cola spp.* – kola		X		X			
7.	*Elaeis guineensis* – oil palm	X	X	X				
8.	*Hevea brasiliensis* – rubber	X	X?				X	
9.	*Mangifera indica* – mango	X	X					
10.	*Musa spp.* – abaca, bananas and plantain	X	X?			X		
11.	*Myristica fragrans* – nutmeg				X			
12.	*Piper nigrum* – black pepper	X			X			X
	Theobroma cacao – cocoa	X					X	
13.	*Vanilla fragrans* – vanilla						X?	X

Table 3. Compatible grasses and legumes and recommended mixtures for three selected legumes used in tropical pastures *

Pasture legume	Compatible grasses and legumes	Recommended mixtures	
Centrosema pubescens	*Grasses*		
(Centro)	*Chloris gayana*	Mixture A	*Sprigs*
	Hyparrhenia rufa	*Panicum maximum*	2-3
	Panicum maximum	*Centrosema*	
	Paspalum dilatatum	*pubescens*	2
	Pennisetum purpureum	*Pueraria*	
	Legumes	*phaseoloides*	1
	Calapogonium mucunoides	*Stylosanthes*	
		guianensis	1
Pueraria phaseoloides	*Hyparrhenia rafa*	Mixture B	*Sprigs*
	Melinis minutiflora	*Digitaria*	
	Panicum maximum	*decumbens*	2
	Pennisetum purpureum	*Stylosanthes*	
	Stenotaphrum	*leucocophala*	1
	secundatum	*Pueraria*	
		phaseoloides	1
	Legumes	Mixture C	*Sprigs*
	Calopogonium mucunoides	*Tripsacum laxum*	*1*
	Centrosema pubescens	*Centrosema pubescens*	4-5
		Pueraria phaseoloides	1
		Calopogonium mucunoides	1
Stylosanthes guianensis	*Grasses*	*Centrosema pubescens*	1
(Stylo)	*Brachiaria ruzuziensis*	Mixture E	
	Chloris gayana	*Stylosanthes guianensis*	
	Hyparrhenia rufa	*Centrosema pubescens*	
	Melinis minutiflora		
	Panicum maximum		
	Legumes		
	Centrosema pubescens		

* *Source:* Food and Agriculture Organisation of the United Nations: Skerman (1977)

Table 4. Advantages and disadvantages of growing crops in mixed stands *

A. *Advantages*

(a) Total yields are often higher than the sole crop yields even if yields of individual components are reduced.

(b) Mixtures result in more efficient utilization of resources (light, water, nutrients) by plants of different height, canopy structure and nutrient requirements;

(c) Diseases and pests may not spread as rapidly in mixtures because of differential susceptibility to the pests and pathogens;

(d) They provide insurance against crop failure;

(e) They afford opportunities for crop diversification, thereby ensuring a steady supply of a range of products without much investment in storage;

(f) They provide effective cover to the soil and so minimize erosion and exposure of soil to solar radiation;

(g) Mixtures spread labour costs more evenly throughout the cropping season;

(h) In grass/legume mixtures, fixed nitrogen from the legume is available to the grass and the nutritional quality of the mixture is improved;

(i) Mixtures in compound gardens constitute experimental plots for screening exotic materials and preservation of germplasm;

(j) They may be used to suppress weeds, thereby reducing the cost of weed control and improving the quality of the product.

B. *Disadvantages*

1. Mechanization of planting, harvesting and various practices is more difficult with mixtures than in sole crops.
2. Determining quantities and mode of application of inputs such as fertilizers and pesticides is a problem.
3. Drought stress may be more serious where moisture is limiting than in sole crops.
4. Experimentation with mixtures is very difficult due to lack of widely accepted methodology for their evaluation and the complexity of interactions among components of mixtures.

---

* *Source:* Okigbo and Greenland (1976).

Sometimes there are several reasons for mixing. For example, melon which matures early but is not harvested early is included in a mixture with maize to suppress weeds, and cowpeas which may be planted earlier may be relay-planted in the maize so as to ensure that it is harvested in the drier part of the year when it is easier to harvest, thresh and store (Fig. 2).

In the traditional farming systems of the rain-forest areas of south-eastern Nigeria, mixtures often contain several leaf, fruit and seed vegetables which are harvested within the first two months after planting during the hungry season and they are used to increase the

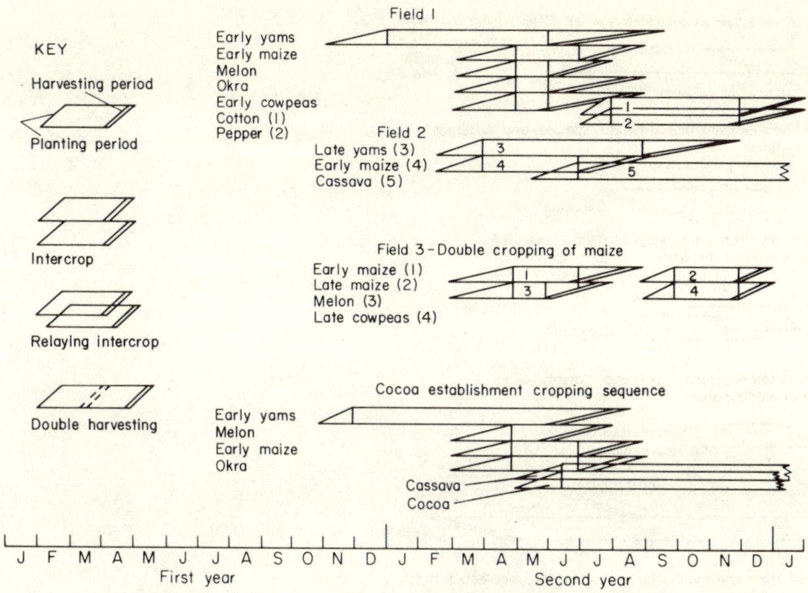

Fig. 1 Cropping schedules in Ibadan area, western Nigeria.

bulk of the yams which become scarce during the hungry season after planting (Fig. 1).

Other mixtures of legumes and cereals take advantage of the complementary relationship between legumes and non-legumes, while at the same time ensuring that the poor nutritive value of the cereal in the diet is compensated by the association with a legume.

Sometimes advantage is taken of peculiar ecological conditions. For example, in hydromorphic soils early maize to be harvested green is intercropped with rice and sugar cane, which are of relatively longer duration and will perform well in soils of high water table at a time when less adapted maize has been harvested (Fig. 3).

A given crop may be planted at different times, because it is less demanding either on soil fertility or on other factors. For example, cassava may be planted early with maize, yams and melon (Fig. 1) but often it may be planted later so as to minimize competition with the other crops which are more sensitive to soil fertility and moisture level. In the middle belt of Nigeria where the length of the dry season may exceed four months, cocoyam, sugar cane and bananas may be intercropped in the valley bottom soil where they have enough moisture for the 12 months that they require to mature (Fig. 3).

Several mixtures involve annuals and perennials. In plantation crops annual catch crops are mixed with tree crops to minimize the cost of

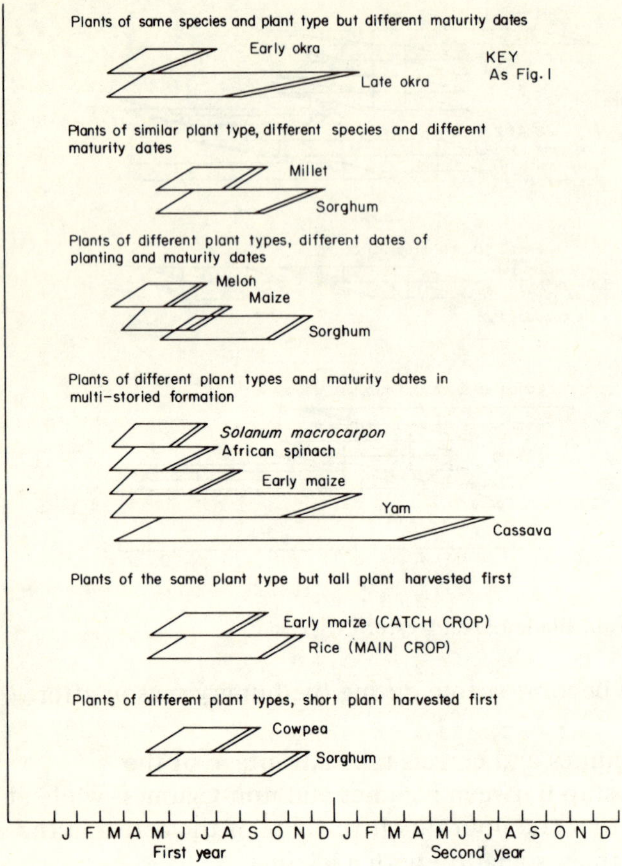

Fig. 2 Mixtures of annuals with annuals or semi-perennials.

plantation establishment. At the same time weeds are controlled and tillage can be charged to all crops in the mixture instead of to the tree crop alone (Fig. 4). Moreover, in the different field systems, protected plants or stumps of fallow crops exist in mixtures with food crops so as to maximize the range of products that can be harvested but also to provide supports for climbers as well as safeguarding the environment by ensuring rapid establishment of vegetation cover. (Fig. 5).

In multi-storied compound farms which may involve over 60 species of plants in the humid tropics, the farmer's status and subsistence requirements are satisfied by growing maize and ware yams. On the other hand, the needs of the housewife for vegetables and condiment plants are satisfied by including a few stands of these vegetables and condiments on the edges of the field where they can be easily obtained

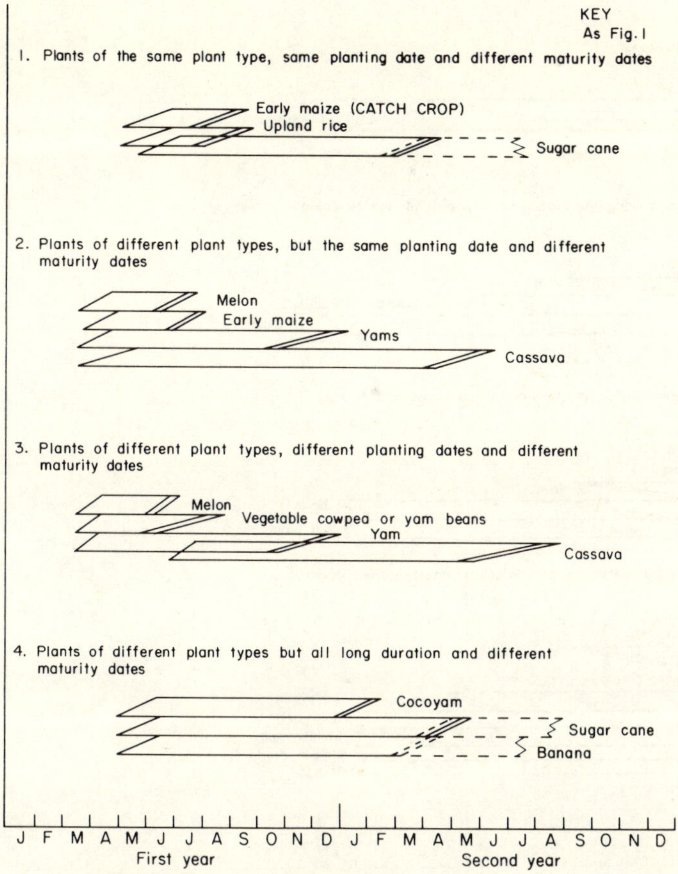

Fig. 3 Mixtures of annuals and long-duration semi-perennials.

without going into the garden or buying from the village market. Fruit trees, drug plants, fibre plants, browse plants etc. are also grown on compound farms to satisfy multifarious needs. It is not surprising therefore to find quite a range of mixtures in relation to the needs of the farmer (Table 5). Another aspect of diversification is in relation to various micro-environmental conditions. Thus Allen (1965) reported that the Bemba of Rhodesia in their chitemene shifting cultivation system maintained different crop combinations and sequences on bush fallow fields, ant-hills, Coleus potato gardens, riverside gardens and hut-site gardens. Miracle (1967) and Schlippe (1956) also reported Azande crop combinations and sequences for maize/oil seeds, gourds, groundnut/finger millet succession, etc.

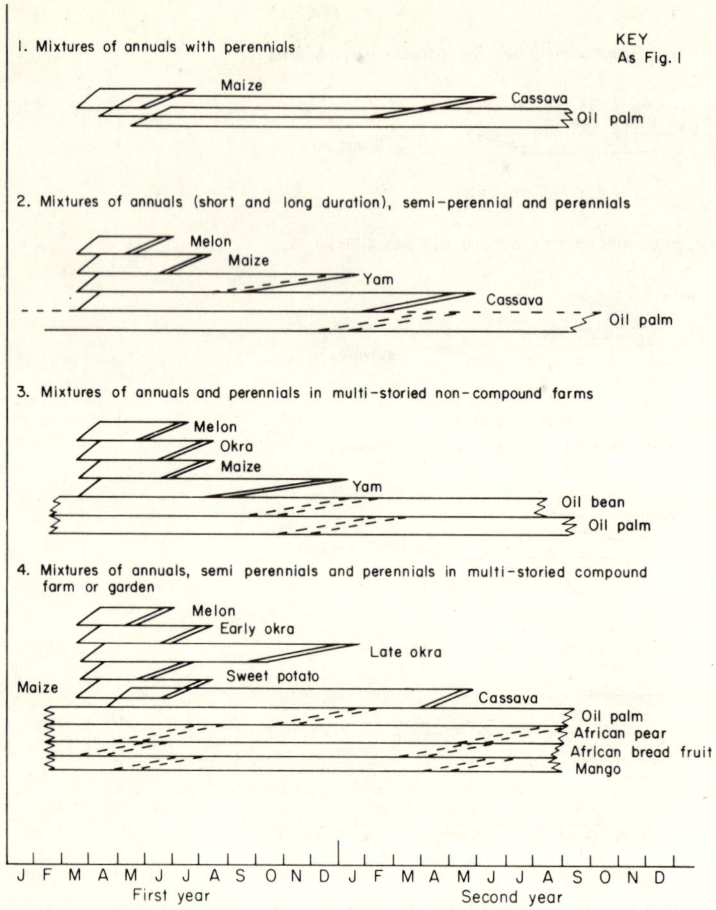

Fig. 4 Mixtures involving perennials.

Probable role of mixtures in the evolution of crops and cropping systems

Technically, all cropping systems of the world are mixtures, and mixtures are important in the evolution of crop plants. This observation is based on the fact that even in sole cropping systems, weeds compete with crops just as components of mixtures interact among themselves. It would appear that the only difference between classical mixtures and crop/weed associations is that all components of mixtures except weeds are purposely planted or utilized by man. The high incidence of mixtures in tropical agro-ecosystems needs to be studied more closely with respect to various interactions among crops (Willey, 1979). In traditional mixed cropping systems, mixtures of species and varieties of

Table 5. Number of food crops and other useful plants found in crop mixtures in compound and outlying farms in parts of south-eastern Nigeria

	Area of sample (ha)							
	A*		B*		C*		D*	
	0·003-0·45		0·04		0·04-0·4		0·04-0·5	
Crop plant group	Range	Mean	Range	Mean	Range	Mean	Range	Mean
1. Roots & tubers	1-12	5	4-6	5	5-8	7	1-8	4
2. Cereals & other starchy staples	0-3	1	1-2	1	2-3	2	0-4	2
3. Leaf vegetables	3-7	4	0-4	2	4-8	6	10-11	6
4. Fruit vegetables	4-6	4	1-3	2	5-6	5	0-6	2
5. Legumes	1-5	3	0-4	2	2-4	2	0-3	1
6. Fruits,nuts & oil plants	1-14	5	0-1	1	10-11	10	2-15	7
7. Spices & beverages	0-3	1	0-1	1	2-6	4	0-9	4
8. Miscellaneous useful plants	0-7	2	0-1	0	10-14	13	1-29	11
Range in total number of species or cultivars	6-52		4-19		40-48		6-62	

* A = Derived Savanna; B = Transition-zone-derived savanna-oil palm bush; C = Oil palm bush-high population density; D = Oil palm bush-medium-high population density.

crop plants interact with each other as well as with protected wild relatives of crop plants in mixtures. The high diversity of traditional polycultural systems has been observed by Harris (1976). Anderson (1954) reported field observations by Harlan about the possible role of weeds in the evolution of crop plants and the transfer of useful characteristics to cultivated plants.

Until the last decade, when concern about the environment and the threat of genetic erosion among cultivated plants increased, the study of crop mixtures was neglected. A study of crop mixtures should receive higher priority than in the past, not only to increase agricultural productivity but also to develop more stable agro-ecosystems that utilize the high diversity index of tropical polycultural systems. Full advantage should be taken of the resurgence of interest in mixtures since the effort made so far to realize their full potential has been minimal.

References

Allen, W. (1965). *The African Husbandman*, Edinburgh: Oliver & Boyd.

Anderson, E. (1954). *Plants, Man and Life.* London: Andrew Melrose.

Andrews, D. J. & Kassam, A. H. (1976). The importance of multiple cropping in increasing world food and supplies. In Papendick, Sanchez and Triplett (1976), pp. 1-10.

Benneh, G. (1972). Systems of agriculture in tropical agriculture. *Economic Geography* **48**, 245-257.

Duckham, A. N. & Masefield, G. B. (1971). *Farming Systems of the World.* London: Chatto and Windus.

Greenland, D. J. (1974). Evolution and development of different types of shifting cultivation. In *Regional Seminar on Shifting Cultivation and Soil Conservation. W. Africa Soils Bulletin* No. 24, pp. 5-13, Rome: FAO.

Grigg, D. G. (1974). *The Agricultural Systems of the World: An Evolutionary Approach.* London: Cambridge University Press.

Harris, D. R. (1976). Traditional systems of plant food production and origins of agriculture in West Africa. In *Origins of African Plant Domestication,* pp. 311-356. Ed. J. R. Harlan *et al.* The Hague: Mouton.

Laut, P. (1971). *Agricultural Geography.* Vol. 1. Sydney: Thomas Nelson.

Miracle, M. P. (1967). *Agriculture in the Congo Basin.* Madison: University of Wisconsin Press.

Norman, D. W. (1974). Rationalizing mixed cropping under indigenous conditions. The example of northern Nigeria. *Journal of Development Studies* **11**, 3-21.

Okigbo, B. N. (1978). Cropping systems and related research in Africa. *10th Anniversary of AAASA Occasional Publications Series OT-1.* April, 1978.

Okigbo, B. N. & Greenland, D. J. (1976). Intercropping systems in tropical Africa. In Papendick, Sanchez and Triplett (1976).

Papendick, R. I., Sanchez, P. A. & Triplett, G. B. Eds. (1976). *Multiple Cropping.* American Society of Agriculture Publication No. 27. Madison: American Society of Agronomy; Crop Science Society of America; Soil Science Society of America.

Ruthenberg, Hans. (1976). *Farming Systems of the Tropics.* London: Oxford University Press.

Sanchez, P. A. (1976). *Properties and Management of Soils in the Tropics.* New York: Wiley.

Schlippe, P. de. (1956). *Shifting Cultivation in Africa. The Land Systems.* London: Routledge and Kegan Paul.

Skerman, P. J. (1977). *Tropical Forage Legumes.* Rome: FAO.

Troll, E. (1966). *Seasonal Climates of the Earth. World Maps of Climatology.* Berlin: Springer.

Webster, C. C. & Wilson, P. N. (1966). *Agriculture in the Tropics.* London: Longmans, Green.

Whittlesey, D. (1962). Major agricultural regions of the earth. In *Readings in Cultural Geography,* pp. 416-444. Eds. P. L. Wagner & M. W. Mikesell. Chicago: University of Chicago Press.

Willey (1979). Intercropping: its importance and research needs: Part I. Competitions and yield advantage. *Field Crops Abstracts* **32**, 1-10.

Williams, C. N. & Joseph, K. T. (1970). *Climate, Soil and Crop Production in the Humid Tropics.* Kuala Lumpur: Oxford University Press.

Wrigley, G. (1969). *Tropical Agriculture.* London: Faber and Faber.

The future of research on mixed cropping in tropical agriculture*

A. H. Bunting
University of Reading

In many tropical countries, mixed cropping is very much the rule. I shall return to the reasons for this, but it is useful first to point out that as agriculture has advanced technically, to become increasingly the agriculture of cities, and as it has become more capital intensive, the requirements imposed by mechanization and the chemical control of weeds have tended to eliminate mixed cropping. It is not profitable to argue that this should be changed. There do not appear to be any sufficiently strong economic or technical trends that will make it advantageous to farmers in the future to produce mixed crops in the industrialized countries.

Ecophysiology of mixed cropping

Producers in the tropics look to their mixed systems for many things besides the yields of the crops they include. To illustrate this, I wish briefly to describe the mixed cropping of the Bamileke region of the north-western Cameroons. This is evergreen forest country which has been cleared for agriculture. It is very completely cultivated, to such an extent that the Roads Department has no need to weed the road verges: crops are planted up to the edge of the tarmac. It is a densely populated region of very industrious people.

The first crop that is sown when the rains begin is a type of melon. When I first encountered experiments with melon at Ibadan I could not imagine what this melon was for. Did people really want to eat all that quantity of melon? They do in fact use it, but the really important thing about it is that it is a cucurbit. Because of its anatomical and morphological structure, a spreading cucurbit of this type can provide more leaf-area index per unit of dry matter accumulated than any other kind of plant known to me. The internodes are essentially annular tubes

*Editors note: The concluding remarks given at the plenary session on 'Mixed Cropping'.

of water. Of course there are some cell walls in amongst the water, but most of the hollow stem is filled with air.

Consequently the melon very rapidly provides leaf cover over the soil surface. This captures some radiation, which can be turned into melon, the seeds of which are a useful source of protein, but it does a lot more than that. It very rapidly protects the surface of the soil against insolation, heat and beating rain, and so against rain splash and subsequent sheet erosion, and it helps the rain water to infiltrate into the profile. It controls the growth of weeds, three weedings being necessary without melon, but only one with melon. Moreover, the crop puts roots down into the soil and begins to cycle nutrients, and (at least as important), it also starts the evaporative part of the seasonal hydrological cycle, without which waterlogging and run-off might lead to later difficulties.

When the melon has grown alone for a while the farmer sows a grain legume into it – usually cowpea. This large-seeded species rapidly pushes itself up through the cover of melon and expands its leaf area above the melon. It no doubt helps to accelerate the death of the melon, as it begins to take over the ecological functions of its predecessor. In due course, but before the melon has completely disappeared, maize is sown. This crop grows rapidly in height and develops its canopy both later in time and higher up in space. Because of the geometry of its canopy, maize allows more light to penetrate to lower levels than its predecessors, so that the cowpea and what is left of the melon can complete their tasks. A number of other crops may then take their turns, but well before the end of the season the farmer plants cocoyam or cassava into the system. As this establishes itself, and the leaf-area index of its predecessors declines, it expands buds and leaves and takes over in its turn. It grows on into and survives the dry season, as best it can, on stored water. When the next season begins, it is immediately ready to develop canopy. In the second season one sees what looks like a pure stand of cassava or of cocoyam, the final phase of the cycle.

All this is a combination of what we would formally call mixed, sequential and relay cropping. The words matter less than the fact that this system of agriculture preserves as nearly as possible a continuous plant cover over the soil, and at the same time produces useful things that people want at different times of the year, so that they can subsist and survive and may also have something to sell.

Above all, it helps to keep the environment stable. The mixed-farming system of the Bamileke region does not resemble the original evergreen forest in any morphological particulars, but in its essential ecological functions it repeats, as well as an arable cropping sequence can, the forest's environmental functions. The evergreen forest maintains a cover over the soil and it restricts the growth of material in the lower stories. It cycles nutrients, and it preserves the hydrological cycle so that local flooding, run off and erosion are avoided. It protects the soil against beating rain, insolation and heat. Taken together, these are the ecological functions of the man-made system, which can consequently be continuous without recourse to shifting cultivation.

As an aside, it may be useful to record two points. First, what is natural is not always most useful to man. Man is in business for his own sake, and it is not often that wild vegetation can be as useful to him as a managed system. Second, wherever we have examined a traditional indigenous farming system in any detail (and we have done this now for a few systems) we have been impressed by the precision and elegance of the simultaneous adjustments to the environment, to the available resources, and to the objectives and requirements of the community. These things can be so finely adjusted that I sometimes wonder at the impertinence of research workers like me, who try to make agricultural systems more productive, without first acquiring any prior understanding of the way they work now.

Many people believe that mixed systems are bound to provide some degree of control over pests and diseases. We have done some work on this question at the International Institute of Tropical Agriculture at Ibadan. Some of the answers are equivocal: the environmental circumstances (perhaps of humidity and temperature) in a mixed crop appear to favour some pests and pathogens while discouraging others. We must also expect to find instances in which the components of a mixture include alternate hosts of common pests and pathogens. For example, the stem borers of sorghum are the bollworms of cotton. In other cases, mixtures can impede the action of predators.

Objects and methods in research

I want next to talk about research and methods of research. Bearing in mind the complexity of the objectives and problems of mixed cropping in the tropics, I was led to wonder about the usefulness of mathematical models of competition. I yield to no-one in my respect for Charles de Wit, but a model which describes competition between two species primarily in respect of the accumulation of biomass and yield under rather strictly defined circumstances seems to me to be rather a long way away from Bamileke agriculture. I see no easy way of developing the model so that it can describe the phenomena in space and time which I have attempted to outline.

Many traditional mixed-cropping systems use a wide range of resources. They fill as much of the year as possible with yield-forming activities, by spreading the use of resources as far as possible round the year. They minimize labour peaks. In short, they are doing many important things that do not enter into the de Wit model at its present stage of development.

This is no criticism of the de Wit model as far as it goes. Conceptually it is undoubtedly valuable, but I do not feel that at this stage it provides an important research tool for studies of mixed cropping.

Phenology and physiology

I have preferred to try to think about the mixture of economic vegetation, as a whole, as a multi-dimensional physiological system using resources to produce products. Of course, in a conventional crop-physiological analysis the sampling problems would be very considerable, and consequently I do not suggest that we should try at the start to make quantitative physiological observations. It would however be useful to have descriptions of the phenology of the systems, the times of planting and harvesting, the sequential history in space and time of the distribution of leaf-area index, and of the products of assimilation between the different species. After the more or less qualitative, descriptive phase one would very soon have to move, because of sampling difficulties, to experiments which arrange the plants in a more regular pattern.

There is one particular piece of equipment which I believe should exist, if it does not exist already, and which would be extremely valuable. If one had a double camera which could take stereoscopic pairs of photographs of the mixed crop from above, it would be possible to use a Wild machine with a floating spot of light to determine the vertical distribution of those parts of the leaf area of different species which receive direct radiation at different times. This could yield, by non-destructive means, a history of the deployment of the physiologically most important part of the leaf-area index of different species in a mixed crop.

Agronomy

I doubt very much whether any precise research on spacing, dates of planting and so on in sole crops or in mixtures is likely to be very useful to farmers who practise mixed cropping. Mixed cropping is seldom as haphazard as it looks, but it is true that as a gap appears in the mixture, for whatever reason, the farmer, or his wife, puts something into it. Consequently the mixed crop may tend to look rather like a jungle. It is a jungle that has plenty of sense in it, as Edgar Anderson pointed out long ago, but it is seldom amenable to precise rules about spacing and timing.

Crop improvement

The most important thing we can do, I suggest, is to find or produce types of plants which are better adapted to mixed cropping than the types that are in the system now. It usually seems possible to improve their resistance to pests and diseases. For example, the types of cassava that we are now putting out from the International Institute of Tropical Agriculture, Ibadan (IITA), at the rate of around a million planting pieces a year, which carry durable resistance to the virus complex and to bacterial wilt, can go straight into the mixed-cropping system and be more productive than the types that are grown now.

I have to confess to a nagging evolutionary question here. If types with superior yield based on disease resistance are not only possible, but can also be found in the indigenous populations and built up by no more than a few years of crossing and selection, why do they not dominate the population already? Is there something wrong with them? Are they likely to collapse for some reason we do not forsee?

I think there is also the possibility that we may be able to produce types which are physiologically marginally better adapted to particular systems. Too often, we set off in quite the wrong direction. At the IITA we started to work with cowpeas on rather short-season, small-statured, upright types, believing that sole crops of these types would maximize protein production per hectare per day. So they well might, but most producers wish to grow cowpeas in mixed systems. We have therefore had to look for types which will withstand shading, and which will survive in mixtures, often by climbing up their more upright neighbours of other species.

Put another way, we need to know the essential attributes of the types which are at present being used in the mixed systems in order to see whether we can find types which are even better. I think we can do that, but this requires that at the start we know what happens in the existing mixed-cropping system, and why.

Commensalism

Speaking very much for myself, I do not feel that I ever got very far in thinking about mixed cropping as long as my thoughts were confined to competition. For me, mixed crops have come to represent a commensal rather than a competitive system. The species are sharing between them both the environmental resources and those which the producers are able to allocate to producing crops. It follows that we have to think about more things than the primary accumulation and distribution of biomass, or the ratios of seeds sown to seeds produced. Mixed cropping is a commensal system which optimizes as best it can the return on all resources available to the producer in order to get for him things that he wants. We are therefore bound to go beyond our disciplinary approaches, and to think from the start in a multidisciplinary way, including the economics of labour use through the year, the nutritional and other needs of the producers, and the social relationships (which affect access to resources) between different members of families and different families in the community.

I conclude that work on mixed cropping, the most complex agricultural system of all, cannot be restricted to any one discipline but requires the active co-operation of the many disciplines of which the Association of Applied Biologists is so harmoniously compounded.

Applying research findings to the cereal crop: the view of a user

R. G. Dawson
King's Farm, Pebmarsh, Halstead, Essex

The objective

My objective, like that of any commercially orientated farmer, is to obtain the maximum profit over a period of years. To achieve this requires the following strategy:

(a) to grow the best possible crop working at a minimum direct cost level;
(b) to examine each and every potential extra input to ensure that the marginal cost of its application is at least covered by the value of the likely increase in yield or quality.

Provided that the latter condition is satisfied I do not mind how high my direct cost structure becomes because maximum profit will be given with maximum yield. While this is my objective, expressed primarily in economic terms, it is also necessary to examine the objective in terms of the crop.

Dry-matter yield

The model

It has been shown that there is a good correlation between total dry-matter production and grain yield (Gallagher and Biscoe, 1978), so I am inclined to try to maximize total dry matter. To achieve this I need to know about those processes controlling dry-matter production so that I can manipulate the inputs available to me and ensure that the process and hence rate of dry-matter production is maximized. The potential dry-matter production for a green, closed crop canopy optimally supplied with water and minerals growing on latitude 52° North is shown by the solid line in Fig. 1 (Sibma 1977). The percentage radiation interception by well-grown winter wheat crops is also shown, which together with the potential production data can be used to calculate the crop growth rate (Fig. 1). During June and July the crop is producing about 250 kg DM ha^{-1} day^{-1} and most of this is going into the grain. A crop performing to this specification produces some 24 t DM ha^{-1} or about 11 t ha^{-1} of saleable grain.

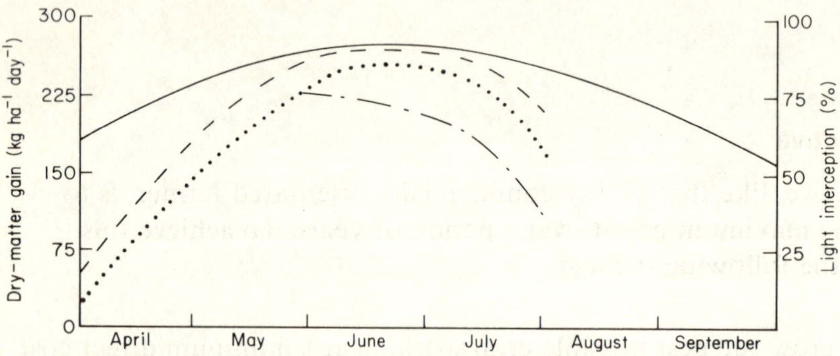

Fig. 1 Change with time of potential rate of dry-matter production (——), percentage light interception (– – –) and calculated rate of dry matter growth for a wheat crop (· · · ·) (after Sibma, 1977). Calculated rate of dry matter growth of the crop modified for water availability in East Anglia (—·—·—).

Unfortunately, in East Anglia we often cannot keep the crop growing until mid-August as water becomes limiting and drought, coupled with higher temperatures, frequently leads to earlier senescence. Figure 2 shows, for my own farm over the last fifteen years, the change in mean air temperature during the season and the ratio between calculated actual transpiration and calculated potential transpiration. Potential transpiration is calculated after the method of Penman (1948). Actual transpiration is calculated from a computer model that I have developed which maintains a daily water balance for 16 layers of soil to a depth of 2 m. Consideration is paid to the release characteristics of the soil type (Hall, Reeve, Thomasson and Wright, 1977), the density of roots in each layer (Gerwitz and Page, 1974), and to the relationships between potential transpiration, soil water potential, temperature and root density (after Cowan 1965). The ratio of actual to potential transpiration indicates the ability of the soil and climate to satisfy the crop's demand for water and I believe that it is a sensitive indicator of drought stress. I also believe, and have experimental evidence, that crop growth (as measured by yield) is directly proportional to this ratio.

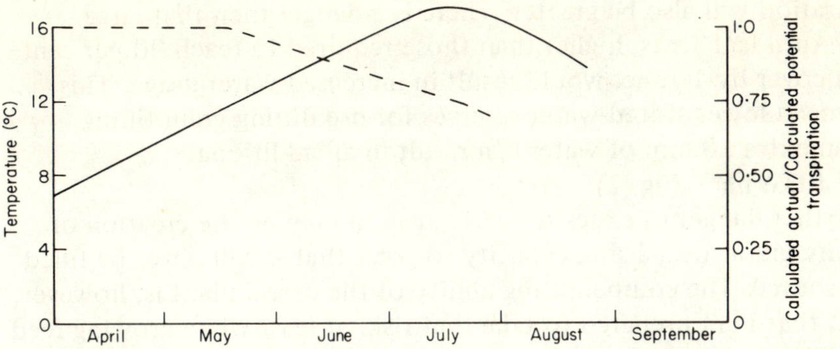

Fig. 2 Change with time of average air temperature (——) and the ratio of calculated actual to calculated potential transpiration (– – –).

Figure 1 shows the reduction in the calculated crop growth rate by applying the ratio.

Therefore, in general in East Anglia I doubt whether we can keep the crop growing past the end of July, giving us a potential of 45 days from anthesis, which is typically around 15 June. However, if we could hasten anthesis by 10 or even 14 days, there would be appreciably less water stress at anthesis, and the critical stage of pollen meiosis would seldom coincide with any drought stress (Fischer, 1973). Anthesis would take place at a lower temperature and there would be a longer potential grain-filling period. Even if we could not maintain fast crop growth during this longer period, the potential dry-matter production is 1·2 t ha^{-1} greater for the 45 days from 1 June than from 15 June (Fig. 1). Sadly we cannot perform this manipulation on the farm and must leave the suggestion to the breeder.

Radiation interception

The rapid increase in crop growth during April is a consequence, as well as a cause, of the rapid increase in radiation interception (Fig. 1). Once the ear has emerged there can be no more increase in green area, and hence, the maximum percentage light interception for the rest of the season is set by 1 June, as is the maximum yield. To reach this position the crop must 'get away' as early as possible in the spring and so should come out of the winter with as large and as healthy a leaf area as is safe.

If one succeeds in having percentage radiation interception values and crop growth rates for April greater than those shown in Fig. 1, then total dry-matter production and possibly grain yields will be larger. However, one consequence of increased radiation interception is that transpiration will also be greater. There is a danger then that large March/April leaf areas, higher than those required to reach 90 per cent ground cover by 1 June, would result in increased water usage. This could mean lower stored-water reserves for use during grain filling, when an extra 10 mm of water can result in an additional 0·5-0·7 t DM ha^{-1} (Fig. 2).

A further danger of excessive early growth may be the creation of too many grain sites, a sink capacity so great that it will never be filled by the source. The compensating ability of the cereal plant is, however, so great that I am prepared to take that risk, at least when growing feed wheat. The decision is harder when growing wheat to achieve a good milling standard, but on balance I believe that the earlier a ground cover can be established the better.

Establishment

Early ground cover must imply winter drilling, indeed drilling by mid-October at the latest. But how early can we start? The earlier we start the more certain we can be of good winter establishment, although if we start in early or mid-September we run into risks of weed problems, of aphid attack introducing barley dwarf yellow virus, and of presenting pests and diseases with a better bridge from last year's crop to this year's. However, two historic objections to early drilling have now been resolved. We know that early drilled crops survived the unusually low temperatures of the 1978-79 winter at least as well as crops drilled at conventional times. This was especially true when autumn fungicides were used. Also, we know that the risk of premature germination followed by death from drought in the autumn can be reduced by direct drilling.

The problem is not only when to start, but also when to finish. It takes me 15 working days to drill my 300 ha, perhaps a total of 20 days allowing for delays caused by the weather. This time could be halved by the purchase of a second drill, involving a capital expenditure equivalent to 200 kg ha^{-1} of wheat and with no increase in annual labour or operating costs. Would it be worthwhile? There have been few trials on this topic—and certainly not enough to establish the critical shape of response curves for September drilling.

The desire for early light interception leads logically to the use of narrow row drills, shown to be beneficial by very many trials over the years. I wonder whether much of the yield advantage that has been claimed from precision drilling may not derive from the use of extremely narrow rows rather than from the accurate spacing of seeds along the row.

It is becoming apparent that the winter is important to the cereal crop. For example, the Weed Research Organization has shown substantial benefits from autumn removal of weed competition as opposed to spring removal (Holroyd, 1977). This may be because there are critical development stages when the crop should be free from competition, or it may be simply a direct benefit of intercepting more light earlier in the spring. Irrespective of the cause, I believe the crop needs looking after as carefully during winter as during spring and summer. This is becoming easier with the advent of light, wide-tyred, vehicles with a capability of spraying or applying fertilizer in almost any ground conditions.

Nutrition

As the objective is to keep the crop growing vigorously throughout the year, it must be adequately supplied with available nutrients at all times. If nutrients removed by the harvested crop are not replaced, then there will be a loss in soil reserves. This obvious fact is the basis for current UK recommendations for phosphate and potash, but why is it usually ignored when considering nitrogen? Surely, this is especially serious as the major soil reservoir of nitrogen is in organic matter and, if this is depleted, it is a long and expensive task to build it up again. Even if the nitrogen is applied elsewhere in the rotation, the cereal crop will still reduce the organic matter if there is a net loss of nitrogen, whereas if sufficient nitrogen had been applied there might have been a build up of organic matter over the years. Although not critical for other elements, the time of application is important for nitrogen. If there is no nitrogen in the surface layer of the soil during the autumn then it must be supplied; but I am also convinced that nitrogen at depth helps encourage deep winter rooting. Fortunately we can get nitrogen to depth by applying it at suitable times to allow percolation with downward movement of rain, even if this means applying the nitrogen to the previous crop. We must apply nitrogen early in the spring to encourage the crop to 'get away', but must avoid excessive supply causing prolific tillering which may occur until approximately the seventh leaf appears.

Phosphate, if applied to the surface, will tend to remain there, especially under direct drilling regimes. The inability of the crop to take up phosphate ions from dry soil (Dunham and Nye, 1976) means a potential deficiency during the summer, which will lead to depletion of phosphate from the leaves and earlier senescence. Foliar applied phosphate to correct this deficiency has sometimes worked (Laloux, 1977), but now some research seems to be concentrating on placing phosphate at about 50 cm depth. While this may help in May I wonder if it is not counter-productive in July as, by then, the soil will usually be quite dry, even at 50 cm, and there would be little phosphate uptake expected from that depth. Certainly in East Anglia there is almost no possibility of getting rain so deep, whereas if the phosphate had been near the surface any appreciable amount of rain would provide sufficient moisture to wet the soil containing the phosphate and allow at least some uptake. Of course this would be of no help in a prolonged drought, but nor would phosphate at 50 cm.

Some clay soils have several centuries of potash reserves, so it seems reasonable to deplete these to some extent. However, most of the potash is unavailable and must undergo a transformation before being taken up by plants. Our ideal crop of 24 t of dry matter requires a maximum rate of potash supply of about 4 kg ha^{-1} day^{-1} (after Knowles and Watkin, 1931; Gregory, Crawford and McGowan, 1979). Can this be met without the addition of fertilizers?

We used to get some of the sulphur requirement free in the rain and it was also a constituent of many fertilizers. Now, since ammonium nitrate is the main nitrogenous fertilizer, little sulphur comes from this source and, with clean air legislation, rain is providing only enough sulphur for a crop yielding 5 t ha^{-1} of grain.

Crop protection

Having established and fertilized the crop, we must protect it from weed competition, pests and diseases. Recent research work has assumed a spraying cost based on a typical contractor's charge, less a 10 per cent 'profit' element, giving a charge of between £5 and £6 ha^{-1} (Elliott, 1978). I believe this is far too high, as almost all serious cereal growers have their own equipment. Table 1 shows my own costs to be about £2 ha^{-1}, or, in real terms, 22 kg wheat ha^{-1}. However, we are usually deciding whether or not to make just one extra application, and so we should be looking not at the total cost, but at the marginal cost. The capital cost is not affected by one extra application and the staff will still be employed even if the application is not made. Therefore,

Table 1. Spray application costs (£ ha^{-1}) for a self-propelled machine covering 3600 ha yr^{-1}, working at 8 ha h^{-1}. Figures in brackets are the cost expressed in terms of kg wheat ha^{-1}.

		Total	Marginal
Machinery ownership			
5 years' depreciation on cost £12 000	0·67		
Interest at 15%	0·25	0·92	
Running			
Repairs	0·42		
Fuel	0·04		
Water	0·02	0·48	0·48
Labour (2 men)			
80% ordinary time	0·44		
20% overtime	0·13	0·57	0·13
		1·97 (22)	0·61 (7)

the marginal cost comprises only machine running costs plus overtime payments, giving a total of £0·61 ha^{-1}, or just 7 kg wheat ha^{-1}.

Agricultural Chemicals Approval Scheme

The armoury we have been given for weed control, together with the data on both its tactical and strategic use, is excellent. The choice of chemical for a given problem is made by reference to its efficiency and to its cost. *But*, the cost we tend to look at is the purchase price rather than the possible cost of crop damage due to phytotoxicity, despite the fact that even at quite low levels of damage this cost could change our choice. Why do we ignore this cost? The experimental work to determine it is beyond the resources of most farmers and, in general, government research undertakings tend to leave it to the manufacturer.

The manufacturer of crop protection chemicals is required to submit the results of his trials before the chemical is given approval under the Agricultural Chemicals Approval Scheme. But there it stops, because the results are given 'in confidence'.

What possible benefit can there be in this secrecy? It might just be bureaucratic habit, it might be to disguise the evidence of the poor levels of crop tolerance that are demanded by the Approval Scheme, or it may be to preserve manufacturers' commercial secrets. If it is because of commercial secrets, why should such secrets, which say 'my product damages the crop but I don't want the customer to know', be tolerated by an approval scheme? If the secrecy is to conceal low standards within the Approval Scheme then the sooner we can find out and improve them the better.

Prevention versus treatment

Farmers are frequently criticized for excessive use of precautionary, or insurance, applications of pesticides and fungicides. Figure 3 shows, on the left, the factors making up the cost of an insurance treatment and, on the right, those for the 'wait and see', or curative, approach. In both cases the direct cost comprises chemical and application costs, together with the costs of mechanical or chemical damage to the crop.

The cost of the insurance approach is the direct cost plus the cost of the risk of having to apply curative treatment as well, less any incidental benefits from, for example, controlling some other problem.

All of the cost of the curative approach is directly proportional to the probability of having to apply the treatment at all, and there are four factors:

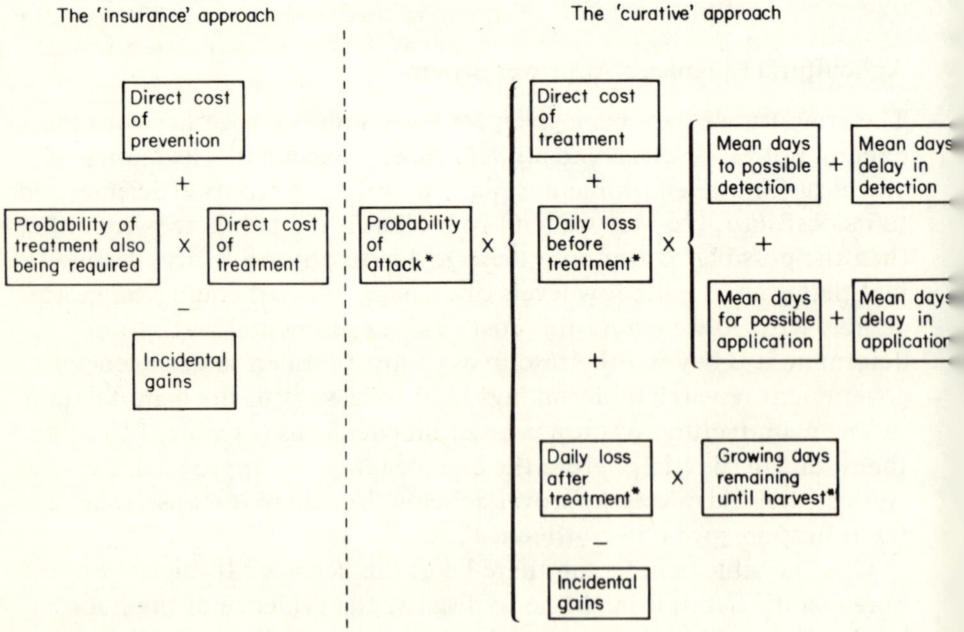

Fig. 3 Comparison between the costs of the 'insurance' approach and the 'curative' approach for the application of chemical sprays.

1. the direct cost;
2. yield losses until the treatment can be applied; delays can be caused by
   (a) the malady not being visible for a period after innoculation;
   (b) a failure (related to the frequency and thoroughness of crop inspection) to detect the malady once it is visible;
   (c) the capacity of the treatment equipment related to the area at risk;
   (d) unsuitable weather delaying or preventing treatment;
3. a continuing loss until harvest if the treatment is incapable of eradicating the disease or of repairing damaged tissue;
4. incidental benefits.

In Fig. 3 the asterisks indicate factors which are substantially weather dependent. The advent of more accurate weather forecasts for 10-14 days ahead would remove many of the uncertainties. However, until then, consider the dangers implicit in a disease such as *Septoria nodorum.* Wet weather will increase the probability of attack, will increase the cost of each day's delay in treatment, and may well delay the actual treatment by making spraying impossible. The adverse effect on the crop is the product of these three factors.

Faced with the simple calculation on the left as compared with the complex and unquantifiable pattern on the right, is it surprising that so many farmers use the insurance approach, even if only to get away for a week's holiday before harvest!

Economics and precision in research

Table 2 shows the marginal spraying costs, again expressed in kg wheat ha^{-1}, for three representative chemicals. The application cost, as we have seen is 7 kg ha^{-1}. The chemical costs are for competitively purchased chemicals applied at the recommended level (the fungicide) or at lower rates which are known to be effective (the insecticide and herbicide). With the use of 'tramlines' I consider there is no additional or marginal loss due to mechanical damage from one extra application. There may be chemical damage but, in the absence of published data, we have to ignore this. The calculation gives total spraying costs between 15 and 85 kg wheat ha^{-1}, or from 0·2 to 1·1 per cent of a typical good crop. Alternatively, as grains grow at about 250 kg ha^{-1} day^{-1}, the spraying cost is only a fraction of one day's grain growth. Therefore, provided that the likely increase in yield is not less than the

Table 2. Costs of spraying a wheat crop with three different chemicals. The costs are expressed as kg wheat ha^{-1}.

	Marginal application	Chemical	Total	Percentage of yield *
Insecticide				
dimethoate	7	8	15	0·2
Herbicide				
CMPP	7	23	30	0·4
Fungicide				
benomyl	7	78	85	1·1

* Assuming a yield of 7·5 t ha^{-1}

spraying cost shown, it is economic to spray the crop.

It follows that in trials assessing the benefits from applying these treatments, the experimental design should be such that yield increases as low as 1 per cent, or even 0·5 per cent, can be determined as statistically significant. This might be possible by increasing the number of replicates, or by reducing the variability between the replicates, or possibly by using a probability higher than the conventional 0·05 for the significance test. If resources are limited, then perhaps fewer but better trials would provide clearer answers.

The very low yields obtained in many experimental trials are of concern. The grain yields from the best commercial farms are now consistently around 7·5 t ha^{-1}, some are frequently over 10 t ha^{-1}, and yet we still see trials with controls yielding around 5 t ha. At that level we know that the yield is being severely restricted by something. If that something happens to be the subject of the trial then there may well be dramatic results. Conversely, if the yield restriction was due to another cause, then there may be no response to the treatment, and this lack of response will often be used as the justification for a recommendation not to use that treatment.

I believe that more information would be obtained if trials were carried out where the only restrictions to crop growth were light and carbon dioxide. This would involve several applications of nutrients, a comprehensive programme for control of pests and diseases, total weed control (preferably by hand removal) and irrigation. The resulting ‘unrestricted’ crop should then be regarded as having received the ‘control’ programme, and the treatments under trial could be removed from, or added to, the control programme. I think this approach would produce high yields, would reduce variability between replicates, would increase variability between different treatments (by ensuring that no other factors were limiting) and so would be more likely to produce significant results.

Finally, in the last five years my yields, and those of many other farmers, have approximately doubled and so, on my own behalf as well as on behalf of my fellow farmers, I would like to thank those members of the Association of Applied Biology whose research has helped to produce this English version of the Green Revolution.

References

Cowan, I. R. (1965). Transport of water in the soil-plant-atmosphere system. *Journal Applied Ecology* **2**, 221-239.

Dunham, R. J. & Nye, P. H. (1976). The influence of soil water content on the uptake of ions by roots. III. Phosphate, potassium, calcium and magnesium uptake and concentration gradients in soil. *Journal of Applied Ecology* **13** 967-984.

Elliott, J. G. (1978). The economic objectives of weed control in cereals. *Proceedings 1978 British Crop Protection Conference, Brighton.*

Fischer, R. A. (1973). The effect of water stress at various stages of development on yield processes in wheat. In *Plant Response to Climatic Factors,* pp. 233-241. Ed. R. O. Slatyer. Paris: UNESCO.

Gallagher, J. N. & Biscoe, P. V. (1978). Assimilate partitioning and high yield in cereals. *Proceedings First British Plant Growth Regulator Symposium,* Reading, pp. 113-124.

Gerwitz, A. & Page, E. R. (1974). An empirical mathematical model to describe plant root systems. *Journal of Applied Ecology* **11**, 773-781.

Gregory, P. J., Crawford, O. V. & McGowan, M. (1979). Nutrient relations of winter wheat. I. Accumulation and distribution of Na, K, Ca, Mg, P, S and N. *Journal of Agricultural Science, Cambridge* **93**, 485-494.

Hall, D. G. M., Reeve, M. J., Thomasson, A. J. & Wright, V. F. (1977). Water retention, porosity and density of field soils. *Technical Monograph No. 9.* Soil Survey, Harpenden.

Holroyd, J. (1977). Weed control in winter cereals. *Technical Leaflet No. 5.* Weed Research Organization, Yarnton, England.

Knowles, F. & Watkin, J. E. (1931). The assimilation and translocation of plant nutrients in wheat during growth. *Journal of Agricultural Science, Cambridge* **21**, 612-637.

Laloux, R. (1977). Ten years of applying phosphate and potash to the foliage of winter barley (Personal communication).

Penman, H. L. (1948). Natural evaporation from open water, bare soil and grass. *Proceedings of the Royal Society, London,* Series A **193**, 120-145.

Sibma, L. (1977). Maximisation of arable crop yields in the Netherlands. *Netherlands Journal of Agricultural Science* **25**, 278-287.

Crop production in the UK — recent changes and future prospects: the role of research

J. J. North
Agricultural Development and Advisory Service, MAFF, London

Introduction

British agriculture is an amalgam of many types of farming with considerable variation in size and operating under widely varying conditions in various parts of the UK. In the past, change has been slow in that for almost 150 years much of the farming of England was based on the well-known and praised four-course rotation, finally established at the end of the eighteenth century following the experiments of Townsend and Coke from Norfolk.

It had technical advantages of producing reasonable yields permanently, but depended on cheap labour and good producer prices for wheat and meat. Grain imports, refrigeration and finally the establishment of the Wages Board produced the depression in the 1920s and 1930s and the rapid demise of this system. In contrast, changes in arable farming over the last 25 years have been rapid, substantial and far reaching. Economic circumstances have encouraged changes, with farmers seeking more efficient and cheaper methods of production, whilst the profitability of farming generally has enabled farmers to invest in necessary new technology.

Recent changes

Much of this improved productivity has arisen from improved yields (Table 1) arising from the introduction of new varieties, the increased use of fertilizers, improved harvesting and storage and by the increasing use of effective pesticides. Additionally the substitution of capital in the form of machinery for labour has very substantially reduced the numbers of workers employed: on average there is only one worker employed on every 81 ha of arable land today, compared with one man to 16 ha 20 years ago (see also Pascal in this volume, p. 272). Machinery investment in volume terms has not been as great as Table 2 might indicate, due to inflation. Much of this machinery has become increasingly specialized and with large capacities leading to a move to a larger scale of operations (Table 3). Additionally cropping farms have become increasingly specialized, to optimize the use of specialist equipment and labour and to ensure more effective management and improved efficiency. In many situations the number of individual enterprises has been more than halved.

Table 1. Average yields of crops over the past two decades.

	Average yields (tonnes/hectare)			
	Wheat	Barley	Potatoes	Sugar beet
1956/58	3·2	3·0	18·8	28·1
1966/68	4·0	3·7	24·9	37·2
1976/78	4·7	4·0	28·1	32·7

Sources: Ministry of Agriculture, Fisheries and Food, *Agricultural Statistics* for 1959/60 and 1969/70, London: HMSO, and *Annual Review of Agriculture*, Command papers for 1977, 1978 and 1979.

Table 2. Capital investment in farm machinery in the UK

	Machinery investment (£ million)
1958	204
1968	231
1978	925

Sources: As Table 1

Table 3. The average area allocated to different crops on UK farms.

	Average crop area per farm (ha)		
	Cereals	Potatoes	Sugar beet
1967	22·2	2·7	7·6
1972	28·8	3·4	10·0
1978	33·7	3·8	14·6

Sources: As Table 1.

Table 4. Number of farms operating in the UK since 1958.

	Number of farms
1958	495 198
1968	369 053
1978	258 200

Sources: As Table 1.

Associated with this there has been a substantial drop in the number of holdings (Table 4). The changes shown by these figures must not be treated as absolute as during this period the definition of a farm for statistical purposes has changed, 'eliminating' approximately 2000 statistically insignificant holdings between 1968 and 1978.

The present

British agriculture supplies around 66 per cent of indigenous food requirements (Table 5) compared with about 60 per cent 20 years ago. This is achieved within essentially the same area of farm land, though increases in areas of some of the major crops have been quite substantial (Table 6).

Since Britain joined the EEC, producer prices have increased substantially in money terms and have encouraged the development of high-yield–high-cost cropping systems. With the prospect of only small price increases in the future, interest is likely to return to more efficient

Table 5. The degree of self-sufficiency in 1977 in the UK and in the EEC with respect to some major vegetable crops.

	Self-sufficiency (per cent) UK	EEC
Wheat	59	105
Barley	96	94
Maize	0·1	40
Potatoes	81	96
Sugar beet	37	104
Oilseed rape	49	112

Source: *Annual Review of Agriculture,* 1979. Command paper 7436.

Table 6. Area of land in the UK devoted to major agricultural crops.

	Cropping (k ha) 1958	1968	1978
Wheat	894	978	1263
Barley	1105	2401	2360
Potatoes	332	279	212
Sugar beet	178	187	210
Oilseed rape	0·2	6	65
Temporary grass	2570	2377	2102
Other	2010	1154	833
Total cropping area	7089	7382	7045

systems, on the basis of reduced inputs per unit of output, only part of which can be achieved from improved yields. Yields approaching the predicted biological maximum have already been achieved on individual fields. The cost of energy as reflected in fertilizer, pesticides and fuel prices, the anticipated higher costs of labour in relative terms, the high prices paid for land and the increased cost of capital have already exerted an influence and some rationalization.

The future

The future role of British agriculture and, associated with this, changes in crop production have recently been considered in a Government White Paper (Command Paper 7459, February 1979) entitled *Farming and the Nation.* It concluded that changes in the demand for food would involve a broadly stable total protein intake with a small reduction in the intake of carbohydrates. In commodity terms this represents an increase in demand for beef, pork and poultry meat, fruit and vegetables and a reduced consumption of sugar, sugar-based products, cereal-based foods, mutton and lamb. Milk and milk products would show little change.

It also suggested that it makes sense to continue to produce those foodstuffs in this country in which we have a competitive edge, either for home consumption or for export. Advantages of soil and climate where they exist should be realized, whilst the overall structure of the UK industry provides potential economies. The proximity of a large consuming market is particularly important where fresh produce is concerned. It also seems to be widely held that UK farmers will continue to operate in an international market similar to that existing today. There is therefore obvious scope for some increases in production, providing an economic advantage can be maintained.

Much of the recent progress in crop production has relied heavily on fossil fuels; recent price and future price changes will have a very important effect on crop production costs and practices. Traditional cultivation involving ploughing is expensive in fuel and labour and this has led to the developments in minimum and modified cultivations and direct drilling. ICI reported that 220 000 hectares were direct drilled in 1977, but many factors including soil type, farming system, cropping and seasonal factors will affect the appropriateness of these changes.

The production of fertilizers, principally nitrogen, is another major consumer of fossil fuel, accounting for 23 per cent of agriculture's total use. Economies in the use of fertilizer nitrogen can sometimes be made by the use of legumes. Since, in marginal energy terms, many fodder crops show advantage over grass as silage, in an energy-limiting situation their use could increase, although they do produce less protein. The marginal energy output when the grain and straw of cereals is used is also greater than from a three-cut grass silage system. These examples are options which may have to be followed if energy becomes expensive or limiting.

The use of pesticides, made largely from fossil fuels, is increasingly under scrutiny by the environmentalists. Reduced pesticide inputs either by reduced or more efficient use will be essential. It remains desirable to use wherever possible integrated methods of control, but the circumstances under which they must operate will become more stringent. Farming will increasingly require reliability of production, not only to service the investment but to meet the market requirements. The level of returns to the producer will be determined by the prices he obtains from the market, which in part depend on repeatable quality standards and continuity of supplies.

Agriculture must make optimum use of available solar energy. Research has identified opportunities for improvement, and the translation of these opportunities into practicality is of vital importance. In doing this attempts must be made to reduce the use of fossil fuels and the timeliness of operation which is often identified as being critical in ensuring better use of solar energy. Modern large-scale operating units require systems in which timing is not critical and which allow considerable flexibility. At present, failure to carry out operations at the right time has serious financial implications for these businesses.

Generally, research should continue trying to remove or minimize the controllable restraints to crop production. In particular, attention should be given to reducing the large-yield fluctuations that occur from year to year, between farms and fields (see, for example, the report by Jackson in this volume, p. 149). The solutions will have to apply in a highly mechanized situation with few traditional 'rotational' advantages. The scale and level of investment, the farmer's assessment and reaction to risk and uncertainty and the prevailing economic conditions will influence his motivation to take up new technology. To remain efficient and competitive he must take them into account, although innovations involving large capital investments are never particularly attractive.

The resources for technology in agriculture

J. A. Pascal
Scottish Institute of Agricultural Engineering, Bush Estate, Penicuik, Midlothian

Introduction

The use of past, present and future technology in agriculture depends on the interaction between food requirements, resource availability and scientific development. Today a world population of 4·3 billion (1000 million) people produce 1·35 million tonnes of cereals which is technically enough to feed 5 billion people (Stevens, 1978). Some of this output is lost as a direct energy food source through poor distribution, wastage, industrial use and for animal feeding. Exponential growth of world population continues, some Latin American countries having had a recent population 'explosion', although population growth has diminished in other countries, namely western Europe and Japan (United Nations, 1966). This will result in demand for food continuing to exceed supply in more and more areas of the world until population growth is controlled by voluntary methods or until Malthusian principles intervene (Malthus, 1798).

In the short term increased food production to meet demand can be obtained by the following methods:

1. Bring more land into production;
2. control pests;
3. breed high-yielding cultivars;
4. increase irrigation;
5. optimize fertilizer use;
6. achieve better tillage;
7. aim at direct consumption of arable crops.

By using these methods, total world food production could be increased by a factor of 8 according to one estimate (Hendricks, 1969) provided an energy- and resource-subsidized agriculture is also available to third-world nations. Present western or first-world agriculture has a significant input of non-renewable resources derived from mineral ores and fossil fuels exemplified in the manufacture and use of machinery and tractors, diesel fuels, pesticide sprays, fertilizers, etc. Much has already been written about the future limited availability of these non-renewable resources through their depletion rate at present and projected high rates of consumption (Cloud, 1969; Meadows *et al.*, 1972); also included have been the fears of increased environmental pollution. However, it has also been suggested that such prognostications are incorrect (Beckerman, 1972; Maddox, 1972; Pavitt, 1973). It is therefore necessary to make an assessment of the validity of such arguments to ascertain whether or not a serious depletion of non-renewable resources will occur and, if so, what effect this will have on agriculture. Such an answer is of particular interest to plant breeders because most modern cultivars have a requirement for a resource-subsidized agriculture to give conditions needed to produce a high crop yield. Indeed it should be noted that all seven conditions previously listed for increasing world food production rely on a subsidy from non-renewable resources.

Resource use

The greatest change in first-world agriculture has been the replacement of man and horses by mechanical power and the subsequent increase in power applied per unit of land area. Examples are given in Fig. 1 showing the decline in the total number of workers in British agriculture since 1923 (Ministry of Agriculture, Fisheries and Food, 1975) and Fig. 2 shows the present trends for tractor numbers and unit power outputs in Britain during the past decade (Ministry of Agriculture, Fisheries and Food, 1978). It will be noted that by 1974 the number of power units had levelled off at just over one per worker, but the ratio of tractors of less than 50 metric horsepower to those of greater than 50 metric horsepower has dropped from 6:1 to approximately 1:1 during the last decade. There has therefore been a sixfold increase in power applied to the land during this period, since total tractor numbers fell only slightly. Commensurate with this has been the increase in machinery for applying the power to the land.

World fertilizer consumption (Fig. 3) shows a sharp increase in consumption, particularly in nitrogen, during the last decade, about 85 per cent being used by the first world in 1976 (Food and Agriculture Organisation, 1977). Present artificial nitrogen manufacture is based on oil-feed stocks, whereas phosphate and potash are found in mineral deposits, and phosphate to a lesser extent in some biological deposits.

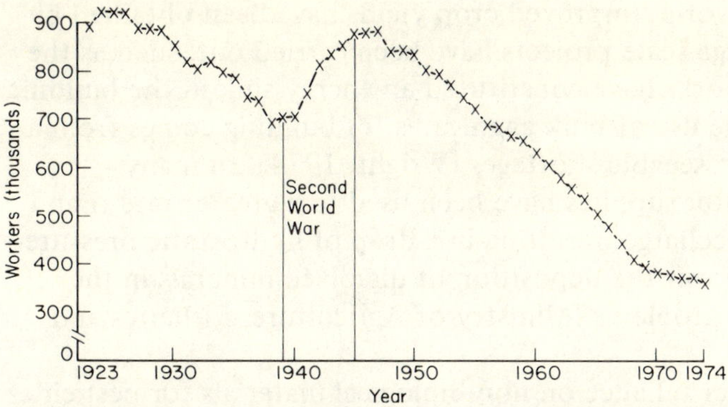

Fig. 1 Number of agricultural workers in Great Britain, 1923-74.

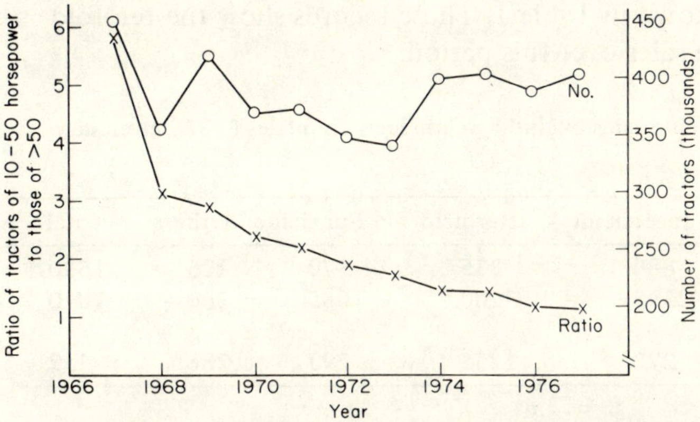

Fig. 2 Number of tractors (including tracklayers) and ratio of the number of those of 10-50 horsepower to the number of those of more than 50 horsepower.

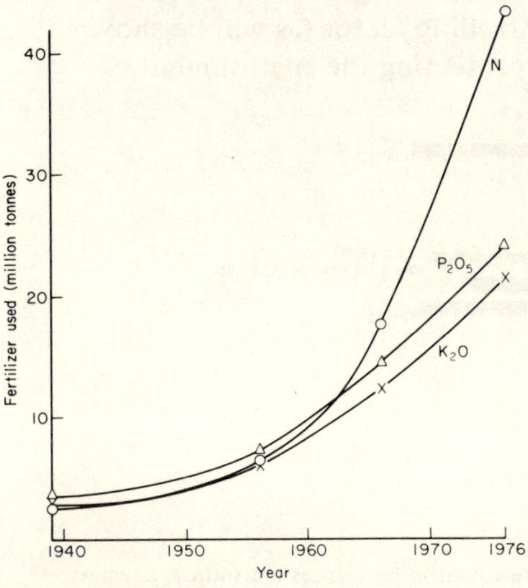

Fig. 3 Fertilizer usage, 1939-76.

Throughout the world, improved crop yields have been obtained by irrigation. Where large-scale projects have been carried out, such as the Aswan Dam, such works have constituted an energy sink in the building process, although the use of bulk aggregates for building comes from a resource with no foreseeable shortages (Wright, 1974). In many instances ground-water supplies have been used to a greater rate than that of the aquifer recharge, resulting in a drop of hydrostatic pressure and of supply; moreover the deposition of dissolved minerals in the water can cause soil problems (Ministry of Agriculture, Fisheries and Food, 1974).

The world's greater reliance on non-biological materials for pesticides has increased markedly in the last 20 years (Chapman, 1978) and total pesticide sales are shown in Table 1. These records show the ten-fold increase in herbicide sales over this period.

Table 1. World pesticide sales excluding Comecon countries (1976 monetary values in $ mill.)

Year	Insecticide	Herbicide	Fungicide	Others	Total
1957	880	255	370	125	1630
1976	2395	3200	1085	360	7040
Percentage increase over 20 years	272	1255	293	288	432

Much concern has been shown over the pollution hazards from the use of pesticide (Carson, 1963), but in the long term resource availability may prove a greater controlling factor (as will be shown later) than protective legislation in protecting the environment.

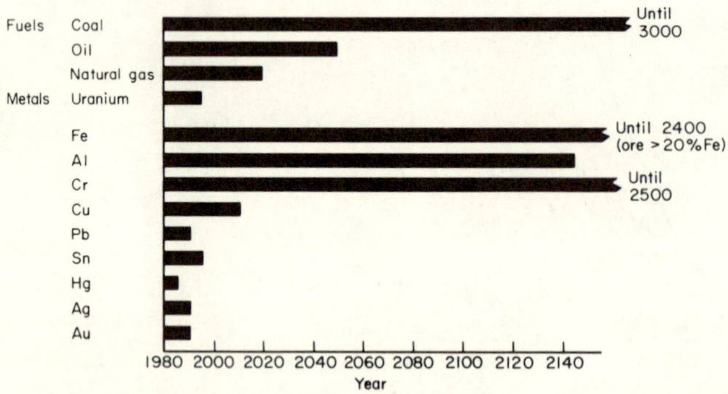

Fig. 4 Projected lifetime of some non-renewable resources (based on present world rates of consumption).

Resource availability

A projection of the 'lifetimes' of a representative number of fuels and metallic ores is given in Fig. 4 (Cloud (Jr), 1968; Goldsmith, 1972; Paul, 1974). Not included in the figures are estimated reserves of phosphate at 50 billion tonnes with a lifetime of 500 years at current use rates and potash at over 40 billion tonnes with a lifetime of over 2000 years at current usage rates (Wright, 1974). Initial examination of Fig. 4 suggests an immediate shortage of some resources, particularly uranium, lead, tin, mercury and precious metals, but further perusal does prompt a number of queries, viz:

1. Do ores ever become exhausted? i.e. why not mine iron ore at <20 per cent Fe or mercuric ores at < 0·2 per cent?
2. What is the effect of uneven resource distribution throughout the world?
3. What are the differences between known and potential reserves?
4. What substitute materials can be used for exhausted minerals?
5. What effect will changing rates of consumption have upon resource lifetimes?
6. In considering points 1-5, how accurate are resource lifetime estimates?

One general principle which operates for many mineral ores where there is a gradation from relatively rich to poor material is that ore tonnage increases at a constant geometric rate as the grade decreases (Lasky, 1950). This principle, if correct, would provide vast reserves of ores until arrested by scale and cost of extraction. In practice, this theory is only applicable over a limited range of ore grade (Lovering, 1969). It is not necessarily true for rich and lean ores or some specific resources, e.g. oil, coal, mercury, where deposits tend to have finite limits. Mercury, used as a cereal-seed dressing fungicide, has only limited reserves remaining, mainly in Spain and Italy. Unlike silver, which is often extracted in commercial quantities from other complex ores, mercury occurs only in discrete ore bodies.

Even those ores that conform to 'Lasky's Law' have economic cut-off points when the grade is too low for exploitation. But in succeeding years, lower- and lower-grade ores become exploitable with corresponding increases in input of energy and subsequent relatively higher prices for the mineral resource. An example of this procedure is given in Fig. 5 (US Bureau of Mines, 1965; Lovering, 1969). Thus resources tend to become relatively more expensive in the long term as exploitation proceeds.

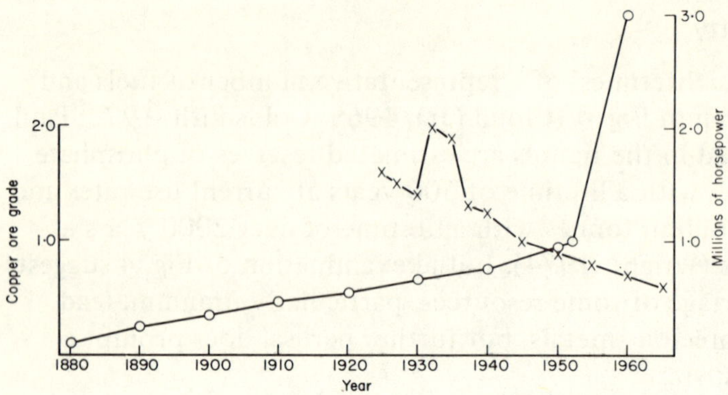

Fig. 5 Changes in the use of power (○) and in the grade of ore exploited (x) in US mines for the extraction of copper.

The uneven distribution of non-renewable resources is due to the irregular distribution of geological processes that have caused the formation or deposition of the resource in the first instance. Thus coal is found only at mid and low latitudes (at time of formation) in shallow deltaic sedimentary basins and certain metal ores occur preferentially at orogenic zones. Unfortunately, geological boundaries are often expressed in landscape relief, which also provides natural national boundaries, and this can give nations exclusive ownership of a resource. Such exclusive or major ownership results in the operation of cartels and also affects the foreign policies and trade agreements of many nations. Some states may be denied a resource through political and geographical accident.

Large differences occur in estimating potential reserves of a resource from proved reserves because new or a previously unexploitable reserve become available or technically feasible to develop. Examples of this are manganese nodules of the deep ocean containing manganese, copper, nickel and other metals (Mero, 1965) which are not included in the Fig. 4 estimates. However, in some circumstances, a large increase in reserves does not necessarily increase a resource lifetime to any great extent, especially where demand and availability are closely linked. World oil production, as shown in Fig. 6 (Cloud (Jr), 1968; Paul, 1974), is an example of this linking, with consumption reaching a peak in answer to demand and then falling away as reserves are exhausted. The recent Mexican oil discoveries may be equivalent to Middle East reserves (Dixon, 1979a), which represent one-third of world production (Cloud, 1968). But when such a quantity is integrated in Fig. 6, oil reserves are extended by only 20-25 years.

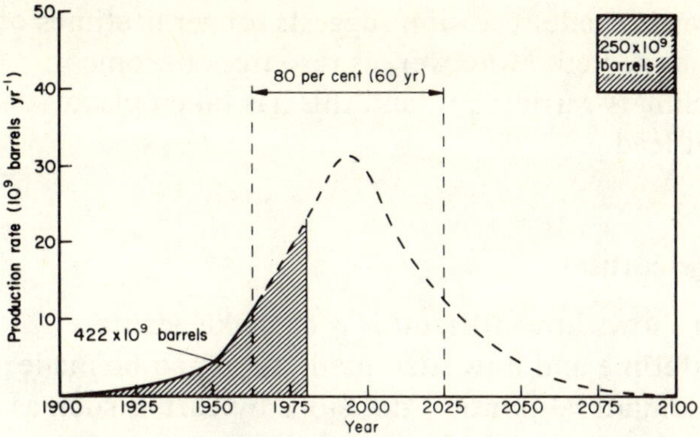

Fig. 6 World oil production.

Other sources of oil, which have been only slightly exploited, are tar sands and oil shales. These are not included in Fig. 4 and contain four times as much oil as already extracted and as remains in liquid reserves (World Oil, 1966 and 1967). However, these deposits have energy and environmental penalties in extraction. The disfiguring shale bings of West Lothian in Scotland bear witness to these penalties.

Short supply of a resource often results in substitution by an alternative resource, and present interest in alternative energy sources to oil is an example. Shortage of oil appears to be due to political pressures and geographical distribution rather than an overall world exploitation of reserves. Substitution results in prolonging one resource lifetime at the expense of the replacement resource (if that is also non-renewable). Some alternative energy sources for oil have been methane production (Shiach, 1978), solar energy (University of Nebraska and other groups, 1979a), synthesizing petrol from coal (Paul, 1974), nuclear power generation and growing crops for alcohol production (Small Farm Energy Report, 1979). These alternatives put additional exploitation pressures on uranium ore and depletion of crop area for food production to alcohol production. With so much under-nourishment prevalent in the world, it has been suggested that using land for 'energy' alcohol is morally indefensible (Meyer, 1979); presumably, crop production for alcoholic drinks would be in the same category? It must be recognized that substitutes for oil as an energy source are generally less convenient and have a relatively higher cost per unit energy produced.

Changing rates of consumption affect the calculated lifetime of a resource. Since the Second World War, industrial expansion initially resulted in an increase in resource exploitation which was then extrapolated to estimate resource lifetime (Hubbert, 1969; Meadows *et al.*, 1972). The present world trade recession suggests longer lifetimes of resources than originally expected. Moreover, as resources become in short supply, more recycling is carried out, and this has taken place particularly in the case of lead.

Future resource use in agriculture

The preceding discussion shows how difficult it is to make accurate predictions of resource lifetime and how such predictions can be made even more inaccurate by unilateral political decisions by cartels such as OPEC. On balance it is probable that depletion of resources is less than that envisaged by some reports, but it would be foolhardy not to expect energy and mineral resources to become comparatively more expensive.

The effect of this on agriculture will be a gradual one. For a number of decades there will be energy for prime movers, heaters etc., but increasing costs will turn farmers gradually to other forms of energy. Perhaps tillage will become more rationalized, with shallow cultivations and/or direct drilling for some crops becoming standard procedures as new systems are developed (Scottish Institute of Agricultural Engineering, 1976-8).

Recent advances in silicon production (Dixon, 1979b) have made the use of solar photovoltaic driven pumps for irrigation a feasible proposition (University of Nebraska and other groups, 1979a) even in higher latitudes, since solar energy is plentiful in dry summers. Similarly photovoltaic cells may well be used to fix atmospheric nitrogen (University of Nebraska and other groups, 1979b) as well as being produced as a by-product from the Lurgi process to produce SNG from coal (Gibson, 1979). The use of these methods does not preclude increasing nitrogen fixation by biological methods (Cooke, 1978) to subsidize industrial fixation and provide a cheaper source for underdeveloped nations.

Deposits of potash in evaporites are adequate and easily obtained for the distant future. Sedimentary phosphate deposits are also plentiful, but most of these are in marine environments and will be more costly to exploit. Land-based deposits are few and therefore subject to being marketed through cartels.

Certain pesticides may disappear from the market, and those containing mercury may well be difficult to obtain in the future. This suggests that new fungicides for cereal seed dressings will be required. Those based on hydrocarbons will remain for many years, especially if new spraying techniques reduce the amount of active ingredient required for pest control (Harris, 1977; Takenaga *et al.*, 1978). Biological control of pests has application in certain situations, but is difficult to operate effectively in many others.

The economic benefits of weed-control methods (Nalawaja, 1974) is given in Table 2. These results indicate the value of hand labour as a renewable resource and perhaps more persons should be returned to the land as other industrial activity declines. Improved ergonomics with agricultural machines and the use of microprocessors can make farm work a pleasant occupation (Reece, 1979). But to carry this out, a new price structure within all industrial activities will be required.

Plant breeders have in the past produced cultivars suitable for an energy-subsidized agriculture. It would appear that this aspect of work should continue alongside breeding programmes for the more distant future to provide greater biological nitrogen fixation, and greater pest resistance. It does not seem worthwhile to improve crop plant competitiveness by sacrificing yield potential.

Table 2. Energy use in weed control

Method of weed control	Energy input for weed control	Yield of corn	Net profit due to weed control	
	(kJ ha^{-1})	(MJ ha^{-1})	(MJ ha^{-1})	(\$ ha^{-1})
None	–	56·3	–	–
Cultivated	580	84·5	27·6	151
Herbicide	393	93·9	37·2	194
Hand labour	342	96·0	39·4	163

References

Beckerman, W. (1972). Economists, scientists and environmental catastrophe. *The Times,* 31 May.

Carson, R. (1963). *Silent Spring.* Hamish Hamilton: London.

Chapman, T. (1978). Pesticides, a period of progress, *Span* **21**, 62-63.

Cloud, P. (1969). Editorial. *Resources and Man.* San Francisco: Freeman.

Cloud, P. (Jr). (1968). Realities of mineral distribution. *Texas Quarterly* (2).

Cooke, G. W. (1978). Advances in crop nutrition. *Span* **21**, 54-55.

Dixon, B. (1979a). Geology of Mexican oil boom. *New Scientist* **81**, 556.

Dixon, B. (1979b). Brighter outlook for cheap solar cells. *New Scientist* **83**, 522.

Food and Agriculture Organisation (1977). *Annual Fertilizer Review* 1956-76. Rome: FAO, UN.

Gibson, J. (1979). The present scene and future of coal utilization. *Conference on Future Energy Concepts.* London: IEE.

Goldsmith, E. R. D. (Ed). (1972). *A Blueprint for Survival.* Harmandsworth: Penguin.

Harris, P. B. (1977). Mixed results with CDA spraying. *Arable Farming* **4**, 62-65. November.

Hendriks, S. B. (1969). Food from the land. In *Resources and Man.* Ed. P. Cloud. San Francisco: Freeman.

Hubbert, M. K. (1969). Energy resources. In *Resources and Man.* Ed. P. Cloud. San Francisco: Freeman.

Jorro, M. A. A. & Paul, P. F. M. (1978). Liquid fuels from coal. *Symposium on Energy and Road Transport.* Crowthorn: Transport Road Research Laboratory.

Lasky, S. G. (1950). How tonnage-grade relations predict ore reserve. *Engineering Mining Journal,* **151** (4), 81-85.

Lovering, T. S. (1969). Mineral resources from the land. In *Resources and Man.* Ed. P. Cloud. San Francisco: Freeman.

Maddox, J. (1972). False prophets of calamity. *The Times,* 3 February.

Malthus, T. R. (1798). *An Essay on the Principle of Population as it Affects the Future Improvement of Mankind.* London.

Meadows, D. H., Meadows, D. L., Randers, J., and Behrens, W. (1972). *Limits to Growth.* London: Pan Books.

Mero, J. L. (1965). *Mineral Resources of the Sea.* New York: Elsevier.

Meyer, J. (1979). World nutrition and international development. *9th Commission Internationale Génie Rural Conference, Michigan.*

Ministry of Agriculture, Fisheries and Food, (1974). Irrigation. *Bulletin 138.* London: HMSO.

Ministry of Agriculture, Fisheries and Food, (1975). UK Manpower 1935-74, *Agricultural Statistics.* London: HMSO.

Ministry of Agriculture, Fisheries and Food, (1978). UK tractors 1967-77, *Agricultural Statistics*, London: HMSO.

Nalawaja, J. D. (1974). Energy requirements for various weed control practices. *Proceedings North Central Weed Control Conference, USA.*

Paul, P. F. M. (1974). Substitute fuels. *Symposium on Re-assessment of Energy Sources*, National Engineering Laboratory, East Kilbride.

Pavitt, K. (1973). Some consequences of physics based research. *Physics Bulletin,* **24**, 22-25.

Reece, A. R. (1979). Future of traditional agriculture. *Mechanization Advisors Conference, Scottish Colleges of Agriculture, Edinburgh.*

Scottish Institute of Agricultural Engineering (1976-8). Reports on Cultivation and Soil Section Work. *Biennial report,* Penicuik.

Shiach, J. G. (1978). *Annual Report*, North of Scotland Agricultural College.

Small Farm Energy Report, (1979). *An Analysis of Farm Alcohol Fuel,* University of Nebraska – Lincoln.

Stevens, J. G. R. (1978). Agriculture: prospects of a better age. *Span* **21**, 49.

Stout, W. A. (1977). Energy sources. *Energy Use in Agriculture.* Committee for Agriculture Science and Technology, Report No. 68, USA.

Takenaga, T., Hashimoto, M., Tsuga, K., Kajiyama, M., & Hashimoto, Y. (1978). Study of ultra low volume sprayers for pest control. *Journal of the Society of Agricultural Machinery, Japan,* **40** (2), 169-178.

United Nations (1966). World population prospect. *UN Population Studies* No. 41, St/SOA/Ser A 41.

University of Nebraska – Lincoln, Department of Energy and Massachussetts Institute of Technology (1979a). Solar photo voltaic powered irrigation system. *Leaflet on working plant at Mead, Nebraska.*

University of Nebraska – Lincoln and C.F. Kettering Research Laboratory, Massachussetts Institute of Technology (1979b). Fertilizer from solar energy. *Leaflet on working plant at Mead, Nebraska.*

US Bureau of Mines (1965). Metals and minerals. *Minerals Year Book* **1**, Washington D.C.

World Oil (1966). *Proved Oil Reserves.* January.

World Oil (1967). *Proved and Probable Reserves.* December.

Wright, J. B. (Ed). (1974). Fertilizer and chemical minerals. *Earth's Physical Resources – Construction and Other Bulk Materials.* Milton Keynes: Open University Press.

Section 4 **Weed Control**

Weed control: past, present and future — a historical perspective

J. G. Elliott
ARC Weed Research Organization, Begbroke Hill, Yarnton, Oxford OX5 1PF

Introduction

There are those who enjoy the reading of history purely for the pleasure and insight that it brings. Usually a research worker is concerned to understand the past in order that he may better analyse the present as a basis for logical predictions of the future. The distant future must be speculation but the immediate future is susceptible to fair prediction because it stems from events and trends apparent today.

In weed control there is little purpose in delving back through our medieval past towards Roman times, interesting though this may be, because the agricultural methods including weed control of the nineteenth century represented a high point of technology based on men and horses which remained intact until 1940. It is the purpose of this paper to show that the principal methods of weed control employed at that time are still with us today, are still relevant to our agriculture to a greater or lesser extent and with one exception will be required to serve the needs of agriculture in the future alongside the single new factor, the chemical herbicide.

For the farmers of the nineteenth century the main weapons of weed control were:

1. prevention of movement of seeds and roots from farm to farm or field to field;
2. the employment of a sequence of crops in rotation;
3. the inversion and disturbance of the soil by cultivation;
4. selective weed control be men with hoes.

Let me describe each of these in modern terms.

The movement of seeds, rhizomes, tubers and roots

The most important need in this respect has long been the prevention of the movement of weed seeds. Vegetative structures also merit mention because of the possibility of moving rhizomes of couch-grass *Agropyron repens* or tubers of potato.

In the latter half of the nineteenth century there was increasing pressure on the seed trade to purify the samples of corn seed being sold to farmers. The Official Seed Testing Station and the Scottish Seed Testing Station were set up during the First World War to improve seed quality, of which analysis of weed content was a part. The object was straightforward: there could be no worse way of spreading weed infestations than actually to transport and sow weeds with the crop seed: this must be prevented. Since farmers had long been accustomed to retaining and sowing their own seed perhaps with only doubtful attempts to clean it, there was associated propaganda to encourage farmers to have all their seed cleansed.

The first Seeds Act was in 1920, after which there was a long gap in the introduction of regulatory procedures. In 1947 the NIAB Cereal Field Approval Scheme contained provisions to limit weed content. In 1964 the Plant Varieties and Seeds Act, and in 1974 the EEC standard system of quality control, updated early legislation on weed content.

A survey in 1970 provided proof in respect of wild-oats, *Avena* spp., that crop seed is still being sown contaminated with weeds (Elliott and Attwood, 1970) (Table 1). Recent evidence on blackgrass, *Alopecurus*

Table 1. Cereal seed drill survey spring 1970.

Percentage of samples containing wild-oat seed from:	merchant	same or other farm
England & Wales	11	41
Scotland	10	24
N Ireland	5	2

Source: Elliott & Attwood, 1970.

Table 2. Official seed testing 1968.

Percentage occurrence of weeds in imported perennial ryegrass	Eire	Denmark	New Zealand
Avena fatua	–	–	1
Bromus mollis	90	–	63
Poa annua	62	55	3
Vulpia spp.	34	–	26

Source: Tonkin, 1968.

myosuroides, and sterile brome, *Bromus sterilis,* suggests that cereal seed is still a source of spread of these weeds. Evidence from the Official Seed Testing Station confirms the importance in England of foreign weeds in ryegrass (Table 2).

While official activities have concentrated on the purification of crop seed, they have overlooked the capacity for contamination of machinery moving on and between farms. The combine harvester is a carrier of weed seeds and the large number of seeds that may be involved is documented (Ministry of Agriculture, Fisheries and Food, 1975). Other carriers are straw bales and the farmyard manure which is made from straw (Kirk and Courtney, 1972). Research by Wilson 1970) provided evidence of the large numbers of wild-oat seeds that may come to rest in a single bale (Table 3) and from this it can be

Table 3. Wild-oat seeds in cereal straw 1970

	seeds found per m^2	seeds per bale	seeds per kg straw
Oxford winter barley	6942	1165	670
Boxworth winter barley	2790	950	62
Boxworth spring barley	1386	143	9

Source: Wilson, 1970.

calculated how many thousands of wild-oat seeds may be transported in a tonne of straw. Since there is a national trade in straw from the cereal-growing areas of the east and south to many livestock and mixed-farming areas, the movement of seed must be considerable. On a more international level the entry into Northern Ireland of wild-oat from North America has been documented (Courtney and Kirk, 1975). In this case the carrier was feed barley from which the wild-oat seeds escaped. Whether or not this means of entry is the reason for Northern Ireland ceasing to be a wild-oat-free area or whether it was contaminated seed corn from England is uncertain.

A conclusion from the general evidence is that crop seed and straw are the most likely means of long-distance spread, and combine harvesters are an additional means of spread within the farm. Prevention of seed movement is as important today as it ever was. It is probably more difficult to control than it was in the nineteenth century because of our ability to handle and move large quantities of produce rapidly.

The sequential rotation of crops

An appreciation of the very detailed and important nature of crop rotations in the nineteenth century can be obtained from Sanders (1949). Much value was attached to well-planned and regular rotations which took into account the climate and soil environment of the farm. Although rotations were altered in the face of changes in crop and stock prices, they continued to be regularly followed until the 1950s. The importance attached to them by landlords can be judged by strictures put into tenancy agreements requiring occupants to follow particular rotations. Although such requirements began to be discarded in the late nineteenth century the attitude of mind remained: for example it continued to be regarded as bad farming to grow more than two straw crops in succession.

The function of rotation in weed control was to avoid a regular pattern of land use which might allow one or several weeds to exploit a regular form of cultivation. Cereal crops were regarded as fouling crops because they were grown in close rows which prevented hoeing, while turnips and the like were regarded as cleaning crops because they allowed regular horse hoeing between the rows and occasional hand hoeing within the rows. Rotation struck a balance between fouling crops so far as weed control was concerned. It had other functions, of which the two most important were the maintenance of soil fertility and a spread of the work of the labour force and horses on the farm.

A major change in twentieth century agriculture has been the movement away from man power with its flexibility in changing from job to job and its limited endurance, to machines, which could work longer without a break and so get over the human endurance problem, but at the expense of an inflexibility of use: a combine harvester can only harvest cereals and like crops, it can do nothing else.

There was a time in the 1960s when rotation was considered an old-fashioned word. The power of chemical herbicides was considered to be such that many believed that farmers could have an almost complete freedom of cropping so far as weed control was concerned. However, more sober counsels now prevail. It can be seen that the abandonment of rotation on cereal farms in favour of long runs, particularly of autumn-sown cereals, has provided the conditions for weeds to exploit a regular form of husbandry. Couch-grass, wild-oat, blackgrass, sterile brome are living examples today of what our forefathers sought to avoid by rotation (Table 4). It is arguable today that cereal farmers would be advised to retain a proportion of their land in spring barley as a means of reducing the difficulties, complications and costs of a complete reliance on selective herbicides inherent in the winter-cereal situation (Table 5). A virtue of oil-seed rape lies in the opportunities that the crop offers for the use of chemicals active on the

Table 4. Effect of past cropping on wild-oat in England 1977

	log (1 + seedheads ha^{-1}) Soil: light	medium	heavy
Past cropping:			
mainly cereals	1·8	2·0	2·6
mixed	1·4	1·8	2·1
mainly grass	0·7	1·0	1·3

Source: Elliot *et al,* 1979.

Table 5. Occurrence of black-grass in cereal crops 1977

	Cereal area infested (percentage) autumn	spring
England	33	13
S.E.	38	20
S.W.	22	4
E.	44	27
Yorks/Lancs	17	5
N.	3	0

Source: Elliott *et al.* 1979.

Table 6. Rhizome buds of couch-grass in a ley

96 buds/plot planted in March 1969	New buds/plot 1969	1970	1971
Nitrogen X cutting			
high-lax	140	291	942
high-intensive	157	127	240
low-lax	138	235	275
low-intensive	110	150	196

Source: Cussans, 1973.

weed grasses that plague the cereal farmer. For farmers who have livestock requiring grass, cereal/ley farming remains one of the forms of cereal growing most free of annual weeds, but not of couch-grass (Table 6). Similarly farmers in eastern England involved in mixed rotations including peas, sugar beet and potatoes have few of the weed problems associated with cereal mono-cropping. These experiences are evidence of the relevance of rotation in our agriculture today. This is a lesson which should not be forgotten as we look forward to the 1980s with seemingly inevitable increases in chemical costs relative to crop returns.

Soil cultivation

Cultivation has traditionally consisted of mouldboard ploughing, tine and disc cultivation, harrowing and hoeing (Fryer and Makepeace, 1977).

The inversion of soil by mouldboard ploughing has for centuries been a mjaor act of weed control causing the weeds to be bruised and buried so that they died from mechanical injury and the exclusion of light. Ploughing had a further function of inverting the soil surface upon which were many weed seeds, placing them to a depth from which they could not successfuly germinate. Inversion of annual plants often caused their death, but it was less effective with perennials which were able to re-sprout from roots or rhizomes in the topsoil. Similarly the burial of weed seeds was effective with small seeds, but less so with large seeds from which plants could emerge from a depth of 10 cm or more. Ploughing was the primary cultivation which not only left a clean surface but disrupted the soil which could be subsequently broken up to form a seedbed. The tine or disc cultivation that followed continued the process of weed control by disrupting germinating weeds.

To this day, soil cultivation is a major factor in weed control. Research in the 1960s demonstrated how tine and rotary cultivation could be used with and without chemicals for the control of couch-grass (Cussans and Elliott, 1969). Ayres (1977) demonstrated the effect of inversion on onion couch *Arrhenatherum elatius* in suppressing its growth. More recent research has shown how differences in soil disturbance affect the seedling population of blackgrass: ploughing has the greatest effect in limiting the subsequent population of the weed and direct drilling (no cultivation) has the least (Moss, 1979) (Tables 7, 8 and 9). However, not all weeds follow this trend, as is evidenced by knotgrass, *Polygonum aviculare* (Table 9). Soil cultivation is a major modifier of weed populations, and particularly when it is associated with rotation, which may allow the time as well as the type of cultivation to be varied.

Ploughing and deep cultivation are now unpopular in British farming because they are time and energy consuming, they require heavy capital investment in machinery and they limit the area of land which can be sown at peak periods, particularly in the autumn. In land devoted to surface harvested crops such as cereals, oil-seed rape and grasses, the trends of reduced cultivation and direct drilling seem likely to continue, urged on by the increased cost of energy and the rising price of tractors (Nix, 1978).

Since ploughing and other cultivations have had a suppressing effect

Table 7. Black-grass in winter wheat in November 1975

Cultural treatment	Plants m^{-2} Buckland	Compton	Begbroke
Ploughed	4	28	8
Tine cultivation	78	130	88
Direct drill	177	159	91

Source: Moss, 1979.

Table 8. Annual meadow grass in winter cereals

	Seedlings m^{-2} 1974	1975	1976
Direct drilled	22	90	106
Shallow-tine cultivation	24	50	52
Deep-tine cultivation	22	56	38
Plough	24	17	8

Source: Cussans *et al.*, 1979.

Table 9. Knotgrass in spring barley

	Seedlings m^{-2} 1969	1971	1973
Direct drilled	1	5	1
Shallow-tine cultivation	2	8	3
Deep-tine cultivation	3	12	8
Plough	11	22	14

Source: Cussans *et al.*, 1979.

on certain weeds, the discarding of these practices must be expected to lead to increased numbers of such weeds. Other means of weed control, principally chemicals, must be that much more effective. Elsewhere in our agriculture where root crops are grown, ploughing and deep cultivation are usual, thus aiding weed control.

Selective weed control

In the nineteenth century the only widespread form of selective weed control was hand hoeing: the distinguishing abilities of a human could select and chop the weed, leaving the crop plant unharmed. As long as the cost of agricultural workers was low in relation to the value of the products, hoeing made economic sense. The Second World War caused a collapse of the hoeing system because the economic balance was destroyed. It was a lucky accident that the new discoveries of selective herbicides in the 1940s coincided with an escalation of labour costs which made the hoeing of traditional row crops uneconomic. The combination of these two factors can be seen clearly in the changes in cropping practices in Britain during the 1950s. Cereals became the cleaning crops and expanded in area rapidly while row crops (turnips, swedes, mangolds, kale) ceased to be grown in many areas of England. Hoeing continued only in those high-value cash crops such as sugar beet, where the cost could be justified. Today hoeing is virtually unused except in very high-value horticultural crops. It is, however, worth remarking that a slightly different form of selective weed control is the hand roguing of wild-oats in cereals. For low infestations this method continues to be cheaper than the use of chemicals.

The growth and the use of chemical weed control has been well chronicled elsewhere (Fryer and Makepeace, 1977) and so I do not propose to describe the events of the past 30 years in detail. Let it be simply stated that during that time some 80 chemicals have come into regular use in British agriculture providing more than 460 herbicidal products. The majority of agricultural crops are sprayed for weed control at least once, and many several times (Table 10). Although slow

Table 10. Cereal area receiving herbicides 1977

	Percentage of area treated against:		
	broad-leaved weeds	wild-oat	black-grass
England	84	30	9
Wales	84	5	0
			couch-grass
Scotland	91	2	4
N. Ireland	96	1	2

Source: Elliott *et al.*, 1979.

to start in grassland, the use of chemicals is now increasing. Chemical weed control is widely regarded as essential for the efficiency and well-being of our agriculture.

The manufacture of herbicides consumes energy, but the products are used in such small quantities on the land that chemical weed control does not constitute an energy-consuming operation. Since herbicides are the key to reduced cultivation and direct drilling with their saving of energy, the chemicals must be seen as right for times of expensive energy. In spite of often-voiced fears of possible contamination of food and the environment, the overwhelming evidence is that herbicides have been safe and I see no reason why this should not continue.

In view of the dependence that British agriculture and horticulture now have on herbicides, what of the future? There can be little doubt that we will continue to be adequately supplied with chemical materials, indeed were no more chemicals to be discovered we could probably subsist well in the 1980s on those that are to hand. However this is not quite the point. The extraordinarily successful developments of chemicals have occurred because of the satisfactory way in which chemical companies could patent their discoveries and proceed to sell and make profits from those patents. The development of the next discovery was financed out of the profits of the last. Now there are fewer niches for a new chemical; health and environmental considerations dictate vastly increased development costs, manufacturing plant is more expensive. A figure of £10m to develop a new herbicide is commonly mentioned. For how much longer can the chemical industry service new development out of patent profits?

Conclusion

Of the traditional methods of weed control only hoeing has fallen into disuse and even that is still used in certain high-value crops. All the other traditional forms of weed control can be detected playing a role in husbandry to this day. All have their virtues and vices, and given proper understanding individual farmers can use them to their advantage.

Although soil cultivation can still be seen as serving a useful purpose purely in terms of weed control, other more demanding pressures will cause its decline in those areas where surface-harvested crops are grown. Thus, in general if not specific instances, the decline of cultivation causes a reduction in efficiency of weed control. This is a fact that should not be overlooked or taken lightly by those whose responsibility it is to construe successful cropping systems for our agriculture.

The future efficiency of weed control would appear to place reliance on three methods, namely preventing weed movement, rotation of crops and chemicals. If agriculture is to continue its present trend of high yields and increased efficiency, much more thought should be given to the complementary use of these three. We should cast aside the attitude that everything can be done by chemicals. Weed seeds are moving around our farms and should be stopped. Farmers could make more use of cropping sequences than they do. However, these are indirect forms of weed control: until there is some new breakthrough, selective control by chemicals is bound to be the main direct means of eliminating weeds growing in crops.

As so much reliance is being placed on chemicals, are we using them to best effect? In the UK we have a system whereby manufacturers produce recommendations which are approved by the Ministry of Agriculture, Fisheries and Food and there is a moral obligation on distributors and farmers to follow these approved recommendations and only these. Can such a rigid approach allow the flexibility in use essential to a major input in crop production? When one considers the many and varied ways in which fertilizers or animal feeds are used, and how much innovation in their use there has been by farmers, can one believe that these two could have developed in all the different ways to suit individual farming environments had they been subject to rigid recommendations? We have a challenge before us. How can we achieve the flexibility of use of herbicides which is essential if they are to play an increasing role in the cropping systems of the future and yet maintain the basically sound performance and safety which have been a feature of the last three decades?

References

Ayres, P. (1977). The growth of *Arrhenatherum elatius* var. *bulbosum* (Willd.) sperm in spring barley, as influenced by cultivation. *Weed Research* **17**, 423-428.

Courtney, A. D. & Kirk, J. (1975). The significance of imported feeding barley as a source for the introduction of wild-oats (*Avena fatua* and *A. ludoviciana*) in Northern Ireland. *Symposium on Status, Biology and Control of Grass Weeds in Europe, EWRS & COLUMA, Paris* **1**, 24-32.

Cussans, G. W. (1973). A study of the growth of *Agropyron repens* (L.) Beauv. in a ryegrass ley. *Weed Research* **13**, 283-291.

Cussans, G. W. & Elliott, J. G. (1969). Perennial grass weeds of arable land. *ARC Weed Research Organization 3rd Report,* pp. 15-19.

Cussans, G. W., Moss, S. R., Pollard, F. & Wilson, B. J. (1979). Studies of the effects of tillage on annual weed populations. *Proceedings EWRS Symposium: The Influence of Different Factors on the Development and Control of Weeds 1979* (in press).

Elliot, J. G. & Attwood, P. J. (1970). Report on a joint survey of the presence of wild-oat seeds in cereal seed drills in the United Kingdom during spring 1970. *ARC Weed Research Organization Technical Report* No. 16.

Elliott, J. G., Church, B. M., Harvey, J. J., Holroyd, J., Hulls, R. H. & Waterson, H. A. (1979). Survey of the presence and methods of control of wild oats, black grass and couch grass in cereal crops in the United Kingdom during 1977. *Journal of Agricultural Science, Cambridge* **92**, 617-634.

Fryer, J. D. & Makepeace, R. J. (1977). *Weed Control Handbook* Vol. 1, 6th edn., pp. 31-36. Oxford: Blackwell Scientific Publications.

Kirk, J. & Courtney, A. D. (1972). A study of the survival of wild-oats (*Avena fatua*) seeds buried in farmyard manure and fed to bullocks. *Proceedings 11th British Weed Control Conference*, pp. 226-234.

Ministry of Agriculture, Fisheries and Food (1975). Guide to the cleaning of combine harvesters. *National Wild-oat Advisory Programme leaflet,* May 1975. London: HMSO.

Moss, S. R. (1979). The influence of tillage and method of straw disposal on the survival and growth of black-grass, *Alopecurus myosuroides,* and its control by chlortoluron and isoproturon. *Annals of Applied Biology* **91**, 91-100

Nix, J. S. (1978). Costs in labour and machinery. *Proceedings 32nd Oxford Farming Conference,* pp. 41-66.

Sanders, H. G. (1949). *An Outline of British Crop Husbandry,* pp. 1-33, Cambridge: University Press.

Tonkin, J. H. B. (1969). The occurrence of some annual grass weed seeds in samples tested by the Official Seed Testing Station, Cambridge. *Proceedings 9th British Weed Control Conference,* pp. 1-6.

Wilson, B. J. (1970). Studies of the shedding of seed of *Avena fatua* in various cereal crops and the presence of this seed in the harvested material. *Proceedings 10th British Weed Control Conference,* pp. 831-837.

The impact of herbicides on crop production

D. W. Robinson
Kinsealy Research Centre, Agricultural Institute, Malahide Road, Dublin, 5.

The past

Chemicals were already being used against weeds, albeit in a small way, at the time the Association of Applied Biologists was founded in 1904. Common salt was probably the first herbicide and was recommended in Germany in 1854 (Timmons, 1970). An important discovery was made in 1882 when a vine grower in Bordeaux used lime water and copper sulphate to produce an unpleasant-looking mixture to discourage the village boys from pilfering his grapes. The unexpected control of mildew provided by the mixture encouraged its wider use and an observant person noticed that charlock (*Sinapis arvensis*) was killed where some of the mixture had been accidentally spilled (Russell, 1956).

The breakthrough to selective weed control occurred when Bonnet (1896) demonstrated that a dilute solution of copper sulphate could kill *Sinapis arvensis* without damaging oats. This discovery prompted widespread testing of other inorganic compounds and by 1904 it was known that ferrous sulphate, sulphuric acid, copper nitrate, sodium nitrate, ammonium sulphate, calcium cyanamide and potassium salts had some possibilities as selective herbicides.

Other compounds including sodium chlorate and sodium arsenate became available as weedkillers during the next two decades. By the 1930s most of the available inorganic compounds had been evaluated but only a few had proved useful and each had some limitations. Dilute sulphuric acid, for example, was effective against a wide range of broadleaved annual weeds but was corrosive and hazardous to handle; copper sulphate was more convenient but was effective only at high volumes, and sodium chlorate, although giving good control of deep-rooted perennial weeds in non-crop areas, was expensive, inflammable and non-selective.

As the limitations of inorganic substances became increasingly obvious the chemical industries in many countries began to test organic compounds on a large scale. The important advantage of these compounds was soon realized; extensive series of related compounds could be made and, as their properties depended on their constitution, these could be modified in various directions once the details of the relationship had been discovered (Russell, 1956). DNOC, dinitrophenols and pentachlorophenol were introduced during the 1930s, but there was little demand for herbicides during this period because labour was plentiful and cheap. The situation changed abruptly with the outbreak of the Second World War. Manpower on the land was scarce and supplies of natural pesticides from overseas were reduced. The search for synthetic organic compounds rapidly expanded and the inorganic and first organic herbicides were soon to be replaced by more effective chemicals.

The revolution in herbicides and weed control practice began with the discovery of the selective herbicidal action of MCPA and 2, 4-D in 1941-42 (Templeman, 1955). These herbicides had many advantages over all others previously produced; they were cheap, non-poisonous, simple to manufacture and they could be easily applied, requiring less water than inorganic salts.

The last 25 years (1954-1979)

The period around the Jubilee Meeting of the Association of Applied Biologists in 1954 saw the start of an important era in the history of weed control. The discovery of 2, 4-D and MCPA 12 years earlier had stimulated the interest of many scientists, chemical companies and governments in selective herbicides. In the 1950s the research activity of the previous decade began to show results. The trickle of new chemicals that started with MCPA and 2, 4-D developed into a steady stream and our knowledge of the science of weed control increased rapidly in the late 1950s. Some of the exciting possibilities for increasing the range of selective herbicides for use in agriculture were described in the Proceedings of the last Jubilee Meeting (Wain, 1955).

Despite the breakthrough with MCPA and 2, 4-D and the progress made in the following decade, there was still in the 1950s a general lack of awareness of the impact that herbicides were to have on crop-production practice during the next 25 years. There were at that time few full-time research workers on weed control and, apart from cereals, chemical weed control was a very minor aspect of agronomy, horticulture, forestry, botany, agricultural engineering and plant physiology. For these reasons few of the delegates at the Jubilee Meeting in 1954 could have anticipated the large number of new compounds that were to become available and the far-reaching changes in weed-control practice that were imminent.

The decade following the 1954 Meeting saw the discovery of a number of important new series of herbicides including the triazines, ureas, acylanilides and uracils. The statistics on the change in the market for herbicides during the last 25 years are impressive. Apart from the greater number of active ingredients, the quantities used are also larger and the monetary value of herbicides has increased fifteen-fold in real terms between 1953 and 1978 (Table 1). But the statistics do not tell

Table 1. Pesticide market at end-user level for the world excluding Comecon countries (1978 monetary values in $mill.)

	1953		1978	
	Market value	%	Market value	%
Insecticides	965	57	3052	34
Herbicides	225	13	4068	45
Fungicides	428	25	1522	17
Others	88	5	380	4
Total	1706	100	9022	100

Source: Shell International Chemical Company.

the full story. In addition to the change in the number and quantities of chemicals used, herbicides have done more than merely control weeds. Their major impact has been their influence on crop-production methods. Weeds are present in all crops throughout the entire growing season and they can be troublesome in overwintered crops throughout the year, particularly in mild areas. There was in the past an effective alternative means of weed control in most crops other than the use of pesticides – notably cultivation. Consequently because of the fundamental importance of weed control and the availability of cultivation aided by rotation as a practical means of control, crop-production practices grew up around the need to control weeds by cultivation. The development of effective weed killers gave farmers a new freedom in crop-production methods; herbicides are now an integral part of production processes and methods of growing practically all crops have changed as a result.

Cereals

In contrast to other crops the use of herbicides in cereals was already well established in 1954. The introduction of 2, 4-D and MCPA was followed in the early 1950s by 2, 3, 6-TBA and in the late 1950s by mecoprop, dichlorprop and 2, 4-DB. The range of herbicides that became available facilitated the control of a wide range of weeds and turned a 'fouling' crop into a 'cleaning' crop, allowed row widths to be narrowed, shorter-strawed varieties to be grown and ploughing to be reduced in depth or eliminated. Because long-season crops could now be kept clean there was a swing towards winter sowing and increasingly cereals are grown in succession. There was a marked increase in the acreage of crops direct drilled without prior cultivation, the area treated in this way in the UK rising from 6915 ha in 1971 to 51 895 ha in 1975 (R. G. Hughes, personal communication).

Cereals are sprayed not only to prevent weed competition with the growing crop but also to ensure weed-free conditions at harvest. This enables combine harvesters to work efficiently and facilitates the conditioning of grain for storage or sale. Elliott (1979) estimated that cereals in the UK were probably treated with £40-50m worth of herbicides in 1978.

Sugar beet and potatoes

Change has been particularly far-reaching with root crops which in the 1950s had a high labour requirement. The acreage of sugar beet treated with herbicides in Britain increased from zero in 1954 and 5 per cent in 1961 to 99 per cent in 1977. Crops today receive at least a single application of herbicides but more usually treatments consist of a planned sequence of pre- and post-emergence herbicides. The sugar beet crop exemplifies the problem of evaluating the impact of any one branch of technology; from being a highly labour-intensive crop 25 years ago sugar beet can now be grown without any hand labour, due in part to the introduction of pelleted monogerm seed and effective precision drills and harvesting equipment. The full value of these developments was realized only by the introduction of effective herbicides and, in turn, the usefulness of chemical methods of weed control was enhanced by the possibility of sowing a spaced crop.

Compared with most other agricultural crops the adoption of herbicides for potatoes has been less rapid, partly because the weed problem in this crop was not so acute and partly because of the opportunities for post-planting, mechanical weed control. Nevertheless the acreage treated increased from 1 per cent in 1963 to 68 per cent in

1976 (Gunn, 1977). Over 30 per cent of the acreage is also treated with haulm desiccants.

Vegetables

The development of a range of effective herbicides for many vegetables has also changed the way these crops are grown. Now that cultivation equipment does not have to pass along every crop row, growers have much greater freedom in selecting optimum plant spacings. Many vegetables such as onions, celery and carrots are now grown in beds and all cultivation is done before sowing. The bed system of producing root crops has given growers more precise control over the proportions of their crop in the desired market grades. The widespread adoption of the bed system has encouraged the development of new machinery for drilling the crops at the required spacing and for harvesting. More specialized mechanization has resulted in more efficient production. Higher plant populations have in turn necessitated higher fertilizer requirements and have resulted in higher yields. Again it was chemical weed control that provided the key to these developments.

Fruit

Herbicides were readily taken up by fruit growers during the 1960s. For example the results of a survey carried out by Rendell (1961) and M. M. Shore, (personal communication) showed that the acreage of black currants treated with herbicides rose from zero in 1957 to over 90 per cent in 1964. Soft-fruit growers quickly saw the advantage of eliminating cultivation through the use of herbicide programmes and many apple growers with modern orchards in the Republic of Ireland have eliminated grass from their plantations. The adoption of complete herbicide management by fruit growers in a number of countries, resulting in the elimination of tillage, was probably the first major advance in a practice of soil management that dates back thousands of years.

When herbicides are introduced into a crop-production system and soil management methods are modified, a cycle of change begins that is not just confined to the way weeds are killed. These changes can affect many factors including the growth and morphology of the crop plant, the composition of the harvested product, soil conditions and microclimate. The many inter-related changes that can follow the introduction of herbicide management are well illustrated in perennial crops such as soft and top fruits.

When soil cultivation is discontinued in soft fruits, root growth is affected in several ways. Mechanical destruction of shallow roots is avoided and this permits higher crop-root densities and greater exploitation of the surface soil which tends to be the most fertile (Robinson, 1965). Similarly, when a grass sward is replaced by herbicide management in apple orchards, higher root densities occur in the surface soil (Atkinson and White, 1976). Roots of apple trees grown in a herbicide row with grassed alleyways are presented with two different environments and concentrate in the herbicide-treated zone (Atkinson and White, 1976; Atkinson, 1977).

There is no evidence that the replacement of soil cultivation or a grass sward by herbicide management has caused a yield decrease in any temperate fruit, although strawberries do not show a positive response. The growth and yield of apples is usually increased substantially where soil cultivation or a grass sward is eliminated either completely or from the tree row. In a comparison of six methods of soil management for apple cultivar Golden Delicious between 1965 and 1976, overall herbicides increased tree vigour and gave a higher yield than all other treatments. Yield was increased by 20 per cent compared with well-mown grass (Robinson and O'Kennedy, 1978). Black currants and gooseberries also appear to be highly responsive to herbicide management and increased plant vigour and yields have been reported from a number of experiments (Robinson, 1964; Allott, Robinson and Uprichard, 1971; Hughes 1972).

Herbicide management results in significant changes in many soil properties including bulk density, pore volume, number and orientation of pores, organic matter, infiltration capacity, soil strength and pH. The significance of some of these changes has been discussed by Robinson (1967 and 1975), Atkinson and White (1976) and Atkinson and Herbert (1979).

Bulk density of herbicide-treated soil is greater than cultivated or grassed soils (Bulfin and Gleeson, 1967; Jelley, O'Kennedy and Robinson, 1974) but the increase in density is largely confined to the top few centimetres and there is no evidence of levels of compaction in fruit plantations that interfere with root growth. In contrast to compaction due to plough pans, such compaction at the surface will not get progressively worse since the soil in this position will be subjected to structure-forming influences, e.g. frost action, cycles of wetting and drying and the activities of soil fauna that are not effective at lower depths.

The increase in bulk density at the surface (0-5 cm) may be

accompanied by a reduction in non-capillary pores compared with cultivated or grassed plots (Jelley *et al.*, 1974), but this does not occur on all soils (Atkinson and Herbert, 1979). Even where a reduction in non-capillary pores occurs in herbicide-treated soils, aeration conditions remain satisfactory (Jelley and Gormley, 1979), due probably to the greater continuity of pores under herbicide management. Soil strength, important for wheeled and foot traffic in orchards, can be up to ten times greater than herbicide management compared with cultivation (Jelley, 1976).

In general less nutrients are required when herbicide systems of soil management are used because of more vigorous plant growth and less competition from grass or weeds. Major nutrients, N, P and K are not normally required in established apple orchards in Ireland where grass has been eliminated. Baxter and Newman (1971) also report that herbicides to keep tree rows free from weeds cost \$30 ha^{-1}, which could be paid for by the saving in 150 kg ha^{-1} of N. Herbicide management often results in a reduction in soil pH in the top 0-3 cm (Roach, 1966). Soil phosphorus tends to increase under herbicide management (Clay and Davison, 1976) but variable results have been obtained with soil potassium.

Fruit crops on many soil types have benefited greatly from herbicide management. The high yields are probably due to a number of factors but especially to the greater volume of soil, including the fertile surface layers, that can be exploited by plant roots. Crops under herbicide management are also able to make better use of natural and supplementary moisture because of the absence of competition by grass and weeds.

The risk of frost damage at flowering is reduced because the insulating effect of a grass sward or loose cultivated soil is removed. In Co. Waterford temperatures averaging 1·6 °C higher at 1·2m above the soil surface were recorded on radiation frost nights in herbicide-managed compared with grassed orchards in 1977.

To obtain full benefits from herbicides, effects on other soil and production factors need to be watched carefully. The reduction of the pH of the surface layer can cause manganese toxicity (Roach, 1966). Run-off and erosion, previously retarded by grass or weed growth, may be severe on sloping sites. Crops that depend on wild insects for pollination may suffer poor fruit set if the insect numbers are reduced by the absence of weeds. However, the advantages of herbicide management of fruit crops are so great that present experience suggests strongly that appropriate action should be taken to overcome these and other problems that may arise.

The future

While the last 25 years have shown remarkable progress, the chemical industry is now facing many difficult problems. The discovery, development and registration of new herbicides is very costly and will become increasingly so as regulatory requirements become more stringent. In 1976 the cost of developing a new product was around £10m and the number of new compounds being introduced is falling at present. This problem could be increased by the decline in patent protection as fewer firms will have the benefit of earlier successful patents to finance present and future searches for new compounds. There is also increasing public interest in, and often unfounded concern about, the use of herbicides in agriculture. This will result in environmental interests being more to the fore politically during the next 25 years.

There have been suggestions that the dynamic period for weed science is now over and that the rate of progress achieved during the last quarter century cannot be continued into the next. This view is unnecessarily pessimistic; there are many areas where significant advances are probable and further substantial progress in weed science is inevitable. Chemical methods of weed control are likely to be enhanced through advances in many different areas of activity including herbicide technology, engineering research, plant breeding and computer technology.

With present progress towards a better understanding of the physiological and biochemical mechanisms of plant growth, we can expect improvements in methods of developing new herbicides. We will have greater opportunities to design specific herbicides to utilize small biochemical differences between crop and weed. The need for safer herbicides both environmentally and for crops is likely to be met in several ways. Screening methods to find new herbicides will become more sophisticated (Hess, 1979) and the traditional pre- and post-emergence greenhouse tests will be supplemented with carefully controlled laboratory procedures, to reveal any intrinsic chemical activity relating to plant processes. As our data base expands, the potential of the computer to help scientists design better herbicides will increase.

Some limited progress has been made with herbicide protectants and antidotes which increase the tolerance of crops to various herbicides (Blair and Dean, 1976). It seems likely that, as our understanding of the mode of action of antidotes increases, other substances will be developed and the principle of protection will be extended to a wider range of crops and herbicides. In particular, herbicide antidotes will be valuable for increasing selectivity and for overcoming the very difficult problem of freeing crops from closely related wild species.

Other developments can be confidently expected which will enable existing herbicides to be used more effectively and with less risk to the environment. With soil-acting herbicides we aim to control weeds over a long period, yet prolonged chemical stability is environmentally undesirable and can lead to problems with succeeding crops. Controlled-release formulations of some herbicides could provide season-long control of many weeds as, after application, the chemical could be released gradually into the environment at a rate more closely matching requirements than is possible with existing formulations. Controlled-release formulations could make useful some compounds that never passed beyond initial screening trials because they were too transient. Some of these chemicals may now be re-assessed because of improvements in screening methods and advances in controlled-release formulations.

There is scope for the significant improvement of equipment for the application of herbicides. New applicators will encourage the development of new formulations of herbicides and conversely new herbicides will stimulate production of novel application methods. We can expect the introduction of many types of applicators designed to enable non-selective herbicides such as glyphosate to be applied selectively and economically, e.g. the recirculating sprayer (McWhorter, 1977). Concurrent with the development of controlled-droplet application (CDA), improvements in conventional sprayers and nozzles can be expected. Highly efficient application equipment will be more urgently needed as herbicides increase in price and must be used with maximum efficiency.

The presence of genetic variation for tolerance to certain herbicides is widespread in many crops and has led to the selection of strains of crops resistant to specific herbicides e.g. *Lolium perenne* to paraquat (Faulkner, 1976). Just as we will see herbicides designed to suit major crop plants, we are likely to see greater emphasis placed on selecting cultivars to suit wide-spectrum herbicides. As increasing development costs reduce the number of new herbicides becoming available, we will need to assess more thoroughly the potential of existing herbicides with their known toxicology. For certain weed problems it may be cheaper to breed cultivars resistant to existing herbicides than to attempt to develop new herbicides for existing cultivars.

Advances in tissue culture and genetic engineering could stimulate the development of herbicide-resistant cultivars. Cultures of a crop plant could be treated chemically or with radiation to induce maximum variation. The cultures could then be grown in a medium containing the herbicide and survivors propagated and tested for agronomic qualities. As more is learned about the genetic control of resistance to herbicides it may be possible to produce herbicide-tolerant plants by transferring the gene or genes which control resistance to a crop plant using genetic-engineering techniques.

As well as helping in the research for new herbicides, computer technology will have significant impact on the way specialist advisers and farmers receive their information and spray their crops in the future. Just as scientists can now research several million abstracts in a few minutes and have printouts on those on which information is required, the same techniques will be used extensively for getting quickly to farmers reliable information on appropriate weed control measures.

Energy

The impact of herbicides on crop production will be increased further as energy costs continue to rise. Even after allowance has been made for the energy required to produce and apply herbicides, the total energy inputs for many weed-control operations will be substantially less with herbicides than with mechanized cultivation (Green, 1979). The concept of energy budgeting at farm level will become more important. All operations will be increasingly scrutinized for their energy inputs and scientists will be more involved in the identification and adoption of energy-saving techniques. Compared with many other agricultural operations, weed-control practices have a favourable energy input-output ratio because of the large increase in crop yields possible as a result of weed control.

Improvements in spraying equipment, both CDA and conventional, will also help to reduce energy requirements. For example, CDA enables the application of herbicides in greatly reduced water volumes without undue risk of drift (Cussans and Taylor, 1976). Just as herbicides were the key to new methods of production of cereals and vegetables, so CDA, with its low spray volumes and low pay-load requirements, could be the key to a new concept of crop production involving low energy inputs.

Excellent weed control has been achieved in cereals at the Weed Research Organization with light-weight, low-ground-pressure vehicles travelling at 20 km h^{-1} and applying herbicides in volumes of 60 l ha^{-1}. Autumn-sown cereals can be sprayed during October to March even when the soil surface is damp (Elliott, 1978). Now that many farmers direct-drill their crops and no longer cultivate the soil to control weeds, the need for major traction has gone. This could result in large tractors with high fuel consumption becoming obsolete on many farms. Because of rapid changes in engineering technology it is impossible to predict accurately the future of application equipment. Not all compounds and formulations are suitable for CDA and the system is still capable of improvement in its present form. Nevertheless the need to conserve fossil fuel and the reduced energy requirements of low-volume spraying suggest that we will see many innovations in application technology during the next 25 years.

Towards a weed-free environment

There has been much discussion lately on the extent to which herbicides should be used for controlling weeds and on the degree of weed control that should be attempted. This debate arises for a variety of reasons including the increasing cost of herbicides and the need to reduce environmental pollution. Stryckers (1979), for example, states that efforts often made in cereals to reach control levels of 90-95 per cent are not economically justified with regard to yield and he suggests that a lower dependence on herbicides could be achieved by reducing the demands on the level of weed control.

In perennial crops, however, the economics of weed control should not be considered merely on an annual basis. If it were possible to achieve a level of control so that no weeds seeded and there was no return of seeds to the soil, the weed problem would fall substantially over a period of years, particularly in non-tillage systems. This would enable the number of applications or amount of herbicides applied to be reduced with economic and environmental benefits. Obviously in extensively grown agricultural crops it is not possible to achieve almost complete weed control, but some horticultural crops such as nursery stock, black currants, raspberries and apples are being grown in a virtually weed-free environment today in some areas. Almost complete weed control in these crops has been achieved by chemicals supplemented by only a small amount of hand work to deal mainly with weeds disseminated by wind or birds. Control has been achieved because the high value of the crop makes it possible to apply sequential treatments adjusted to the specific weed problem.

Complete weed control in a crop will not be achieved easily. But the present range of effective herbicides for woody perennials is still expanding. In addition, greater use is likely to be made in the future of herbicide mixtures and of fumigants to eradicate weed propagules in the soil. These and other advances along with better application equipment and with the development of increasing skills by growers suggest that a virtually weed-free environment is a realistic and commendable target for growers of intensive perennial crops.

In contrast, advice to aim for less than complete weed control in these crops has associated dangers. Weed growth will reduce the persistence of soil-acting herbicides and could encourage the development of populations of species that are more difficult to control. Because weed growth is possible throughout the year in perennial crops, it is easier to aim for complete freedom than to maintain weed cover at a certain level. This approach, particularly in a perennial crop, would carry with it grave risks of the development of genetically resistant weed species. Obviously in a weed-free situation, resistance to herbicides cannot evolve.

In intensive areas there will be much in favour of attempting to achieve a weed-free environment even if this requires a period of what is apparently cosmetic weed control. For a number of reasons complete eradication of weeds in woody perennials should not be as difficult to achieve as insect or disease control. Firstly the weed-control programme is likely to be based on a persistent herbicide applied to the soil. Thus the target, in contrast to crop foliage, is flat, static and remains protected throughout the season. If weed seedlings become established, they will usually be clearly visible; weeks and probably months will elapse before further viable seeds are produced giving time for appropriate control measures, cultural or chemical. Growers will need to develop a different philosophy of weed control. They will need to think more about long-term strategies for reducing the weed-seed population in the soil by chemical, cultural and other means.

At the same time as high-value crops are being managed more and more intensively there will be an even greater need to manage parts of the landscape more positively for nature conservation. Conservationists are justified in asking the farming community not to reclaim the last remnants of good wildlife habitat, the small woods, old hedges and ponds that still exist (Moore, 1977).

Landscape management

The next Jubilee Conference of the Association of Applied Biologists will be held in the next century when earning a living is likely to constitute a much smaller part of a person's activities than at present. Increasing leisure will give people opportunities to develop new skills and arts and time to appreciate new delights. Many of today's outdoor leisure pursuits will become more important to larger numbers of people.

Herbicides are likely to become a more valuable tool in the preservation and enhancement of the environment. Just as we are now learning to manipulate agricultural grassland with selective herbicides and seed introduction, the scope for managing amenity grassland is equally great. Herbicides may well be used in the future to increase species diversity, for example along roadsides. As we accumulate knowledge of the life history of wild flowers and their response to herbicides we will be able to create and manage for recreational purposes large areas of grassland, bluebells, primroses, cowslips and other attractive native species.

During this century western Europe has lost many magnificant gardens due to the high cost of upkeep, mainly for weed control. Herbicides can be very useful not merely to maintain amenity areas, sites of special scientific interest and nature reserves but their use can result in better quality plants since woody ornamentals respond positively to herbicides in the same way as closely related fruit crops.

The last 25 years have seen remarkable progress with herbicides. Much has still to be learned about their long-term use, particularly with regard to their influence on the environment. The effects of herbicides on plants, soil organisms and soil-management techniques are very complex and inter-related. There is much potential for interactions between plants, micro-organisms, soils and herbicides and it is impossible to predict precisely the consequences of such interactions for plant and soil health.

Herbicide usage has been particularly intensive in fruit growing. If the results obtained in this area can be extrapolated to other sectors of agriculture, then there are grounds for optimism. After over twenty years experience of complete herbicide management in fruit crops in Ireland, there is no evidence on a range of soil types of any adverse effects on soil stability, structure or fertility resulting from the long-term use of herbicides. On the contrary, present experience indicates that compared with traditional cultural methods, the judicious use of herbicide management can result in significant increases in the yield of many crops and result in long-term improvements to soil structure. This suggests that the present trend from crop and soil management by machinery towards chemical management with herbicides will continue during the next 25 years.

References

Allott, D. J., Robinson, D. W. & Uprichard, S. D. (1971). The response of gooseberries to non-tillage systems of management. *Horticultural Research* **11**, 166-176.

Atkinson, D. (1977). Some observations on the root growth of young apple trees and their uptake of nutrients when grown in herbicide strips in grassed orchards. *Plant and Soil* **49**, 459-471

Atkinson, D. & Herbert, R. F. (1979). A review of long-term effects of herbicides on the soil with particular reference to orchard crops. *Annals of Applied Biology* **91**, 125-146.

Atkinson, D. & White, G. C. (1976). Soil management with herbicides. The response of soils and plants. *Proceedings 1976 British Crop Protection Conference-Weeds* **3**, 873-884.

Baxter, R. & Newman, B. J. (1971). Orchard soil management trials No. 2. Effect of herbicides and nitrogen on growth and yield of young apple trees in permanent pasture. *Australian Journal of Experimental Agriculture and Animal Husbandry* **11**, 105-112.

Blair, A. M. & Dean, M. L. (1976). Improvements in selectivity of perfluidone against *Rottboellia exaltata* in maize with herbicide protectants. *Weed Research* **16**, 47-52.

Bonnet, L. (1896). Le Cuivre contre les ravenelles et les sénés. *Communications dans la séance du Comice de Reims du 12 Décembre 1896.* Société Nationale d'Agriculture de France.

Bulfin, M., & Gleeson, T. (1967). A study of surface soil conditions under a non-cultivation management system. 1: Physical and chemical properties. *Irish Journal of Agricultural Research* **6**, 177-188.

Clay, D. V. & Davison, J. G. (1976). The effect of 8 years of different soil management treatments in raspberries on weed seed and worm populations, soil microflora and on chemical and physical properties of the soil. *Proceedings 1976 British Crop Protection Conference – Weeds* **1**, 249-258.

Cussans, G. W. & Taylor, W. A. (1976). A review of research on controlled drop application at the ARC Weed Research Organization. *Proceedings 1976 British Crop Protection Conference – Weeds* **2**, 885-894.

Elliott, J. G. (1978). Fast low-ground-pressure sprayers. *Big Farm Management,* June 1978, 16-23.

Elliott, J. G. (1979). The economic objectives of weed control in cereals. *Proceedings 1978 British Crop Protection Conference – Weeds* **3**, 829-836.

Faulkner, J. S. (1976). A paraquat resistant variety of *Lolium perenne* under field conditions. *Proceedings 1976 British Crop Protection Conference – Weeds* **2**, 485-490.

Green, M. B. (1979). The energy balance of pesticide use. *Biologist* **26**, 123-126.

Gunn, J. S. (1977). The role of herbicides in highly mechanised cash root crop production. *Proceedings 1976 British Crop Protection Conference – Weeds* **3**, 831-858.

Hess, D. (1979). Detection of intrinsic activity of growth inhibitor herbicides using an algal bioassay. *Proceedings Weed Science Society America, San Francisco, 1979,* Abstract 199:94.

Hughes, H. M. (1972). Soil and bush management studies on gooseberries. *Experimental Horticulture* **24**, 43-49.

Jelley, R. M. (1976). Methods of soil management. Effects on soil strength. *Research Report, An Foras Taluntais, Horticulture 1976,* pp. 62-63.

Jelley, R. M. & Gormley, T. R. (1979). Carbon dioxide levels in soils. *Research Report, An foras Taluntais, Soils, 1978*, pp. 14-15.

Jelley, R. M., O'Kennedy, N. D. & Robinson, D. W. (1974). Effects of soil management on soil structure, moisture and organic matter. *Research Report, An Foras Taluntais, Soils, 1973,* p. 12.

McWhorter, C. G. (1977). Weed control in soybeans with glyphosate applied in the recirculating sprayer. *Weed Science* **25**, 135-141.

Moore, N. W. (1977). Agriculture and nature conservation. *Bulletin British Ecological Society* **8**, 2-4.

Rendell, J. (1961). Black currant survey. *3-year report 1957-1959,* University of Bristol.

Roach, F. A. (1966). The use of herbicides in fruit production. *British Fruit Growing Symposium of the British Weed Control Council,* pp. 5-21.

Robinson, D. W. (1964). Investigations on the use of herbicides for eliminating cultivation in soft fruit. *Scientific Horticulture, 1962-1963* **16**, 52-62.

Robinson, D. W. (1965). The integration of chemical weed control in horticultural practice. *Mededelingen Directeur van de Tuinbouw* **28**, 401-412.

Robinson, D. W. (1967). Production of soft fruits without cultivation. *Proceedings, 17th International Horticultural Congress* **3**, 145-160.

Robinson, D. W. (1975). Some long-term effects of non-cultivation methods of soil management on temperate fruit crops. *Proceedings, 19th International Horticultural Congress* **3**, 79-91.

Robinson, D. W. & O'Kennedy, N. D. (1978). The effect of overall herbicide systems of soil management on the growth and yield of apple trees 'Golden Delicious'. *Scientia Horticulturae* **9**, 127-136.

Russell, J. E. (1956). Weed control: a record of striking progress. *Third British Weed Control Conference* **1**, 3-12.

Stryckers, J. (1979). Problems in chemical weed control. *EPPO Bulletin* **9**, 83-93.

Templeman, W. G. (1955). The uses of plant growth substances. *Annals of Applied Biology* **42**, 162-173.

Timmons, F. L. (1970). A history of weed control in the United States and Canada. *Weed Science* 18, 294-307.
Wain, R. L. (1955). A new approach to selective weed control. *Annals of Applied Biology* 42, 151-157.

New weeds for old in annual crops

R. J. Chancellor
Agricultural Research Council, Weed Research Organization, Begbroke Hill, Yarnton, Oxford OX5 1PF

Vegetation is never static. There is always a continuous process of change, as plant communities adapt to the changes in their environment. The rate of change may well depend both upon the life span of the dominants in the community and upon the degree of change in important environmental factors. Thus in undisturbed forests little change in the species present may occur over centuries or even thousands of years, but in annual crops the changes in the weed flora, which is mostly composed of short-lived annual species, can be relatively very quick indeed.

Types of change

Changes in vegetation can be of two sorts, either fluctuating or successional. In weed communities of annual crops, changes in the short term are more likely to be fluctuating. Variation in climate is known to alter the proportions of individual weed species occurring in a field, often quite outstandingly. A good example is *Stellaria media*, which is reported to be most abundant in wet years (Warington, 1958). Similarly, Thurston (1969) has recorded that *Myosotis arvensis* is more frequent after wet summers, while *Tripleurospermum maritimum* subspecies *inodorum* is more frequent after dry ones because rain has an adverse effect on seed set. Dry summers favour the dispersal of seeds of weedy *Compositae* which are wind-dispersed. The increase in number and distribution of these species in the year after the 1976 drought was very noticeable. Fluctuations can also be due to changes in management, which of course change the environment in various ways.

The other type of change, succession, is probably less frequent in the ephemeral communities in annual crops although it could occur under prolonged systems of monocropping. Succession in vegetation implies certain species giving way to others which are more competitive or better adapted to that particular habitat. This continues until a more or less stable equilibrium is reached between the plants available to colonize the area and the factors regulating the habitat conditions. This adaptation of the composition of vegetation to fit a particular habitat situation is called autogenic succession (Tansley, 1946). In contrast, a succession in vegetation can be brought about by changes occurring in the habitat, which make the existing vegetation less suitable to occupy that particular situation than other better-suited species. These better-suited species then gradually (depending upon life span and the rate of change of the factors) replace the original species. This is known as allogenic succession.

The Broadbalk Wilderness

The classic example of a complete autogenic succession in arable land is the Broadbalk Wilderness at what is now the Rothamsted Research Station. This was an area of winter wheat which was completely abandoned before harvest in 1882. That crop of wheat shed its seed naturally, but the selfsown crop in the following year was suppressed by a bad infestation of *Agropyron repens*. Continuing and increasing competition from weeds resulted in the wheat becoming extinct in the area in 1886 just four years after the field was abandoned (Garner, 1965). The first survey in 1886, the last year of the wheat, showed 40 species of herbs with *Agrostis tenuis* and *Medicago lupulina* as co-dominants. Eight years later (1894) there were 49 herbs with the same two dominants, but with *Lathyrus pratensis* becoming abundant. After a further nine years (1903), there were 57 species present with *Dactylis glomerata* and *Lathyrus pratensis* as co-dominants (Witts, 1965). During the first 31 years the flora changed from the original weed flora to other species not normally occurring in cereal crops. Thus in 1913 it was described as a dense thicket of trees and shrubs composed of *Quercus robur, Corylus avellana* and *Rubus* sp. with a carpet of *Hedera helix* (Brenchley and Adam, 1915). By 1960 the taller trees including a greater number of species, had reached 18m (Witts, 1965).

Examples of allogenic succession in annual crops are less easy to give, for permanent and natural changes in the habitat over the short term are difficult to find. Man's activities, such as liming, adding fertilizers, using herbicides or improving drainage, are not irreversible and so can almost be regarded as temporary alterations in the habitat, but nonetheless have probably the greatest effect on weed flora in annual crops.

Long-term survival of weed species

Having reviewed the types of changes that can occur in vegetation, let us now consider the evidence of any long-term persistence of species in the weed flora of annual crops. Archaeological evidence shows that many of our currently worst weeds have been present in crops for hundreds or even thousands of years. *Avena fatua,* for example, has been found as carbonized grains in excavation of Iron Age, Roman and Anglo-Saxon sites (Godwin, 1956). It probably first appeared in central Europe in the Bronze age, but did not reach Britain until somewhat later. It is frequently found mingled with seeds of other oat species and presumably behaved as a weed at least in some instances, especially in other crops. Another weed of very considerably importance today is *Polygonum aviculare,* which has also frequently been found in Iron Age, Roman and later sites where it undoubtedly was a weed of crops. Much earlier records of this species from glacial and post-glacial deposits, however, show that it occurred in natural plant communities in open habitats during those periods.

One of the most interesting and indeed poignant pieces of evidence comes from the discovery of well-preserved human bodies from the Iron Age in raised bogs in Denmark. From one of these, Tollund Man, the stomach contents reveal that his last meal consisted of a gruel of roughly ground up seeds consisting mainly of barley, but with considerable admixture of linseed (*Linum usitatissimum*), *Polygonum lapathifolium* and *Camelina linicola,* and lesser quantities of *Chenopodium album, Polygonum convolvulus, Spergula arvensis, Brassica campestris, Viola arvensis, Galeopsis* sp. (Godwin, 1956), most of which are still today well-known and important weeds of barley and other crops. There is some evidence which suggests that primitive man, although not deliberately growing weed species as crops, nonetheless occasionally did collect seeds of them for food. For example, it has been concluded that *Chenopodium album* was occasionally used as a grain by prehistoric man in Europe (Helbaek, 1960), like another species, *C. quinoa,* which is still used as a food plant in South America (Willis, 1951). More recent and more reliable evidence from books, floras and other documents confirm this earlier story, that many of the most serious weeds today have been present since very early times.

The decline or disappearance of weeds

Thus far we have considered evidence for stability and persistence in the weed flora; now the evidence for change should be considered, for changes have undoubtedly occurred. Again both archaeological and historical evidence should be examined. One of the most informative early pieces of evidence is from an excavation of a Roman fortress at Caerleon in Monmouthshire (Helbaek, 1964). Here some carbonized grain, probably rye, was discovered. In the debris surrounding it were seeds of various weeds including species common in Britain today such as *Polygonum convolvulus, Rumex crispus,* and *Raphanus raphanistrum.* In addition, there were seeds of *Agrostemma githago* and more surprisingly some currently more unusual species not normally weeds, such as *Lathyrus aphaca* and *Lathyrus nissolia.* Here is possibly some evidence of change.

Agrostemma githago is a good example of a species which is known to have changed greatly in occurrence. It was probably introduced to these islands in rye along with *Raphanus raphanistrum*, another common contaminant of seed of that crop (Helbaek, 1964). It obviously flourished, for by the sixteenth century it was regarded as one of the worst weeds (Salisbury, 1961). This abundance continued until the end of the nineteenth century. In one flora published in 1877 it states that this plant occurred in cornfields everywhere (De Crespigny, 1877). Thereafter its decline began and lasted about half a century. Described as a common plant in 1910 (Long, 1910), and in 1920 as sometimes plentiful in wheat (Brenchley, 1920), it is today rarely found (Perring and Walters, 1962; Chancellor, 1977).

The decline was due mainly to improved seed-cleaning methods combined with the short life span of the seed. Survival of the seed is roughly one year in the soil and only slightly longer in dry storage (Salzmann, 1954). Survival of the plant consequently came to depend heavily on its seed being planted each year with the crop for maintaining its populations, so that when improved methods of cleaning crop seeds were introduced, which removed this contaminant, populations of this weed inevitably declined.

The two vetch species, *Lathyrus aphaca* and *L. nissolia,* found with the rye seed in Monmouthshire, occur locally today in the south of England, but not as arable weeds. *Lathyrus aphaca* does occur, however, as a contaminant of clover and lucerne seed from southern France and Italy (Salisbury, 1961). Those two species may then have originally been weeds in cereals.

Another example of the decline of weeds in annual crops is *Chrysanthemum segetum.* This species was formerly a very serious weed, indeed, Henry II issued an ordinance against the species.

Salisbury, (1961) observes pithily that 'this is perhaps the earliest recorded enactment requiring the destruction of a pernicious weed'. Nowadays, although widely distributed, it is greatly diminished in density, a diminution attributed by Salisbury to improved seed cleaning and a decline in the practice of 'home-saved seed'.

Other important weeds which have declined for various reasons are *Lolium temulentum, Adonis annua, Centaurea cyanus* and *Bromus secalinus.* Many minor weeds must have declined or even disappeared over centuries of agriculture as improvements in crop husbandry made annual crops ever more competitive. Those weeds unable to survive have disappeared altogether while small plants such as *Kickxia spuria* and *K. elatine* must have been greatly diminished, although these last were still present in 7 per cent of tetrads recently assessed (Chancellor, 1977).

Improved seed cleaning has undoubtedly meant considerable changes in weed floras. *Agrostemma githago*, as reported earlier, has virtually disappeared in consequence. In addition, legislation enforcing a declaration of weed seed content of crop seed has completely removed the incidence of *Cuscuta trifolii* from English red clover seed at least (Wellington, 1960) and probably kept the injurious weeds of the 1961 seeds regulations at a low level in other crops (MacKay, 1964).

Liming of soil may have contributed greatly to a reduction of species preferring acid soils such as *Chrysanthemum segetum* and *Spergula arvensis.* Improved drainage in many areas may similarly have reduced species liking damp situations such as *Juncus bufonius* and *Polygonum hydropiper,* which is still sometimes an arable weed.

The advent of selective herbicides has had a more profound effect on the composition of weed populations than any other factor since the very beginning of agriculture. A review of the effects of herbicides on weeds some ten years ago showed on the evidence then available that the main effect of herbicides was to reduce rather than to eliminate weed populations (Fryer and Chancellor, 1970a). It was found that one of the most important weeds, *Sinapis arvensis,* which is very susceptible, although greatly reduced in density, occurred just as widely as before the introduction of herbicides. However, there were signs that certain species could become extinct locally if herbicides to which they were susceptible were used every year. Evidence from Rothamsted showed that *Ranunculus arvensis* and *Vicia sativa* could possibly be locally extirpated (Thurston, 1969). Another piece of evidence from sugar-cane plantations in Trinidad showed that continuous usage of the same herbicide (2,4-D) could induce a deflected plant succession in the weed population (Tate and Lyle, 1967). Thus dicotyledonous species gave way to grasses, which in turn gave way to sedges. Monocropping can therefore in some instances give rise to a herbicide-induced plagiosere.

Very recently there have been several reports from phytosociologists on the continent that a number of weed species have shown considerable diminution and even eradication locally. They are concerned that this is becoming so widespread that it is a serious ecological problem which could lead to a pseudo-desert (Aymonin, 1976). Others take a less extreme view and consider that a majority of the declining species are in fact retreating to areas which constitute their climatic optimum where they should be able to survive in vegetation outside farmland or even within it (Holzner, 1978). It is suggested that only the most specialized weeds are likely to be eradicated completely for their environmental requirements are likely to disappear in the levelling of ecological conditions that is occurring in farmland. Holzner (1978) suggests that the often quoted statement that weed communities are no longer indicators of environmental conditions is not true, for the increasing uniformity of weed communities indicate increasingly uniform farmland conditions.

New weed introductions

Another important type of change in weed flora is of course new introductions, which are due almost invariably to man's activities. The history of British agriculture is full of examples of this. The earliest weed introductions were brought about by the waves of early warlike invaders from the European continent bringing their weed-contaminated seed corn with them. Later there were the more peaceful although equally accidental introductions of recent times, either in exotic crops or new crop varieties or even possibly through the increased transport of goods and materials.

In Neolithic times a number of important weeds such as *Sinapis arvensis, Galium aparine* and *Spergula arvensis* are thought to have been introduced by man. Similarly, those great colonisers, the Romans, introduced various crops and unwittingly their associated weeds such as *Agrostemma githago, Chrysanthemum segetum* and *Fumaria officinalis* (Godwin, 1956). In more recent times a continuing trickle of new introductions has become established, some of which have become problems in annual crops.

Flax is one of the oldest crops. Weeds associated with this crop derived from linseed samples from the La Tène period (the second Iron Age culture of central and western Europe) to the twelfth century have been investigated (Lange, 1978). Evidence was found of a gradual development in the weeds typical of flax crops. Thus a linseed sample from the La Tène period contained *Spergula arvensis* and *Camelina* sp. and these are also found from a twelfth century sample. They are still today considered as typical weeds of flax, However, two species which are frequent in flax today, *Lolium remotum* and *Silene linicola,* have never been found in any of the early samples and so are assumed to have been introduced only recently.

Cardaria draba first occurred in this country in 1809. It is believed to have been introduced by soldiers involved in the invasion of the island of Walcheren off the coast of the Netherlands. Wounded soldiers were apparently brought home on mattresses stuffed with hay which is thought to have contained seeds of the plant. The hay was subsequently ploughed into arable land by a farmer. This very effectively started off a weed problem which is still spreading in this country.

Two other introduced weeds of annual crops are *Veronica persica,* which was recorded first in Berkshire in 1825 having come from western Asia via Europe, and *Matricaria matricarioides*, which first arrived in 1871 from north-east Asia via North America. Both these species have spread very rapidly indeed. The spread of *M. matricarioides* is thought to have been due to transport in mud on motor car tyres in the days before roads had tarmac surfaces (Salisbury, 1961).

This century has seen the introduction of *Avena ludoviciana* (in 1917) possibly through movement of war material as well as crop seed during time of war. Another similar instance is *Amsinckia intermedia,* which was possibly introduced from western North America during the last war, although one report suggests that Canadian linseed imported in 1947 was the source (Chater, 1975). A recent survey of selected arable weeds has indicated that this species is already widespread and well established in annual crops and even on roadsides (Chancellor, 1977).

Increases in individual species

Of recent years the most obvious cause of increase in individual weeds has probably been through resistance to herbicides. The changes thought to have been and to be occurring in the British arable weed flora over four decades was figured in a review in 1970 (Fryer and Chancellor, 1970b). In this it was postulated that much of the increase in annual grass weeds has been undoubtedly due to the effective control of dicotyledonous weeds and that their increase would continue until 1980 at least and probably longer. This, however, has not proved to be the case. The phenomenal increase in *Avena fatua* at least has been checked, by herbicides!

Species such as *Anisantha sterilis* and *Lactuca virosa* are apparently now increasing their range of habitats by invading annual crops. The former is associated with reduced cultivations in agriculture, which allow its seeds to germinate under conditions suitable for establishment of the seedlings. This is a species currently being studied at the Weed Research Organization. The latter is a species which is becoming increasingly aggressive with the increase in disturbed ground suitable for colonization, such as presented by the extensive construction of arterial roads (Salisbury, 1961). It also appears to have been encouraged by the drought in 1976 which reduced competition and helped spread the windborne seeds. We should be careful of this species for it has been introduced into the USA and Australia, in both of which it has become a serious pest.

The future

There are so many factors to be considered when making forecasts of the future that it is easy to be wrong. The most obvious possibilities are, first, that weeds will remain a characteristic of annual crops for the foreseeable future, unless some effective and cheap method of sterilizing soil or stimulating dormant weed seeds to germinate and thus reduce seeds in the soil is discovered. Second, weed populations will comprise species relatively resistant to herbicides and will consequently be present in considerable density. This has been called the 'compensation effect' (Holzner, 1978) in which resulting communities will be poor in species, but high in density. Thirdly, many of these species will be very similar to the annual crops that are grown. Crop mimics have always been very successful. The ultimate of these possibilities is likely to be that the only remaining important weeds in the future will be those which are close relatives of the crop, and crop plants themselves which have persisted from previous years.

References

Aymonin, G. G. (1976). Observations sur le processus de regression des adventices de culture ('mauvaises herbes') et consequences biocoenologiques. *IVe Colloque International sur l'Ecologie et la Biologie des Mauvaises Herbes, Marseille, 1973,* pp. 105-114.

Brenchley, W. E. (1920). *Weeds of Farm Land.* London: Longmans Green.

Brenchley, W. E. & Adam, H. (1915). Recolonisation of cultivated land allowed to revert to natural conditions. *Journal of Ecology* **3**, 193-210.

Chancellor, R. J. (1977). A preliminary survey of arable weeds in Britain. *Weed Research* **17**, 283-287.

Chater, O. A. (1975). Amsinckia in Britain. *Botanical Society of the British Isles News, March 1975,* pp. 12-14.

De Crespigny, E. C. (1877). *A New London Flora.* London.

Fryer, J. D. & Chancellor, R. J. (1970a). Herbicides and our changing weeds. In *The Flora of a Changing Britain,* pp. 105-117. Ed. F. Perring. Hampton: Classey.

Fryer, J. D. & Chancellor, R. T. (1970b). Evidence of changing weed populations in arable land. *Proceedings of the 10th British Weed Control Conference,* pp. 958-964.

Fryer, J. D. & Makepeace, R. J. (1977). *Weed Control Handbook Volume I Principles.* Oxford: Blackwell Scientific Publications.

Garner, H. V. (1965). History of Broadbalk Wilderness. *Report of Rothamsted Experimental Station for 1964,* pp. 218-219.

Godwin, H. (1956). *The History of the British Flora.* Cambridge: University press.

Helbaek, H. (1960). Use of *Chenopodium album* as a food plant in prehistory. *Bericht über das Geobotanische Forschungsinstitute Rübel in Zurich* **31**, 16-19.

Helbaek, H. (1964). The Isca grain, a Roman plant introduction to Britain. *New Phytologist* **63**, 158-164.

Holzner, W. (1978). Weed species and weed communities. *Vegetatio* **38**, 13-20.

Lange, E. (1978). Unkräuter in Leinfunden von der Spätlatenzzeit bis zum 12. Jahrhundert. *Bericht der Deutschen botanischen Gesellschaft, Berlin* **91**, S. 197-204.

Long, H. C. (1910). *Common Weeds of the Farm and Garden.* London: Smith, Elder.

MacKay, D. B. (1964). The incidence and significance of injurious weed seeds in crop seed. *Proceedings of the 7th British Weed Control Conference*, pp. 583-591.

Perring, F. H. & Walters, S. M. (1962). *Atlas of the British Flora.* London: Nelson.

Salisbury, E. (1961). *Weeds and Aliens.* London: Collins.

Salzmann, R. (1954). Untersuchungen über die Lebensdauer von Unkrautsamen im Boden. *Mitteilungen für die schweizerische Landwirtschaft, Zurich.* **2**, 170-176.

Tansley, A. G. (1946). *Introduction to Plant Ecology.* London: Allen & Unwin.

Tate & Lyle Ltd (1967). *Report of Tate & Lyle Research Centre 1967,* pp. 270-271.

Thurston, J. M. (1969). Weed Studies in Broadbalk. *Report of Rothamsted Experimental Station for 1968,* pp. 186-208.

Warington, K. (1958). Changes in the weed flora on Broadbalk permanent wheat field during the period 1930-1955. *Journal of Ecology* **46**, 101-113.

Wellington, P. S. (1960). Assessment and control of the dissemination of weeds by crop seeds. In *Biology of Weeds*, pp. 94-107. Ed. J. L. Harper. Oxford: Blackwell Scientific Publications.

Willis, J. C. (1951). *A Dictionary of the Flowering Plants and Ferns.* Cambridge: University Press.
Witts, K. J. (1965). Flora of Broadbalk Wilderness. *Report of Rothamsted Experimental Station for 1964,* pp. 219-222.

Developments in herbicide application

E. S. E. Southcombe
Fisons Limited – Agrochemical Division, Chesterford Park Research Station, Saffron Walden, Essex.

The beginning

This paper reviews pesticide application, and that of herbicides in particular, over the last 75 years or so, and offers some thoughts for the future. I will concentrate on spraying as it is the most important technique. Going back a little further takes us to the latter part of the nineteenth century, when it all began.

Many of the earliest attempts at pest control were of course non-chemical (Long and Brenchley, 1946). Various washes for insect control had been used throughout the nineteenth century, particularly by gardeners and horticulturists. Weed control was by crop rotation, various cultivation techniques or by weed-seed removal. The first systematic, scientifically based control seems to date from 18 August 1883, when a mixture of copper sulphate and lime was used in the Bordeaux region of France to control downy mildew on vines (Lodeman, 1896). This discovery had been made accidentally in 1882 when vines treated with an unpleasant brew containing copper sulphate as a deterrent against children and casual thieves were noticed to remain clear of the disease.

The Bordeaux mixture, as it became to be known, was applied with whisks or brooms which were dipped in the mixture and flicked over the crop. The considerable beneficial effect of Bordeaux mixture was immediately obvious and prompted a rapid development programme. Of more interest to this paper is that it also prompted the first specific developments in application technology. The broom was modified from a simple bundle of bristles to a device having an oilcloth arrangement for even distribution to the bristles with a feed tube through the handle. This was fitted with a stop-cock and fed from a knapsack device.

The important feature of this story is that from the very start there was an appreciation of correct design and use of the applicator. A strong and positive link was immediately established between formulation, application equipment and biological efficiency.

Early applicator development

The brooms were even by 1883 rather outdated as steam-driven 'garden engines' and hand syringes were well established. Their continued use with Bordeaux mixture was probably due to the thick nature of the brew. As early as 1874 a mobile sprinkler device for applying Paris Green to two potato rows was described. Barrel sprayers drawn by horses and with a wheel-driven pump were available by 1887, though hand lances rather than booms were used. Nozzles are described as early as 1858. By 1886 both flat-fan and swirl or hollow-cone types, often of rather bizarre design, were in use.

The years 1880-1900 were the founding years and France was the most influential country, exporting equipment and knowledge to the USA, UK and many other countries.

Dusting equipment was also under rapid development during this period and in many cases made for easier application.

The transient, contact action of the inorganic salts used necessitated high volumes, between 1000 and 6000 $l\ ha^{-1}$, and as a consequence large, high-capacity equipment.

The contracting era

The main characteristic of crop protection in the early days was that it was clearly a job for contractors (Bates, 1948). The chemicals were far too unpleasant and the equipment far too large and complex for the average farmer to handle. Farmer application was restricted to dust formulations using fertilizer equipment or to fruit spraying.

The introduction of DNOC in the 1930s as the first selective herbicide in cereals, boosted the development in the UK of the boom sprayer, though these had been in use for quite a time.

During the 1939-45 war the contracting companies grew large and highly organized. In a few years the industry had moved from horse-drawn barrel sprayers to 500 gallon tractor-trailed monsters with 40 foot booms and high capacity piston pumps. These were often based on John Bean sprayers, the supply of which from America had ceased at the beginning of the war. Nozzles were either various hollow-cone designs of tee-jet flat-fans from Spraying Systems of America.

Contracting reached a peak in the mid 1950s. The larger companies were active in many fields. Pest Control Ltd., for example, were by then manufacturing most of their own chemicals and their own spraying equipment, selling smaller sprayers to farmers, operating over most of the UK and in many overseas territories on the ground and by air. They had research facilities and could provide a full technical backing to the use of their products, both chemical and equipment. In fact a total pesticide package.

The hormones

The demise of this apparently sound situation started in 1945 with the introduction of the hormone herbicides MCPA, discovered by ICI during the war, and 2,4-D and 2,4,5-T from the USA. These products were so easy to use, so tolerant of misuse and could be applied at such relatively low volumes, that a fundamental change in application technology was inevitable. Spraying herbicides need no longer be a contractor's job; the farmer could do his own application exactly when he wanted to and be free of his reliance on the availability of contractors.

MCPA in fact appeared first as a dust formulation for use in the fertilizer/dusting equipment already on many farms. The aqueous salt solutions, however, proved much easier to apply and gave better control. The first spray applications were at the traditional high volumes of 1000 l ha^{-1}. Soon, however, volume rates had been dramatically reduced to 50 to 100 l ha^{-1}.

Clearly, the massive equipment then being built was totally out of scale to the new products and to farm use, and it was not very long before the small, low-volume, tractor-mounted sprayer appeared. The 'model T' of these sprayers was the Pest Control 'Weedmaster' and its derivatives which first appeared in 1948. It had a 45 gallon tank, 19 ft 6 in boom, gear pump and fan nozzles, and it cost just £49.10.0d.

The numbers of sprayers on farms increased dramatically; from 4200 in 1950 to 49 100 in 1959 (BCPC, 1968). Farmers reaped the benefits of the experiences gained by the contractors and soon adapted to the new task. Besides, the tolerant products made that task so easy. And so by 1950 the sprayer had moved firmly into the hands of the farmer.

The only main feature of the next two decades in the UK was the subtle change in style of application caused by the introduction of less systemic chemical products. The effect was a gradual increase in volume rate from 100 to 300 l ha^{-1} or even higher. The chemical companies felt more secure recommending these higher rates, which allowed more tolerance, to the frequent misuse already becoming apparent on many farms. Sprayers gradually increased again in size in response not only to increase in volume rates but also to improved work rates by using larger tanks and booms. The general layout and construction of the equipment followed the pattern established during the 1940s. Sprayers were generally simple, cheap and not very imaginative in design (Rose, 1963).

Crop protection in UK agriculture was influenced mainly by herbicide application, whilst in Europe the need to apply fungicides and insecticides to vines and fruit maintained an interest in high-volume, high-pressure sprayers. In the USA, crop protection had largely taken to the air and except for certain applications in a few crops where special equipment was designed, general farm sprayers did not advance much technically.

One development in the UK during this period was that of band spraying, particularly in the sugar-beet crop. Although early sprayers suffered from the general malaise of the time of low grade design and construction, the art of accurately applying a restricted band of selective, expensive product over row crops at an early growth stage in conjunction with inter-row hoeing became well developed.

The modern sprayer

During the early 1970s significant developments were made in European pesticide application by means of special metering and control devices. Conventional sprayers rely on constant pressure and forward speed to maintain a constant application rate. Whilst pressure is easily controlled accurately, speed is not. The new sprayer designs overcome this by adjusting nozzle output to speed changes. These systems have become known in France as DPA (Débit Proportional à l'Avancement). In the UK the best accepted term is AVR (Automatic Volume Regulation).

All AVR systems currently in use achieve constant application rate by changing pressure at the nozzle in relationship to the square of a change in forward speed. The pressure change may be caused either by the operation of a valve or by increasing pump output. It was this feature which caused many of the earlier doubts about AVR systems. Clearly, large pressure changes could significantly alter spray characteristics – droplet size in particular. A study of volume median diameter (VMD) change with pressure, however, showed that the often arbitrary selection between hollow-cone or flat-fan nozzles can have a bigger effect than quite large pressure changes. As nozzle selection has never been regarded as a major problem area in agricultural spraying the use of AVR systems with suitable restraint in forward speed changes (± 25 per cent is advised) is therefore unlikely seriously to affect spray

Table 1. Drop-size change with pressure change, relevant to automatic volume regulation systems

Pressure (bar)		Pressure (% change)	Equiv. speed ($km\ h^{-1}$)		Equiv. speed (% change)	Flat fan – 8006 (VMD)		Flat fan – 8006 (% change)	Hollow cone – D10·25 (VMD)		Hollow cone – D10·25 (% change)
2·08	↑	−49·0	5	↑	−28·6	540	↑	+7·7	460	↑	+13·6
4·08			7			505			405		
6·75	↓	+65·4	9	↓	+28·6	468	↓	−7·3	344	↓	−15·1

at 7 $km\ h^{-1}$: Flat fan (468) → Hollow cone (344): −19·8

Drop size data adapted from figures supplied by Spraying Systems Inc., USA quality, and is to be recommended (Table 1).

Many different AVR systems have been developed, mainly in France or the USA (Matthews, 1979). Some use electronic computers into which it is possible to feed, and to be given in return, all relevant data affecting sprayer performance. Application rate can now be automatically controlled with great precision, and often great cost.

The 'ideal' system whereby neat chemical is metered in proportion to forward speed and injected into a constant-pressure water supply at the nozzles has been used for many years by Fisons Ltd in their spray trains, but has so far proved extremely difficult to adapt to the much smaller agricultural sprayer. When such a system is successfully developed it will be most welcome, and will rapidly lead to the adoption of the plug-in chemical container concept.

Conventional sprayers have benefitted in recent years from electronic instrumentation which monitors speed and output, thus indicating the degree of error to the operator who can then take corrective action.

Many other design improvements have taken place particularly in the use of plastic and rubber materials to construct tanks, pumps and pipework, sources of rust and contamination in the past. However, the susceptibility of some plastics and most rubbers to chemicals containing organic solvents is a cause for concern and care. The importance of boom stability has been well demonstrated by Nation (1978a) and rapid improvements in this area are being made. The size of sprayers has increased dramatically, due partly to the doubtful idea that bigger sprayers must give better work output, and partly to the increased capacity of the much larger tractors.

But underneath the gloss we find that the modern sprayer is still the same machine it was in 1939. Forty years have only modified the details, not the fundamental design, nor the fundamental manner in which herbicides are applied.

Atomizers

If spraying equipment has not changed much in the last 40 years, nozzles have not changed at all. Nearly all agricultural spraying is still done with hydraulic flat-fan or hollow-cone nozzles of identical design to those developed by the Spraying Systems Company and others before the war. The atomizer is the only really critical component on a sprayer, yet it is the least understood by the user.

It is interesting that although a very complete knowledge of the spray characteristics from hydraulic nozzles has been built up over many years and many inadequencies shown, their dominance in the farm sprayer has remained unchallenged. This is because of their simplicity.

The hydraulic nozzle is really a unique component. Small, with a machined hole that meters the output, it determines spray angle, drop-size spectrum and produces the required distribution on the ground; it is also cheap and simply fitted to the sprayer.

The potential of hydraulic nozzles to produce very fine drops, with consequent risk of drift, particularly when used for low-volume application, has for long been of concern to research workers and farmers. The recent development of laser-operated equipment has enabled a rapid analysis of drop spectra to be made, confirming this problem. As much as 69 per cent of the number of drops in a spray of VMD 192 μm can have a diameter of less than 20 μm (Bals, 1978). The volume they contain is negligible, however. As these drops contribute little to the action of the chemical, research on their prevention or removal would seem to be very desirable; a plea first made by Frazer (1958).

Controlled droplet application and low volume

The first major challenge to the hydraulic nozzle has been the concept of Controlled Droplet Application (CDA).

The term CDA was first coined by J. D. Fryer, but the father of the concept and of most of the rotary atomizer equipment used is E. J. Bals, (1969) who defined the concept in the following manner: 'The efficiency of a spraying machine is inversely proportional to the range of droplet size it emits whilst the suitability for a specific problem depends on the actual size of droplets produced'.

Rotary atomizers can produce the narrow range of drop sizes required for CDA. The Micron 'Herbi' disc working in direct-drop mode can produce a mono-size spray with the drop size proportional to disc speed – typically 250 μm at around 2000 r.p.m. The more recent Micron 'Micromax' disc has been designed to operate in the ligament mode of drop production and by matching chemical feed rate to disc speed, closely controlled sprays of between 100 and 300 μm VMD can be obtained.

CDA as applied to Ultra-Low Volume (ULV) insecticide spraying has produced some dramatic improvements in overall efficiency.

ULV application has been recently proposed for use with certain insecticides and fungicides in the UK, using a rotary atomizer producing a drift spray of 70 μm VMD. However, the proposal to convert the whole chemical dose into driftable components has caused much concern and no product has yet received the necessary safety clearance.

An extension of this work of more interest to herbicide application is to charge the spray electrostatically in order to encourage fine sprays to impact efficiently onto crops. Work has been done for some years at Rothamsted and by FMC in the USA, but publication of results is still awaited. If successful, this concept will be most influential in crop protection in the future.

CDA can of course be applied to any volume rate or any drop size, providing these are scientifically and logically chosen for optimum performance. The pioneering work by Cussans and Taylor (1976) at the Weed Research Organization, and by Bailey, Kitchen, Smartt and Roebuck (1978) and many others has all been done at very low-volume rates of 5-50 l ha^{-1} using equipment supplied by Micron Sprayers and by Horstine Farmery.

The results of CDA herbicide applications seem so far to be of the same order as those given by conventional sprayers. The cost and effort needed to use successfully the more complex rotary atomizer equipment means for the farmer an outlay with little obvious return, except the valuable acquisition of a means of spraying at low volume with less risk of drift.

It is this feature of CDA that is now seen as the most important and the wheel of history has turned nearly full circle, for there is much current interest in means of using conventional sprayers at around 100 l ha^{-1}, not necessarily with true CDA sprays, but nevertheless restricted drop spectra, 150-350 μm being the most useful range.

So for the time being the conventional sprayer has survived, though now the quality of the spray it produces is recognized as being of vital importance. CDA has profoundly influenced agricultural spraying and it is to be hoped that real improvements in spray quality and the selection of atomizer types relevant to the job being done will be seen in the near future.

Other application considerations

Correct timing of spray applications has always been of great importance, particularly today. There are a number of developments aimed at improving the farmer's ability to organize efficiently and accurately carry out his operations.

Nation (1978b) has been using computer studies to identify the most influential parameters on overall work rate. Interestingly, he has shown the much greater effect of lower volume rates than, say, ever wider booms.

Adams (1978) has studied meteorology aspects and has shown the potential benefits of low-drift sprays for increasing the number of working days available.

The use of low ground-pressure vehicles, often also capable of high speeds, can further increase the time available for spraying, particularly on ground which is near saturation point.

There is a need for more accurate matching of swaths now that chemicals are less tolerant to over- and under-dosing. Whilst the ideal solution still eludes us, the use of tramlines provides an elegant solution to those prepared to adapt to the technique.

Formulation development will continue to improve the chemical products. The change from wettable powders to suspension concentrates has improved operator safety and efficiency. There are problems, however, if new techniques demand special formulations to be made alongside conventional ones. This consideration may limit some low-volume applications.

The future

It is disappointing that a history of spraying has produced so few events of significance – the low-volume sprayer, band spraying, AVR and other control systems, rotary atomizers and CDA, low-volume application (again) – and no real change in standard spraying equipment.

However, a much greater awareness is now apparent of the fringe parameters such as organization of field operations, timeliness, accuracy, drift and the potential benefits, both biological and logistic, which low-volume application can offer.

The modern sprayer is used throughout the year and a small fortune in pesticides passes through it. Yet the cost of the equipment is kept low by customer demand and the standard of use and maintenance is generally very poor. It has been said that combines can be seen to be making money, sprayers only to be spending it. Only when the farmer puts the sprayer high on his purchasing list, and is prepared to demand and pay for high-quality technology and equipment, will real advances take place.

The 90 000 sprayers in the UK are all essentially the same. Any fundamental change in design or use must show real benefits to be successfully adopted.

Farmers cannot be expected to have the same understanding of spray techniques as research workers. It is also difficult for chemical suppliers to make the enormous effort required to develop recommendations for new volume rates or techniques.

Not all chemical products will tolerate very low volumes. This means that the sprayer must either be flexible enough to cover a range from 50 to 500 l ha^{-1} or that the farmer will require different machines for different crop-protection operations and the knowledge of how to use them. This may prove to be an unpopular idea.

There have been more useful developments which could influence UK herbicide application in the last decade than in the previous three. It is hoped that the farming industry and those associated with it will allow new techniques time for proper development and take advantage of their benefits.

References

Adams, R. J. (1978). Meteorological aspects of changes in spraying techniques. *Proceedings of the 1978 British Crop Protection Conference – Weeds* **2**, 625-632.

Bailey, R. J., Kitchen, R., Smartt, A. & Roebuck, J. (1978). A review of ADAS results in 1978 with CDA of herbicides in cereals. *Proceedings of the 1978 British Crop Protection Conference – Weeds* **2**, 695-699.

Bals, E. J. (1969). The principles of and new developments of ultra-low volume spraying. *Proceedings of the Fifth British Insecticide and Fungicide Conference* **1**, 189-193.

Bals, E. J. (1978). The reasons for CDA (controlled droplet application). *Proceedings 1978 British Crop Protection Conference – Weeds* **2**, 659-666.

Bates, G. H. (1948). *Weed Control.* London: Farmer and Stockbreeder.

British Crop Protection Council (1968). *Weed Control Handbook.* Oxford: Blackwell Scientific Publications.

Cussans, G. W. & Taylor, W. A. (1976). A review of research on CDA at the Weed Research Organisation. *Proceedings of the 1978 British Crop Protection Conference – Weeds* **3**, 885-894.

Frazer, R. P. (1958). The fluid kinetics of application of pesticidal chemicals. In *Advanced Pest Control Research* II. pp. 1-106. Ed. R. L. Metcalf. New York: Interscience.

Lodeman, E. G. (1896). *The Spraying of Plants.* New York: Macmillan.

Long, H. C. & Brenchley, W. E. (1946). *Suppression of Weeds.* London: Crosby Lockwood.

Matthews, G. A. (1979). *Pesticide Application Methods.* London: Longmans.

Nation, H. J. (1978a). Developments in sprayer boom design. *Proceedings of the 1978 British Crop Protection Conference – Weeds* **2**, 649-653.

Nation, H. J. (1978b). Logistics of spraying with reduced volumes of spray and higher vehicle speeds. *Proceedings of the 1978 British Crop Protection Conference – Weeds* **2**, 641-648.

Rose, G. J. (1963). *Crop Protection*. London: Leonard Hill.

Herbicides: past, present and future

R. J. Makepeace
Agricultural Chemicals Approval Scheme
A.R.C. Weed Research Organization, Oxford

Introduction

The first chemicals to be used for the control of weeds were inorganic salts such as copper sulphate, ferrous sulphate, sodium chlorate and sulphuric acid. Although both the technique and the chemicals were somewhat crude, they demonstrated the potential of herbicides at a time of changing husbandry techniques.

The origin of modern herbicides, as with all other pesticides, was directly due to the emergence of the fine chemical industry in the early part of this century. Large manufacturing chemists appeared whose trade was based on pharmaceutical chemicals, dyestuffs, paints and other synthetic organic chemicals. Many of our day-to-day drugs such as aspirin were pioneered in these years.

These companies relied heavily on research to synthesize new chemicals and it was natural for them to turn to crop protection, this being a potentially large market. This early research discovered some of the most important pesticides we shall ever know. In 1941 and 1942 DDT was announced in Switzerland, BHC (now HCH) in England and 2,4-D and MCPA in America and England.

The early herbicides

The inorganic herbicides were usually applied by contractors as there were no farm sprayers available in the form known today. The practical problem of application was a delaying factor in the acceptance of the new herbicides. MCPA was initially available as a powder applied by fertilizer applicator. One outcome of the war years was the production of the modern farm sprayer. Most manufacturers and farmers do not appreciate the debt that they owe to these engineers, such as Frank Woodham, who worked in this field for the Ministry of Defence.

As the field sprayer was refined, salt and ester formulations of herbicides were produced and spraying as we know it arrived. It was at once clear that the technique would succeed. The target was easily identified, the benefits were obvious, the selectivity to the crop was good, the application comparatively easy and the cost acceptable. The first herbicides to be used through these early machines were: DNOC (patent granted in the UK in 1932); 2,4-D (announced 1942); MCPA (introduced in the UK in 1945 by ICI); 2,4,5-T (announced in 1944 by Amchem); Dinoseb (announced in 1954 by Dow); Propham (announced in 1945 and introduced in 1948).

The search for new herbicides

The impact of herbicides in cereals led to the search for other compounds. The techniques of chemical synthesis and screening for activity were well established in the chemical industry and rapid strides were made in producing closely related compounds and testing their potential once the originating compound had been discovered. This occurred first with the phenoxy acid group and led to several discoveries (Table 1). Apart from 2,4,5-T these herbicides found their use in selective weed control in cereals, and with the discovery of the benzoic acids found continued use with 2,3,6-TBA (1952) and dicamba (1965).

Table 1. The dates of discovery of the phenoxy-acid herbicides

Derivative	Phenoxyacetic acid	Phenoxyproprionic acid	Phenoxybutyric acid
4-chloro-2-methyl	MCPA 1945	mecoprop 1956	MCPB 1954
2,4-dichloro	2,4-D 1942	dichlorprop 1961	2,4-DB 1957
2,4,5-trichloro	2,4,5-T 1945	fenoprop 1953	2,4,5-TB 195?

Propham was the first carbamate herbicide to be introduced and also the first selective herbicide for grass weed control in broad-leaved crops. It was followed by many other widely used carbamates and thiocarbamates: chlorpropham, 1951; barban, 1958; di-allate, 1960; tri-allate, 1961; carbetamide, 1966; phenmedipham, 1967; asulam, 1968.

Many new groups followed in the 1950s and 1960s. The largest and most important were the substituted ureas (Table 2) and the triazines (Table 3). The first substituted urea herbicide to be described was monuron (1951) which was used as a total herbicide or sterilant as it was described at the time. The rates of use were very high, in the order of 20 kg ha^{-1}, and it was some time before the selective value of reduced doses was discovered in perennial crops such as asparagus. Both monuron and the subsequent members of the group were particularly useful for the control of grass weeds in both broad-leaved and cereal crops.

Table 2. The dates of discovery of the major substituted urea herbicides

monuron	1951	diuron	1951	neburon	1951
fenuron	1951	linuron	1960	chlorbromuron	1961
monolinuron	1958	metobromuron	1963	methabenzthiazuron	1967
chloroxuron	1960	metoxuron	1968	isoproturon	1971
chlortoluron	1969				

The triazines were an outstanding group of herbicides. They illustrate clearly the principle of producing chemical analogies of an active compound and have yielded an amazingly diverse group of herbicides from soil-acting residual total herbicides to contact foliar acting herbicides. Virtually all the compounds in the group were discovered and patented by Geigy Agricultural Chemicals. The development of triazine compounds is still not exhausted and we have seen the introduction in the last ten years of several important triazines, namely hexazinone and the substituted triazines metribuzine and metamitron.

Table 3. The dates of discovery of the major triazine herbicides

Chlorotriazines		Methylthiotriazines		Methoxytriazines	
simazine	1956	ametryne	1960	atratib	1960
atrazine	1958	simetryne	1959	prometon	1960
propazine	1957	prometryne	1962	simeton	196?
trietazine	1958	desmetryne	1964		
terbuthylazine	1966	methoprotryne	1965		
cyanazine	1966	terbutryne	1966		
		aziprotryne	1967		

In addition to these groups many other groups have been successfully introduced such as the amides and bi-pyridyls and are still in use today. Over the past thirty years there must have been 10 000 herbicides screened. The majority never proceeded beyond the laboratory, being eradicated for reasons of cost, manufacture, toxicology or marketability. Nonetheless over this period something in the order of 1000 compounds must have reached the trials stage and at present it is likely that two hundred herbicides are in commercial manufacture world wide. What conclusions can be drawn from the past that can be used to guide us in the future use of herbicides? The first is the differing acceptance of chemicals by growers for use on different crops. The acceptance of cereal herbicides was rapid and by the late 1950s some 65 per cent of the crop was sprayed annually. This is reflected in the List of Approved Products for 1956 issued by the Crop Protection Products Approval Scheme (Table 4).

Table 4. The list of approved products 1956

Herbicide	Number of products	Name
sodium arsenite	1	Weedicide
sodium chlorate	9	
2,4-D	16	
dinoseb	9	
DNOC	8	
MCPA	38	
MCPB	1	Tropotox
2,4,5-T	1	Phortox
Oil herbicides	1	Shell W

It was not until the late 1960s that the use of herbicides became widespread in the majority of crops grown. This is illustrated by comparing the numbers of herbicides and growth regulators approved over the last 23 years (Table 5). In this it can be seen that in the period from 1960 to 1970 the number of herbicides increased fourfold and the number of products threefold. During this period the number of herbicides that have been lost is reflected in the levelling out of the number of herbicides from 1974 onwards in Table 5. These are listed in Table 6. They all can be said to have been minor herbicides in one way

Table 5. The number of herbicides and products approved 1956-1979

Year	Herbicides	Products
1956	8	81
1958	11	104
1960	14	127
1962	23	176
1964	38	241
1966	43	308
1968	46	355
1970	60	407
1972	70	434
1974	77	461
1976	81	453
1978	84	475
1979	85	468

Table 6. Some herbicides no longer approved or available in 1979

ametryne	pentachlorophenol
benzthiazuron	medinoterb acetate
buturon	methoprotryne
cycluron	metobromuron
cyprazine	morfamquat
dimexan	monuron
endothal	noruron
propazine	trimeturon

or another. None of them maintained a dominant position in any crop for long, although both pentachlorophenol and endothal played important roles in the development of selective herbicides for spraying against annual weeds in broad-leaved crops. Their use established the technique that facilitated the introduction of the better herbicides that followed them. Interesting though these losses may appear, the most important lesson to be learned from this backward glance is the large number of herbicides that have survived the last three decades, that are still performing a useful role in weed control, have not been superseded and are still commercially viable. Of the earliest mentioned here, all are still used, although there is now only one DNOC formulation available for purely weed control purposes in the UK. This is further illustrated in Table 7, in which the number of individual herbicide formulations are compared over a period of years.

Table 7. The number of approved products for several herbicides: 1956-1979.

Year	2,4-D	MCPA	dinoseb	mecoprop and mixtures	dichlorprop and mixtures	dalapon
1956	16	38	9	–	–	–
1958	18	41	9	–	–	–
1962	22	51	13	28	2	4
1968	30	42	35	47	36	14
1973	30	33	42	43	43	18
1979	24	29	31	41	40	8

Present herbicides

For the purpose of this paper the present can be defined as the period from 1975 to 1979. This is represented in Table 5 by the levelling out of the number of herbicides and products approved. It is in fact a reflection of a stable situation in which the herbicides lost are equal to new herbicides gained. This is a period in which herbicides form an accepted part of crop husbandry and few weed situations exist where there is no herbicide available, even though it may have serious shortcomings. In this period commercial considerations have become much more important, with many older products taking on the role of commodity materials. The standards of efficacy in this situation are those set by the registration bodies, the main area of competition being that of price.

Development costs

The present period has been one of considerably increasing costs from synthesis to marketing. The greatest increases have been in the cost of toxicological studies for registration purposes and the high capital costs of setting up new plant. The British Agrochemical Association quote the cost of currently taking a pesticide from synthesis through to marketing to be in the order of £10 000 000. The result has been that no product is taken beyond the early screening stages unless it has a potential in one of the major world crops. In the UK herbicide terms this would exclude all fruit and vegetables. The result has been a stream-lining of the synthesizing companies, an increase in the number of chemicals screened and rejected with far fewer being taken through to registration and an increase in registration personnel to deal with differing national clearance and approval requirements. It was forecast that this would result in a steady diminution of new herbicides coming forward for development. Table 8 shows that if this is likely it has not begun to happen.

New uses

Having launched a herbicide onto a major crop considerable potential can then be exploited in minor crops. This has certainly been one feature of development in the last decade. Where a company has only a restricted number of herbicides on which to concentrate, this can be a profitable exercise, greatly increasing the turn-over for that herbicide. This has been particularly important in the truly minor crops where growers have not had new compounds developed for them for many years (Table 9).

The problem of minor uses has been with the horticultural industry for many years. To try to alleviate this situation the Agricultural Chemicals Approval Scheme in conjuction with the Agricultural Development and Advisory Service, Agricultural Research Council and

Table 8. New herbicides approved since 1973

Year	Herbicide	Crop
1973	metribuzin	potatoes
	methazole	onions
1974	cycloate	sugar beet
1975	carbetamide	brassica and clover seed crops
	ethofumesate	sugar beet
	isoproturon	cereals
	propyzamide	fruit, ornamentals, beans and rape
1976	bromofenoxim	cereals
	glyphosate	cereals, fruit, non-crop situations
1977	difenzoquat	cereals
	flamprop-isopropyl	cereals
	flamprop-methyl	cereals
1978	metamitron	sugar beet
	dinitramine	swedes and turnips
	chlorthal-dimethyl	brassicas
1979	3,6-dichloropicolinic acid	cereals
	isocarbamid	sugar beet
	pendimethalin	cereals
	diphenamid	strawberries, forest nurseries
1980 (herbicides that may be approved)		
	hexazinone	forestry
	dichlofop-methyl	cereals, arable broad-leaved crops
	ethephon	cereals
	alloxydim sodium	sugar beet, peas
	allachlor	oil-seed rape
	devrinol	swedes

the British Agrochemical Association have set up a minor use category on approved labels. This enables the manufacturer to obtain clearance, the biological data being provided by acceptable sources at a slightly lower order to that required for full approval. The granting of a use as 'Provisionally Accepted under the Agricultural Chemicals Approval Scheme' is for an initial five-year period. This scheme has certainly assisted the grower and has led in some circumstances to an examination of a minor use revealing sufficient data for full, instead of provisional approval to be granted.

Mixtures and sequences

With the continued mechanization of agricultural husbandry the levels of weed control required has increased in almost direct proportion to

Table 9. The acquisition of new recommendations of three modern herbicides

	Tri-allate		Trifluralin		Propyzamide
1967	wheat	1969	Brussels sprouts	1975	apple
	barley		cabbage		pear
	peas		broccoli		black currant
	beans		cauliflower		red currant
					gooseberry
1970	carrot	1975	calabrese		strawberry
			kale		lettuce
1979	beetroot		swede		oil-seed rape
	Brussels sprouts		turnip		field bean
	cabbage		broad bean		
	broccoli		carrot	1976	blackberry
	cauliflower		parsnip		loganberry
	swede		raspberry		plum
	turnip		strawberry		raspberry
	kale		sugar beet		rose
	mustard				ornamentals
	linseed	1976	dwarf bean		chicory
	onion		French bean		
	parsnip		runner bean	1979	forestry
	sugar beet		navy bean		spring cabbage
	fodder beet		oil-seed rape		sugar-beet seed
	mangold	1977	lettuce		clover seed
	rye		mustard		lucerne
			parsley		brassica
		1979	ornamentals		seedcrops

the decline in the ability of the grower to carry out hand or mechanical weed-control measures. The very success of herbicides has allowed a change over to husbandry methods such as closer drilling and reduced cultivations in cereals and drilling to a final plant stand in sugar beet and vegetables. These systems require complete control which no single herbicide can generally offer. Faced with this situation in cereals, the grower learned to mix herbicides to give differing 'cocktails' for different weed problems. This was overcome, in the main, by the manufacturers introducing wide-spectrum herbicide mixtures that gave good results on almost all weeds likely to be met with by the grower. This situation could only work where the different components were available to a manufacturer. In most crops this was not the case and with patented herbicides tank mixtures of sequences could not be recommended as the separate parties involved could not accept liability for the other's recommendation. The result was that advisers and independent bodies such as the British Sugar Corporation were forced to look into the use of herbicide programmes independently of the manufacturer. The situation was further complicated by the implications of clearance of mixtures and sequences by the Pesticides Safety Precautions Scheme. Faced with this situation the users were able to bring pressure to bear on the manufacturers through such bodies as the Annual Review of Herbicide Use with the result that the manufacturers consulted with each other, were able to agree on joint recommendations, obtain clearance and ultimately make available many programmes and mixtures to the grower. In this exercise the manufacturers have spent considerable amounts of time and money in establishing that in any mixture or sequence there is no incompatibility, either physical or biological.

The lapse of patents

One feature of the last few years has been the ending of a number of patents. This is of such an order that both in the UK and overseas many new manufacturers have started up in the last decade and their products are being marketed in competition with the original products. In the case of chloridazon (pyrazone) the number of products has jumped from one to six and in the case of chlormequat from two versions of the same formulation to some twenty products. The result is a mixed one. Further development has ceased on new uses for unpatented herbicides, and although prices have been reduced to the grower, there is no longer any advice or supervision of the products on the farm.

Herbicides in the future

Turning to the future in weed control one cannot sensibly predict more than one or two years ahead and even so such speculation would not take into account many variables over which we have no control, such as the effect of fuel prices and manufacturing costs. This must be a very personal interpretation and should be taken as being a 'crystal ball' exercise.

What then are the general features of weed control? I think it is clear to say that there will be a continuation of new compounds appearing over the years. We have seen already that they have continued to appear in spite of escalating costs and there seems no reason why they should not do so, especially as this form of income is the life-blood of the main manufacturing companies. We should also continue to expand uses of current herbicides. This may be the biggest area for expansion in weed control over the next decade. To some extent we have been rather crude in our handling of the compounds we have to date in that we have too often looked at narrow uses for compounds and not explored their other uses either alone or in mixture. With modern data-recording procedures it should be possible to look at far more associations of active factors in herbicides in order to produce products for the grower in the future.

The area of new weeds is obviously one which one constantly keeps in mind. New weeds will continue to appear, not at a very fast rate but fast enough to keep our research workers busy, and this of course will be suited to both new compounds and new mixtures.

Individual crops

Cereals

In cereals one can contemplate a continuing change to autumn-sown crops. This will be accompanied by a progressive change to pre-emergence and autumn-applied herbicides following the trend which is already well under way. It can be seen that within 2-3 years there will be a major impact from this new technique of weed control provided we have the machinery, such as the low-ground-pressure vehicle to apply herbicides on wet land. The benfits are obvious, prophylactic weed control preventing the yield loss, far better timeliness of application and considerably reduced problems of drift on suceptible crops.

In the area of grass weeds this will be one situation where new compounds will continue to appear, but we can look forward to more expansion of the programmes and mixtures that we have seen appearing over the last two years. One area of considerable change will be the control of perennial grass weeds, especially couch grass, by the use of glyphosate applied at grain moisture contents of 20-30 per cent. This has shown tremendous promise so far this year, and from previous data it looks extremely reliable. Its greatest impact will come in the areas of northern England and in Scotland where the couch problem is acute. This is confounded by the fact that farmers in these areas experience a late winter and cannot spray before they drill, and they have no autumn in which to get re-growth before respraying. To be able to attack the weed within the crop phase is of considerable importance and will have a great impact on cereal growing over the next 2-5 years.

Arable crops

We have some very promising grass weed herbicides of which the most important is alloxydin-sodium. There are others similar to this and we can see new compounds moving into this area over the next 1-3 years. Control of broad-leaved weeds and perennial weeds in most of these crops is similar to vegetables and looks forward to safe post-emergence herbicides.

Vegetable crops (field and arable scale crops)

In this area the greatest impact we are likely to see over the next 2-6 years will be that of soil sterilization. At the moment we are using conventional soil sterilants (e.g. metham sodium) to achieve weed control. We have an application machine working quite successfully in the fens and silt lands, namely the Rumstedt machine, and from this we can see the large potential of soil sterilization for weed control as costs are tailored and mechanical problems of application ironed out.

The other area of herbicide development which is open to new compounds is that of post-emergence herbicides in vegetable crops. There may be some potential in re-formulation of existing herbicides, but until the industry produces new compounds there looks to be little in view at the moment and one can see no change for the next 3-4 years.

Volunteer crops as weeds

Our immediate answer must be the selective application of contact herbicides and there are many innovations in this line at present. We will also see the introduction of mechanical means of removing these volunteer crop weeds; these may be by physically pulling or by such methods as the electric-discharge machine that is being developed by the British Sugar Corporation. The use of selective herbicide compounds is an area for research into existing materials and possibly new compounds.

Grassland

One can see little change here. With current information it must now be possible to lay down a programme of prophylactic herbicide treatments whereby a short-term ley can be kept clean of grass and broad-leaved weeds over a period of up to 5 years. This will need co-ordinated management from the grower but could revolutionize this area of grassland production over the next 2-5 years. Broad-leaved perennial weeds will continue to be a problem and in this area over the next 1-3 years I think we can see considerable impact of selective applicators such as rope-wicks, horizontal sprayers and roller-sprayers.

Future of the industry

One can foresee over the next decade changes in the structure of the herbicide industry. The effects of registration may restrict development, if for no other reason than cost. One can also see changes occurring from the marketing point of view that will affect the grower. Manufacturers will clearly become more specialized and one can see a divergence between those concentrating on synthesis and development of new compounds and those concerned solely with manufacture. The latter will not be involved in research, barely in development, and may go no further than registration. This will be accompanied by a change in marketing and distribution and one can clearly see that marketing and servicing of agricultural chemical products will move further and further into the hands of the members of the British Agricultural Supply Industry Scheme. They will be the ones with the crop specialization needed to answer the growers' problems. There will be less development by primary manufacturers than we have seen in the past. This will involve a greater dependence on research from the Agricultural Research Council, the Agricultural Development and Advisory Service and independent bodies such as the British Sugar Corporation and the Pea Growing Research Organization. Associated with this will be the self-help aspect seen already in the emergence of crop-protection consultants who contract to take the problems out of the farmers' hands. We may see all these changes occur with time, but what is clearly of importance is that herbicides will continue to be part of crop growing as we know it today and we can continue to look forward to more interesting developments in this field in the next decade.

Section 5 **Post-harvest physiological changes**

Recent advances in the control of physiological changes

E. C. Lougheed and D. P. Murr
Department of Horticultural Science, University of Guelph, Guelph, Ontario, Canada, N1G 2W1

Introduction

Over the past decade several reviews (Halevy and Mayak, 1979; Isenberg, 1979; Lougheed, Murr and Bérard, 1978; Smock, 1979), books (Abeles, 1973; Fidler, Wilkinson, Edney and Sharples, 1973; Hulme, 1970; Hulme, 1971; Ryall and Lipton, 1972; Ryall and Pentzer, 1974), and proceedings of conferences (Dewey, 1977; Dewey, Herner and Dilley, 1969; Isenberg, 1977) cover much of the literature up to 1977-79. This review attempts to cite literature representative of that indicating potential for the control of physiological change and in some cases refers to older and as yet unused theories, results, or practices.

Effect of the gaseous environment

Relative humidity (r.h.)

A recent symposium of the American Society for Horticultural Science (Lipton, 1978) reviewed information on r.h. Consideration of r.h. usually occurs only in relation to reducing desiccation by the use of r.h., and with some crops, such as dried onions, to reducing fungal attack by the use of low r.h. However, there are also physiological effects of humidity.

There may be a higher incidence of breakdown in some apple cultivars at high relative humidities (Porritt and Meheriuk, 1973) which may be related to water loss from the fruit at low humidities acting as a 'carrier' for potentially harmful 'volatiles' (Wills, 1968). Forsyth and Lightfoot (1977) indicated that with McIntosh apples a high humidity acted antagonistically to the beneficial effect of low concentrations of ethylene during storage. Schippers (1975) reported that after storage at 100 per cent r.h. potato chip colour was significantly darker than after storage at 70-80 per cent. It is therefore likely that the physiological effects of r.h. are considerably broader, and its relationship with other factors more complex, than usually considered.

Controlled atmosphere (CA) storage

Much of the recent work on CA storage is summarized in the proceedings of a conference in 1977 (Dewey, 1977) and two recent reviews (Isenberg, 1979; Smock, 1979). CA storage should be potentially useful for those crops which undergo chilling injury, replacing the lower but unusable temperature with atmospheric control (Robinson, Browne and Burton, 1975).

Oxygen

Research (North, Bubb, Cockburn and Gull, 1978) and semi-commercial studies (Sharples, *et al.*, 1978) have established that Cox's Orange Pippin apples may benefit, or at least not be harmed as might be expected, from storage at atmospheres of 1-1·25 per cent O_2. However, the general improvement in quality of fruit held at 1·25 per cent O_2 compared to 2·0 per cent O_2 was marred somewhat by lack of flavour and the occurrence of a 'necrotic disorder'. With 'McIntosh' apples Forsyth (personal communication) has found beneficial results in the use of 1·5 per cent O_2.

As far as we know, O_2 levels as low as 1 per cent have not been used commercially on other commodities, although such atmospheres have been used in research (Isenberg, 1979) and very low partial pressures have been used in low-pressure-storage research (Bubb, 1978).

Carbon dioxide

Treatment with CO_2 before storage is not a new technique (Brooks, Bratly and McColloch, 1936), but there has been considerable recent work with apples on the use of very high levels of CO_2 prior to CA storage. With McIntosh the effect on firmness has been small and the potential for injury very great (Bramlage *et al.*, 1977). Evidently this treatment is used in the USA (Washington and Oregon) on Golden Delicious (Olsen and Bartram, 1978) but is too harmful to be used on this cultivar in British Colombia, Canada (Lau and Looney, 1978), an area not too distant. Recent work by Handwerker (1979) indicates that

CO_2 injury of two apple cultivars was reduced if the treatment was imposed at 10°C and 21°C as compared to lower temperatures. Couey and Wright (1977) have shown that high CO_2 levels are used commercially on pre-cooled strawberries shipped from California (Smock, 1979). High CO_2 levels before storage may ameliorate chilling injury of citrus in subsequent storage (Hatton and Cubbedge, 1977) and may reduce the evidence of chilling injury on chili peppers and okra (Morris and Kader, 1977), although such levels may increase symptoms in tomatoes and bell peppers. Wang (1977a) showed that high CO_2 levels, although retarding some symptoms of senescence, caused subsequent severe calyx injury and decay on bell peppers.

Poapst, Price and Forsyth (1978) have developed a new technique for control of greening in potatoes. They used a food-grade surfactant to coat the surface of the tuber to increase the interal content of CO_2 to a level that controls greening. This could be rather hazardous considering the effect of CO_2 on blackheart of potato.

Ethylene

Dewey (1979) has called ethylene the 'hyaline thread woven through postharvest physiology research'. For several years ethylene production and action have been associated with fruit ripening and senescence, and ethylene is still considered to be the ripening trigger or the fruit-ripening hormone (Burg and Burg, 1962). However, recent experiments cast doubt as to whether ethylene is responsible for the ripening reactions in fruits such as the grape (Coombe and Hale, 1973), the tomato (McGlasson, Dostal and Tigchelaar, 1975) and the pear (Frenkel, 1975; Frenkel and Dyck, 1973; Frenkel, Haddon and Smallhee, 1975). Furthermore, the concept of a single ripening hormone in fruit tissue can be questioned by recent work, which indicates that ripening may be controlled by interactions between ethylene and the levels of other endogenous hormones (Lieberman, Baker and Sloger, 1977; Bruinsma, Knegt and Varga, 1975).

In climacteric fruit ethylene action has been associated with an upsurge in respiratory metabolism at the onset of ripening. This respiratory climacteric appears to be induced by a prior increase in endogenous ethylene production which induces an increase in glycolytic substrates and specific glycolytic enzymes (Solomos and Laties, 1975). However, activation of respiratory metabolism may not be due to a direct regulation of enzymes by ethylene, but may involve the action of other hormones as well as ethylene.

Much of our present technology is limited to the control of ethylene synthesis and action by means of ethylene removal, low temperature, CA storage and low-pressure storage (LPS), but little emphasis has been given to the control of other endogenous effectors. Perhaps when a clearer understanding of hormonal interrelationships in ripening is accomplished, new technology may be aimed at controlling these changes as well.

Over the years there has been considerable controversy whether ethylene had any effect on fruit ripening at storage temperatures of 10°C or below (Smock, 1979). However, there were indications in the early literature that scrubbing of 'volatiles' from apple stores with the objective of controlling scald did improve the firmness of apples (Smock and Southwick, 1948).

There has been renewed interest in the removal of ethylene from CA storage with the work of Forsyth, Eaves and Lightfoot (1969) and LPS which maintains an essentially ethylene-free environment (Dilley, 1972). Much of the work on removal of ethylene from atmospheres in apple stores was discussed by Liu (1977) and Smock (1979). There have been other recent research reports not reviewed here on the control of ethylene levels in the storage of apples and other fruits with accompanying delay of senescence. In most fruits it may be presumed that ethylene removal would be most effective on preclimacteric fruit before the surge in ethylene production coincident with the climacteric rise results in irreversible ripening changes (Smock, 1979).

For some time the function of ethylene in the senescence of flowers has been known (Nichols, 1968) and atmospheric scrubbers have been used in florists' stores for some time, ostensibly to control ethylene levels. Vegetable tissue produces little ethylene and is generally less sensitive to ethylene than many ornamentals. However, there is an ethylene scrubber in a CA cabbage store in New York State (Furry, Isenberg and Jorgensen, 1979) which may indicate future development.

Whatever the method of ethylene removal (Abeles, 1973; Smock, 1979), the difficulty is in removing ethylene down to <1 p.p.m. and indeed removing ethylene from *within* the crop too, which is easiest with LPS.

Some control of ethylene production in fruits (or vegetables if it were desirable) is possible by some chemical treatments such as daminozide (Liu, 1979) which are approved in some jurisdictions by food and drug regulations. Other chemicals which may reduce endogenous ethylene production in fruits (Janes and Frenkel, 1978a) or vegetables (Wang, 1977b) are not approved for food use and are unlikely to be. In ornamental plants, for which regulations may not be as restrictive, chemical control of endogenous ethylene production is more promising. Such treatments appear as senescence inhibitors (Wang and Baker, 1979) and also in opening solutions (Venkatarayappa, Tsujita and Murr, 1979).

Retardation of ripening or senescence by removal of ethylene is important, but there are instances where ethylene applied as ethephon (Liu, 1978a) may be used to promote ripening. Daminozide and other chemicals may increase ethylene production (Ben-Arie and Guelfat-Reich, 1979; Shaybany *et al.,* 1979). In some cases ethylene-inducing treatment may control the occurrence of physiological disorders (Liu, 1976; Padfield, 1977). There is also evidence that some fungicides may promote ripening (Thomas, Gray and Drew, 1977), ethylene production, respiration and senescence (Ben-Arie and Guelfat-Reich, 1975).

Acetaldehyde

There are indications that acetaldehyde may increase respiration in strawberry and blueberry and cause softening in pears (Janes and Frenkel, 1978b). Researchers working with ethylene scrubbers should also consider their effect on acetaldehyde accumulation. On the other hand gaseous acetaldehyde or acetaldehyde dips have been used recently to control undesirable physiological changes in strawberries (Morris, Cawthorn and Buescher, 1979).

Other volatiles

Herregods (1977) has studied the effect of other volatiles on stored products and indicated the desirability of atmospheric scrubbers.

Carbon monoxide

Carbon monoxide has been shown effective in controlling discoloration of several commodities (Kader, Morris and Klaustermeyer, 1977), but in conjuction with elevated CO_2 level, induced injury on lettuce. Symptoms of chilling injury and subsequent decay of tomatoes and peppers were also reduced by carbon monoxide. Woodruff (1977) has described the 'intriguing possibilities of the use of carbon monoxide in modified atmospheres for fruits and vegetables'.

Low-pressure storage (LPS)

Although hardly a recent innovation (patented in 1967; Burg, 1967) it still remains only an experimental procedure to most physiologists. However, the results possible with LPS do at times appear physiologically impossible and better than by any other method of storage (Lougheed *et al.*, 1978).

The most promising results have been increased firmness of apples (Dilley, 1972, 1977) a carry-over effect of LPS on rate of ripening of some fruit after storage (Dilley, 1977), delay in tomato ripening (Dilley, 1977), retention of green colour in leafy crops (Bangerth, 1974), the potential storage of leafy crops with ethylene-producing fruits (McKeown, Lougheed and Murr, 1978), and the anti-senescent effects on some bud-cut flowers (Staby, 1977). Despite the research failures, the high cost and the lack of application for fixed storage, LPS remains one of the most intriguing and potentially widely applicable storage methods.

Temperature

Despite the early indications that growth regulator treatments (Dedolph, Wittwer, Tuli and Gilbart, 1962) or in-storage treatments such as LPS (Kim, Lee, Hong and Sohn, 1969) might control rate of physiological change without temperature control, low temperature remains a fundamental storage treatment. Recently there has been renewed interest in fluctuating temperatures to allow metabolism of potentially harmful substances accumulated at chilling temperatures (Anderson, Penney and Smith, 1977; Furmanski and Buescher, 1977).

Some time ago use of pre-storage high-temperature treatments in air were investigated for control of scald (Hardenburg, 1965) or rotting (Sharples, 1967) in apples. Recent research has shown that interesting and as yet not understood physiological changes result from high-temperature treatment of apples before storage (Liu, 1978b; Porritt and Lidster, 1978).

Calcium

Calcium content has long been related to the incidence of bitter pit and other disorders in apples but the potentially broader implications and applications resulted in the conference at Beltsville, Maryland (Shear, 1979).

The relationship of calcium to pectinaceous compounds (thus texture) is well known (Christiansen and Foy, 1979), although its direct involvement in cell growth appears poorly understood. Calcium appears to be related to control of enzyme activity (Poovaiah, 1979a), respiration and ethylene production (Lougheed, Murr and Miller, 1979) and senescence (Sharples and Johnson, 1977; Poovaiah, 1979a; Levy and Poovaiah, 1979). Indirectly these effects may be related to maintenance of membrane integrity (Christiansen and Foy, 1979; Levy, Poovaiah and Kolattukudy, 1979).

On a practical basis the improvement of firmness with calcium treatment may be marginal (Lougheed *et al.*, 1979) but the control of breakdown in Spartan apples in British Colombia is of major importance (Mason, 1979). Recent attempts to induce high levels of calcium into the fruit without injury involve pressure or vacuum infiltration (Poovaiah, 1979b; Poovaiah, Shekhar and Patterson, 1978; Johnson, 1979), instead of post-harvest dips or tree sprays (Mason, 1979). However, injury still occurs (Johnson, 1979) and the corrosive effect of $CaCl_2$ upon equipment remains an impediment to the general use of such treatments.

Genetics

It hardly needs to be said that endogenous control of physiological change begins at the nuclear level. The most dramatic genetic control of 'ripening' is evident in tomato mutants reviewed by Tigchelaar, McGlasson and Buescher (1978). Some control of the rate of ripening may be achieved through incorporation of certain 'mutant' characteristics into commercial varieties. Hence, shelf life may be lengthened, allowing shipments over longer distances. Furthermore, the use of cell-culture techniques may permit development of cultivars less sensitive to chilling temperatures and permit the use of lower temperatures than presently used to control ripening or senescence.

The development of cabbage cultivars which lose their green colour less rapidly than standard cultivars is another striking example of direct genetic control of post-harvest behaviour.

Physiological stress

Physiological stress resulting from physical contact or deformation (bruises or lacerations) may result in modified patterns of respiration and ethylene production and may also influence fruit quality (McLeod, Kader and Morris, 1976). In addition post-harvest handling causes physiological and physical changes which may result in internal discoloration in many crops and probably physiological changes so subtle as to be immeasurable at any one time but possibly with cumulative harmful effects (Smock and Gross, 1950). Grierson and Soule (1979) on the other hand have indicated that moderate physiological stress (temperature or water-induced) may double the storage life of grapefruits.

Conclusions

Rapid imposition of proper storage temperatures and r.h. remains the most important means of control of post-harvest physiological change. The use of growth regulators and controlled atmospheres is of limited but nevertheless important application. Low-pressure storage has yet to be proven economically feasible for large-scale stationery storages. The greatest possibilities for improvement in current practices appear to be in techniques to overcome chilling injury, through plant breeding or genetic manipulation to improve storage responses, control of levels of ethylene and possibly other volatiles within storage rooms, and proper management conditions to control pre-storage stress.

Acknowledgements. Research reports of the authors cited herein were supported by funds from the Ontario Ministry of Agriculture and Food and from the National Science and Engineering Research Council, Canada (NSERC). The authors also acknowledge the support of the Association of Applied Biologists.

References

Abeles, F. B. (1973). *Ethylene in Plant Biology.* New York: Academic Press.

Andersen, R. E., Penny, R. W. & Smith, W. L. (Jr.) (1977). Peach storage in a controlled atmosphere with intermittent warning: a pilot test using inexpensive flowmeters and plastic bags as CA chambers. *HortScience* **12**, 345-346.

Bangerth, F. (1974). Hypobaric storage of vegetables. *Acta Horticulturae* **38**, 23-32.

Ben-Arie, R. & Guelfat-Reich, S. (1975). Stimulation of internal breakdown, respiration and ethylene evolution of Spadona pears with benzimidazole fungicides. *Canadian Journal of Plant Science* **55**, 593-596.

Ben-Arie, R., & Guelfat-Reich, S. (1979). Advancement of nectarine fruit ripening with daminozide and fenoprop. *Journal of the American Society for Horticultural Science* **104**, 14-17.

Bramlage, W. J., Bareford, P. H., Blanpied, G. D., Dewey, D. H., Taylor, S., Porritt, S. W., Lougheed, E. C., Smith, W. H. & McNicholas, F. S. (1977). Carbon dioxide treatments for 'McIntosh' apples before CA storage. *Journal of the American Society for Horticultural Science* **102**, 658-662.

Brooks, C., Bratley, C. O., & McColloch, L. P. (1936). Transit and storage diseases of fruits and vegetables as affected by initial carbon dioxide treatments. *Technical Bulletin No. 519,* United States Department of Agriculture.

Bruinsma, J., Knegt, E. & Varga, A. (1975). The role of growth-regulating substances in fruit ripening. In *Facteurs et regulation de la maturation des fruits.* pp. 193-199. Colloq. Int. C.N.R.S. No. 238, Paris, 1974.

Bubb, M. (1978). Hypobaric Storage. *Report of the East Malling Research Station for 1977,* p. 150.

Burg, S. P. (1967). Method of storing fruit. *U.S. Patent Office 3,* 333, 967. Filed 26 September 1963.

Burg, S. P. & Burg, E. A. (1962). Role of ethylene in fruit ripening. *Plant Physiology, Lancaster* **37**, 179-189.

Christiansen, M. N. & Foy, C. D. (1979). Fate and function of calcium in tissue. *Communications in Soil Science and Plant Analyses.* **10** (1 & 2), 427-442.

Coombe, G. B. & Hale, C. R. (1973). The hormone content ot ripening grape berries and the effects of growth substance treatments. *Plant Physiology, Lancaster* **51**, 629-634.

Couey, H. M., & Wright, T. R. (1977). Effect of prestorage CO_2 treatment on the quality of 'd'Anjou' pears after regular or controlled atmosphere storage. *HortScience* **12**, 244-245.

Dedolph, R. R., Wittwer, S. H., Tuli, V. & Gilbart, D. (1962). Effect of N^6-benzylaminopurine on respiration and storage behaviour of broccoli *(Brassica oleracea* var. *italica). Plant Physiology, Lancaster* **37**, 509-512.

Dewey, D. H. (Ed.) (1977). Controlled atmospheres for the storage and transport of perishable agricultural commodities. *Horticultural Report No. 28,* Michigan State University.

Dewey, D. H. (1979). Three remarkable generations of postharvest horticulture. *HortScience* **14** (37), Sec. 2, 342-344.

Dewey, D. H., Herner, R. C., & Dilley, D. R. (Ed.) (1969). Controlled atmospheres for the storage and transport of horticultural crops. *Horticultural Report 9,* Michigan State University.

Dilley, D. R. (1972). Hypobaric storage – a new concept for preservation of perishables. *102nd Annual Report of the State Horticultural Society of Michigan for the year 1972,* pp. 82-89.

Dilley, D. R. (1977). The hypobaric concept for controlled atmosphere storage. In *Controlled Atmospheres for the Storage and Transport of Perishable Agricultural Commodities*, pp. 29-37. Ed. D. H. Dewey, Horticultural Report No. 28, Michigan State University.

Fidler, J. C., Wilkinson, B. G., Edney, K. L. & Sharples, R. O. (1973). *The Biology of Apple and Pear Storage.* Research Review No. 3, Commonwealth Agricultural Bureaux.

Forsyth, F. R. & Lightfoot, H. J. (1977). Effect of ethylene and humidity on quality of stored McIntosh apples. In *Controlled Atmospheres for the Storage and Transport of Perishable Agricultural Commodities* pp. 97-107. Ed. D. H. Dewey. Horticultural Report No. 28, Michigan State University.

Forsyth, F. R., Eaves, C. A., & Lightfoot, H. J. (1969). Storage quality of McIntosh apple as affected by removal of ethylene from the storage atmosphere. *Canadian Journal of Plant Science* **49**, 567-572.

Frenkel, C. (1975). Role of oxidative metabolism in the regulation of fruit ripening. In *Facteurs et regulation de la maturation des fruits,* pp. 201-209. Coloq. Int. C.N.R.S. No. 238, Paris, 1974.

Frenkel, C. & Dyck, R. (1973). Auxin inhibition of ripening in Bartlett pears. *Plant Physiology, Lancaster* **51**, 6-9.

Frenkel, C., Haddon, V. R. & Smallhee, J. M. (1975). Promotion of softening and ethylene synthesis in Bartlett pears by 3-methylene oxindole. *Plant Physiology, Lancaster* **56**, 647-649.

Furmanski, R. & Buescher, R. (1977). Wooliness in peach fruits, a serious problem affecting internal quality. *Proceedings of the 98th Annual Meetings. Arkansas State Horticultural Society, 1976,* pp. 116-117.

Furry, R. B., Isenberg, F. M. R. & Jorgensen, M. C. (1979). Post-harvest controlled atmosphere requirements of cabbage intended for long duration storage. *Joint Meeting of American and Canadian Societies for Agricultural Engineering, University of Manitoba, June 1979.*

Grierson, W. & Soule, J. (1979). Making use of physiological stress. *HortScience* **14**, Sec. 3, 425 (abstract).

Halevy, A. H. & Mayak, S. (1979). Senescence and postharvest physiology of cut flowers. In *Horticultural Reviews*. Vol. 1, pp. 204-236. Ed. Jules Janick. Westport, Connecticut: AVI.

Handwerker, T. S. (1979). Effect of high CO_2 treatments on apples when applied at various storage temperatures. *HortScience* **14**, Sec. 3, 471 (abstract).

Hardenburg, R. E. (1965). Scald. In *A Review of Literature and Handling, Storage and Transportation of Apples,* pp. 194-215. USDA, ARD 51-4.

Hatton, T. T. & Cubbedge, R. H. (1977). Status of controlled atmosphere storage and transport of perishable agricultural commodities. In *Controlled Atmospheres for the Storage and Transport of Perishable Agricultural Commodities,* pp. 250-259. Ed. D. H. Dewey. Horticultural Report No. 28, Michigan State University.

Herregods, N. (1977). Disorders caused by some volatiles other than ethylene found in storage rooms *Acta Horticulturae* **62**, 247-256.

Hulme, A. C. (Ed.) (1970). *The Biochemistry of Fruits and their Products.* Vol. 1. London: Academic Press.

Hulme, A. C. (Ed.) (1971). *The Biochemistry of Fruits and their Products.* Vol. 2. London: Academic Press.

Isenberg, F. M. R. (Ed.) (1977). Symposium on vegetable storage. *Acta Horticulturae* **62**.

Isenberg, F. M. R. (1979). Controlled atmosphere storage of vegetables. In *Horticultural Reviews,* Vol. 1 pp. 337-394. Ed. Jules Janick. Westport, Connecticut: AVI.

Janes, H. W. & Frenkel, C. (1978a). Inhibition of ripening processes in pears by inhibitors of cyanide-resistant respiration and by silver. *Journal of the American Society for Horticultural Science* **103**, 394-397..

Janes, H. W. and Frenkel, C. (1978b). Promotion of softening processes in pear by acetaldehyde, independent of ethylene action. *Journal of the American Society for Horticultural Science* **103**, 397-400.

Johnson, D. S. (1979). New techniques in the post-harvest treatment of apple fruits with calcium salts. *Communications in Soil Science and Plant Analyses* **10**, 373-382.

Kader, A. A., Morris, L. L. & Klaustermeyer, J. A. (1977). Physiological responses of some vegetables to carbon monoxide. In *Controlled Atmospheres for the Storage and Transport of Perishable Agricultural Commodities*, pp. 187-202. Ed. D. H. Dewey. Horticultural Report No. 28, Michigan State University.

Kim, K. S., Lee, K. L., Hong, S. Y. & Sohn, T. H. (1969). Studies on the reduced pressure storage of fruits II. Preservation of Jonathan under various storage chamber pressures. *Journal of the Korean Agricultural Chemistry Society* **11**, 77-82.

Lau, O. & Looney, N. E. (1978). Effects of a pre-storage high CO_2 treatment on British Colombia and Washington State 'Golden Delicious' apples. *Journal of the American Society for Horticultural Science* **103**, 341-344.

Levy, D. & Poovaiah, B. W. (1979). Effect of calcium infiltration on senescence in apples. *HortScience* **14**, Sec. 3, 466 (abstract).

Levy, D., Poovaiah, B. W. & Kolattukudy, P. E. (1979). Effects of calcium infiltration on membrane leakage and changes in phospholipid and fatty acid content in apples during senescence. *HortScience* **14** (3), Sec. 3, 466. (abstract).

Lieberman, M., Baker, J. E. & Sloger, M. (1977). Influence of plant hormones on ethylene production in apple, tomato, and avocado slices during maturation and senescence. *Plant Physiology, Lancaster* **60**, 214-217.

Lipton, W. J. (Chairman)(1978). Proceedings of the symposium: Relative humidity – physical realities and horticultural implications. *HortScience* **13**, 550-574.

Liu, F. W. (1976). Ethylene inhibition of senescent spots on ripe bananas. *Journal of the American Society for Horticultural Science* **101**, 684-686.

Liu, F. W. (1977). The ethylene problem in apple storage. In *Controlled Atmospheres for the Storage and Transport of Perishable Agricultural Commodities,* pp. 86-96. Ed. D. H. Dewey. Horticultural Report No. 28, Michigan State University.

Liu, F. W. (1978a). Ripening bananas with ethephon in three polymeric film packages. *Hortscience* **13**, 688-690.

Liu, F. W. (1978b). Modification of apple quality by high temperature. *Journal of the American Society for Horticultural Science* **103**, 730-732.

Liu, F. W. (1979). Interaction of daminozide, harvesting date, and ethylene in CA storage on 'McIntosh apple quality'. *HortScience* **14,** Sec. 3, 421 (abstract).

Lougheed, E. C., Murr, D. P. & Bérard, L. (1978). Low pressure storage for horticultural crops. *HortScience* **13**, 21-27.

Lougheed, E. C., Murr, D. P., & Miller, S. R. (1979). Effects of calcium and daminozide on ethylene production and softening of apple fruits, *Experientia* **35**, 43-44.

McGlasson, W. B., Dostal, H. C. & Tigchelaar, E. C. (1975). Comparison of propylene-induced responses of immature fruit of normal and rin mutant tomatoes. *Plant Physiology, Lancaster* **55**, 218-222.

McKeown, A. W., Lougheed, E. C. & Murr, D. P. (1978). Compatibility of cabbage, carrots, and apple in low pressure storage. *Journal of the American Society for Horticultural Science* **103**, 749-752.

McLeod, R. F., Kader, A. A. & Morris, L. L. (1976). Stimulation of ethylene and CO_2 production by impact bruising. *HortScience* **11**, 604-606.

Mason, J. L. (1979). Increasing calcium content of calcium-sensitive tissues. *Communications in Soil Science and Plant Analysis* **10**, 349-372.

Morris, L. L. & Kader, A. A. (1977). Commodity requirements and recommendations for transport and storage-selected vegetables. In *Controlled Atmospheres for the Storage and Transport of Perishable Agricultural Commodities,* pp. 266-276. Ed. D. H. Dewey. Horticultural Report No. 28, Michigan State University.

Morris, J. R., Cawthon, D. L. & Buescher, R. W. (1979). Effects of acetaldehyde on post-harvest quality of mechanically harvested strawberries for processing. *Journal of the American Society for Horticultural Science* **104**, 262-264.

Nichols, R. (1968). The response of carnations (*Dianthus caryophyllus*) to ethylene. *Journal of Horticultural Science.* **43**, 335-349.

North, C. J., Bubb, M., Cockburn, J. T. & Gull, P. (1978). Extending the storage life of Cox's Orange Pippin apples. Effects of small differences in oxygen concentration. *Report of the East Malling Research Station for 1977,* p. 146.

Olsen, K. L. & Bartram, R. D. (1978). Carbon dioxide treatment of 'Golden Delicious' apples. *USDA, SEA, Advances in agricultural technology,* AAT-W-2/Sept.

Padfield, C. A. S. (1977). Preharvest ethephon-treatment, a possible control of apple scald. *New Zealand Journal of Experimental Agriculture* **5**, 183-184.

Poapst, P. A., Price, I. & Forsyth, F. R. (1978). Controlling post storage greening in table stock potatoes with ethoxylated mono- and diglyceride surfactants and an adjuvant. *American Potato Journal* **55**, 35-42.

Poovaiah, B. W. (1979a). Role of calcium in ripening and senescence. *Communications in Soil Science and Plant Analysis* **10**, 83-88.

Poovaiah, B. W. (1979b). Post-harvest calcium infiltration system for apple fruits. *HortScience* **14**, Sec. 3, 465 (abstract).

Poovaiah, B. W., Shekhar, V. C. & Patterson, M. E. (1978). Postharvest calcium and other solute infiltrations into apple fruits by pressure and vacuum methods. *HortScience* **13**, 357 (abstract).

Porritt, S. W. & Lidster, P. D. (1978). The effect of pre-storage heating on ripening and senescence of apples during cold storage. *Journal of the American Society for Horticultural Science* **103**, 584-591.

Porritt, S. W. & Meheriuk, M. (1973). Influence of storage humidity and temperature on breakdown in Spartan apples. *Canadian Journal of Plant Science* **53**, 597-599.

Robinson, J. E., Browne, K. M. & Burton, W. G. (1975). Storage characteristics of some vegetables and soft fruits. *Annals of Applied Biology* **81**, 399-408.

Ryall, A. L. & Lipton, W. J. (1972). *Handling Transportation and Storage of Fruits and Vegetables. Vol. I. Vegetables and Melons*. Westport, Connecticut: AVI.

Ryall, A. L. & Pentzer, W. T. (1974). *Handling, Transportation and Storage of Fruits and Vegetables. Vol. 2. Fruits and Tree Nuts.* Westport, Connecticut: AVI.

Schippers, P. A. (1975). The influence of storage conditions on chip color of potatoes. *Potato Research* **18**, 479-494.

Sharples, R. O. (1967). The effect of a post-harvest heat treatment on the storage behaviour of 'Cox's Orange Pippin' apple fruits. *Annals of Applied Biology* **59**, 401-406.

Sharples, R. O. & Johnson, D. S. (1977). The influence of calcium on senescence changes in apple. *Annals of Applied Biology* **85**, 450-453.

Sharples, R. O., Chappell, D. J., Sharp, L. E., Thompson, A. E. & Frost, J. (1978). Extending the storage life of Cox's Orange Pippin apples. Semi-commercial scale storage of Cox's Orange Pippin apples in 1·25 per cent oxygen. *Report of the East Malling Research Station for 1977.* p. 147.

Shaybany, B., Costa, G., Brown, S. S., Obenauff, G., Martin, G. C. & Gerdts, M. (1979). Effect of 2-(3-chlorophenoxy)-propionic acid and 2-(3-chlorophenoxy) proprionamide applications on fruit size and maturity of peach. *Journal of the American Society for Horticultural Science* **104**, 34-36.

Shear, C. B. (Ed.) (1979). International symposium on calcium nutrition of economic crops. *Communications in Soil Science and Plant Analyses* **10**.

Smock, R. M. (1979). Controlled atmosphere storage of fruits. In *Horticultural Reviews,* Vol. 1., pp. 301-336. Ed. Jules Janick. Westport, Connecticut: AVI.

Smock, R. M. & Gross, C. R. (1950). Studies on respiration of apples. *Memoir No. 297.* Cornell University Agriculture Experiment Station, Ithaca, N.Y.

Smock, R. M. & Southwick, F. W. (1948). Air purification in apple storage. *Bulletin 843.* Cornell University Agricultural Experiment Station, Ithaca, N.Y.

Solomos, T. & Laties, G. G. (1975). The mechanism of ethylene and cyanide action in triggering the rise in respiration in potato tubers. *Plant Physiology, Lancaster* **55**, 73-78.

Staby, G. L. (1977). Controlled atmosphere and low pressure storage of floral crops – overview. In *Controlled Atmospheres for the Storage and Transport of Perishable Agricultural Commodities*, pp. 60-70. Ed. D. H. Dewey. Horticultural Report No. 28, Michigan State University.

Thomas, T. H., Gray, D. & Drew, R. L. K. (1977). Investigations into outdoor tomato production and storage in the United Kingdom. *Annals of Applied Biology* **87**, 117-121.

Tigchelaar, E. C., McGlasson, W. B. & Buescher, R. W. (1978). Genetic regulation of fruit ripening. *HortScience* **13**, 508-513.

Venkatarayappa, T., Tsujita, M. J. & Murr, D. P. (1979). Effect of cobaltous ion on vase life, water balance, and quality of 'Samantha' roses. *HortScience* **14**, Sec. 3, 433 (abstract).

Wang, C. Y. (1977a). Effect of CO_2 treatment on storage and shelf-life of seet peppers. *Journal of the American Society for Horticultural Science* **102**, 808-812.

Wang, C. Y. (1977b). Effect of aminoethoxy analog of rhizobitoxine and sodium benzoate on senescence of broccoli. *HortScience* **12**, 54-56.

Wang, C. Y. & Baker, J. E. (1979). Vase life of cut flowers treated with rhizobitozine analogs, sodium benzoate, and isopentenyl adenosine. *HortScience* **14**, 59-60.

Wills, R. B. H. (1968). Influence of water loss of volatiles by apples: *Journal of the Science of Food and Agriculture* **19**, 354-356.

Woodruff, R. E. (1977). Use of carbon monoxide in modified atmospheres for fruits and vegetables in transit. In *Controlled Atmospheres for the Storage and Transport of Perishable Agricultural Commodities,* pp. 52-54. Ed. D. H. Dewey. Horticultural Report No. 28, Michigan State University.

Physiological aspects of the storage of cassava and other tropical root crops

J. Marriott, R. A. Plumbley and J. E. Rickard
Tropical Products Institute, 56/62 Gray's Inn Road, London WC1X 8LU, England

Introduction

Root crops are widely adapted to many tropical environments and supply about a third of the staple food of the developing world (Coursey and Haynes, 1970; Goering, 1979). The principal root crops grown exclusively in the tropics are cassava (*Manihot esculenta* Crantz, world production 110 Mt yr^{-1}); yams (*Dioscorea* spp., 19 Mt yr^{-1}) and the aroid root crops (principally taro, *Colocasia esculenta* (L.) Schott and tannia, *Xanthosoma sagattifolium* (L.) Schott, 3 Mt yr^{-1}). Additionally, substantial amounts of sweet potatoes (*Ipomoea batatas* (L.) Lam., 18 Mt yr^{-1}) and potatoes (*Solanum tuberosum* (L.) 24 Mt yr^{-1}) are grown within the tropical world. None of these crops, except potato, are normally stored under refrigeration in tropical countries. Usually tropical root crops are stored under ambient conditions where the produce reaches temperatures which lie between 20° and 30°C, although if exposed to the sun, produce may reach temperatures of 40°C or even higher (Rickard, Burden and Coursey, 1978). Physiological investigations have been conducted largely within normal tropical ambient temperatures. Maximum practicable storage periods for each crop are given in Table 1.

Cassava

Unlike the storage organs of other root crops, that of cassava is not an organ of dormancy and is not adapted to survival when detached from the plant. It is a thickened root which begins to accumulate starch in the parenchyma cells about 4 weeks after root initiation and which can continue growth and starch accumulation for 18 or even 24 months according to cultivar and growth conditions. It is usually harvested for use as a fresh vegetable between 6 and 14 months and for production of

Table 1. Factors limiting storage of root crops at ambient conditions in the tropics

Crop	Factors	Approximate maximum storage period (weeks)
Yam	Wound pathogens, sprouting, nematodes	24
Aroids	Wound pathogens	16
Potato	Wound pathogens, sprouting	12
Sweet potato	Wound pathogens, weevils	12
Cassava	Vascular discoloration, secondary pathogens	1

flour or starch up to 18 months, acceptability being limited mainly by gradual lignification as the root ages. Deterioration usually commences 2-5 days after harvest, so that the crop is not often stored as fresh roots, but is marketed or processed immediately.

The symptoms of deterioration in stored roots have been described and illustrated by Drummond (1953) and Montaldo (1973) and investigated further by Averre (1967). Deterioration initially manifests itself as discolouration of the vascular system which extends with variable intensity along most of the root length. The pigmented material at first appears blue but gradually turns brown and spreads into the storage parenchyma causing the development of dry brown necrotic areas. In a comparison between cultivars, Montaldo (1973) found that these symptoms developed within 7 days in 63 out of 65 cultivars when stored under typical tropical conditions of 24°C and 80 per cent relative humidity. Deterioration of whole roots could be retarded by storage in an oxygen-free atmosphere or by immersion in water (Drummond, 1953; Averre, 1967) or by coating with paraffin wax (Anon., 1972) and it was suggested that enzymatic oxidation processes were involved.

Recent investigations conducted by Tropical Products Institute (TPI) in collaboration with the Centro Internacional de Agricultura Tropical (CIAT) in Colombia (Booth, 1976) showed that these symptoms usually develop from sites of mechanical damage to the roots and that their intensity is strongly correlated with the degree of mechanical damage (Table 2). The symptoms can be largely suppressed by storage packed in moist sawdust or in a clamp at temperatures between 25°C and 40°C. These conditions minimize weight loss and aid suberization and wound periderm formation. Using these packing treatments, cassava roots can be stored for 4 to 8 weeks or even longer before deterioration

Table 2. Effect of severity of mechnical damage on primary deterioration of cassava roots stored at 24°C (Booth, 1976)

Days in storage	Deterioration index (per cent)		
	Slightly damaged	Severely damaged	Significance
1	0	7·5	–
2	12·5	30·0	*
4	7·5	52·5	*
7	12·5	72·5	*
11	12·5	67·5	*

* Differences between slightly and severely damaged treatments significant at $p = 0{\cdot}01$.

through other causes such as microbial rots or excessive dehydration sets in (Booth, 1977). These effects are similar to those that occur in other root crops, where preliminary storage at high temperature causes wound healing and leads to the avoidance of invasion by wound pathogens. However, no micro-organisms could consistently be isolated from freshly discolored areas, and those isolated from decaying tissue did not induce typical discoloration symptoms when re-inocultated (Noon and Booth, 1977). Thus it is concluded that primary injury is probably a physiological response to wounding (Booth, 1976) and that there is no evidence of a causal role for micro-organisms (Noon and Booth, 1977).

Subsequent studies of the physiological changes associated with the development of vascular discoloration were made using cylindrical sections of cassava roots, 100-150 mm long (Marriott, Been and Perkins, 1978). Exposed wounds were either one of the transverse cut ends or an area on the side from which the periderm had been removed, other wounds to the periderm and at the cut ends being covered with PVC film. This made possible examination of the discoloration reaction at a controlled, reproducible wound rather than in the complex situation in whole roots which suffer multiple, variable injuries at harvest.

To investigate the effect of moisture retention in storage systems such as packing in moist sawdust or in clamps, the intensity and depth of discoloration beneath exposed transverse cuts was measured using root pieces held at 20°C, ventilated in an enclosed system at high (90-95 per cent) or low (50-60 per cent) relative humidity and of a third group on an open surface. Discoloration increased parallel to the rate of weight loss (i.e. water loss from the wound) and was almost

Table 3. Deterioration of cassava root pieces stored at 20°C and at different levels of water stress for 3 days (Marriott and Rickard, unpublished observations)

Ventilation conditions	Humidity	Weight loss (% day^{-1})	Percentage discoloration at depths (mm) below wound surface 10	30	Total depth discoloured (mm)
Enclosed	High	1·3	6	4	4
Enclosed	Low	2·9	66	32	56
Open	Low	3·4	84	44	54
Open and fan	Low	4·0	100	56	86

completely suppressed at high humidity (Table 3). Similar results were obtained below an exposed periderm injury. Development of discoloration beneath both types of injury in root pieces kept in the open was already evident after 24 h and was nearly maximal after 48 h. In root pieces with exposed periderm injuries held at 20°C, no respiratory change occurred in pieces ventilated at high humidity after injury. A large, well-defined respiratory increase occurred during ventilation at low humidity (Fig. 1) and similar respiratory changes, superimposed on the wound respiration (which is independent of humidity) occurred at exposed transverse cuts (Marriott *et al.*, 1978). The rapidity of the discoloration reaction at low humidity, the inhibition of this response at high humidity at a temperature which usually is too low for rapid suberization, and the lack of respiratory

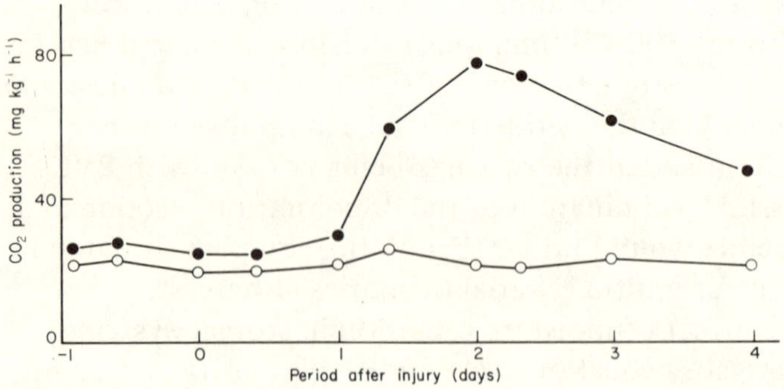

Fig. 1 CO_2 production in cassava root pieces when held at high and low humidity before and after periderm removal. Pieces were held 4 days after cutting and wrapping, before removal of periderm. The flow rate was 20 l h^{-1} ○–○, high humidity; ●–● low humidity. (From Marriott, Been and Perkins, 1978.)

response at high humidity all suggest that an active wound response is unnecessary for the inhibition of vascular discoloration. Rather, it appears that vascular discoloration develops in response to rapid water loss, occurring when the periderm is injured or the tissue is cut, unless the external humidity is extremely high.

Attempts to extract and identify the pigmented material were unsuccessful and cytochemical and cytological investigations were therefore undertaken (Rickard, Marriott and Gahan, 1979). Observations by both light and electron microscopy show that the pigmented material originates from adjacent parenchyma cells, entering the xylem via breaks in the pit membranes (Plate 1A). It appears as gel-like material lining the vessel walls and forming connecting strands or as large granular masses filling the vessel lumen. Transmission electron microsopy observations show the pigmented material as a fine stranded matrix containing electron dense particles with no detectable enclosing membrane (Plate 1B).

Cytochemical investigations show that the major components of the occluding material in fully discoloured tissue are polysaccharides, lignin-like material and acidic lipids but other constituents including free phenols cannot be detected (Table 4). Recent observations (Rickard, 1979) have shown, however, that before the appearance of pigmented occlusions, parenchyma cells adjacent to the xylem vessels give a strongly positive reaction for free phenols. No evidence of fungal

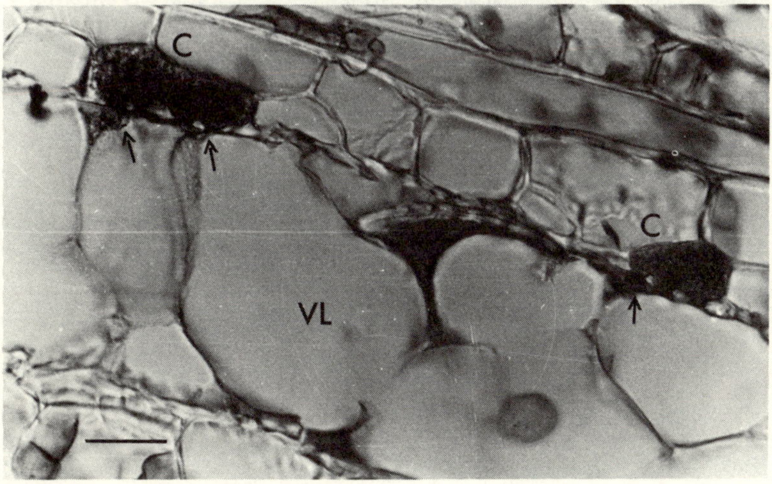

Plate 1A Unfixed, frozen section of cassava root showing pigmented material entering vessel lumen (VL) from adjacent parenchyma cell (C) via pit areas (arrow). Bar = 25 μm.

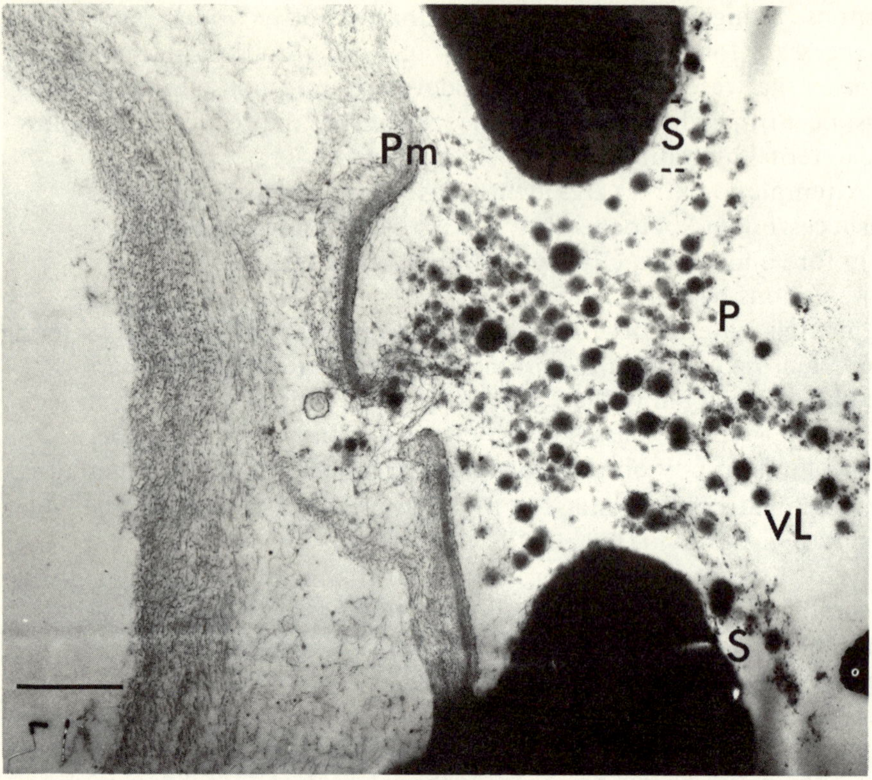

Plate 1B Transmission electron micrograph of pigmented material (P) entering vessel lumen (VL) via ruptured pit membrane (Pm) between secondary wall thickening (S). Bar = 3 μm. (From Rickard, Marriott and Gahan, 1979.)

Table 4. Cytochemical tests on xylem occlusion material in unfixed frozen section of cassava root (Rickard, Marriott and Gahan, 1979)

Compound	Result
Polysaccharides	+
Pectins	–
Proteins	–
Lipids	+
Terpenoid-aldehydes	–
Lignin-like material	+
Phenols	– *

* Reactive after injury but before appearance of pigmented material (Rickard, 1979).
Results are given as + = reactive; or – = non-reactive.

or bacterial infection could be found in any of the material during the examination, confirming the conclusions of Drummond (1953), Averre (1967) and Noon and Booth (1977) that vascular discoloration is of physiological origin.

To investigate possible enzymatic oxidative processes, changes in the activities of peroxidases and polyphenol oxidases in cassava roots undergoing vascular discoloration have been studied. Peroxidase activity is evident as three groups of bands after electrophoresis on polyacrylamide gels, when stained with benzidine dihydrochloride (Wang and Pinckard, 1973). During the development of vascular discoloration, the peroxidase shows changes, including the appearance of a new band in the middle group and the intensification of several others (Fig. 2). These changes are accompanied by an increase in the

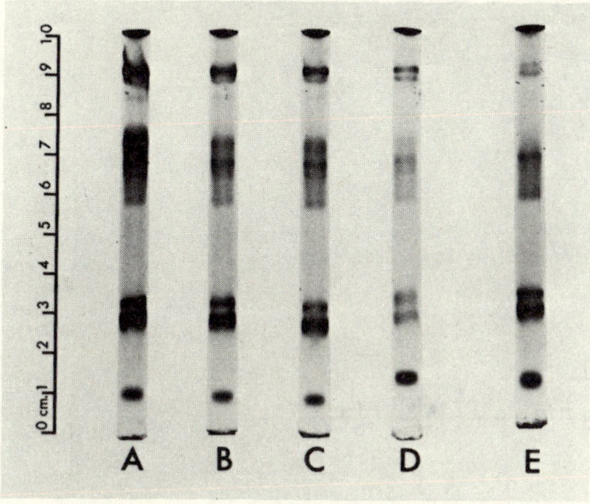

Fig. 2 Peroxidase activity of electrophoretic gels of cassava tuber tissue extracts (cultivar Ankrah). ABCD extracts from tissue 15, 30, 50 and 70 mm from a desiccated surface. Vascular discoloration decreased with distance from the surface. E. Extract from healthy undesiccated tuber tissue (Plumbley and Hughes, unpublished data).

total peroxidase activity, as measured by oxidation of ascorbic acid with a benzidine indicator (Gregory, 1966). Changes are evident not only close to the wound (15 mm depth) but also in discoloured tissue further from the wound (30 and 50 mm depth). When pieces are cut but stored at high humidity to minimize vascular discoloration, some changes in peroxidase bands do occur in tissue at 15 mm depth but not at 30 and 50 mm depths (data not shown). This suggests that the changes observed in discoloured tissue relate to the stress causing vascular discoloration rather than solely to a wounding response.

Polyphenol oxidase (PPO) activity is associated with the upper and lower groups of bands showing peroxidase activity (Fig. 2) but not with the middle group, and no difference in the electrophoretic pattern of PPO activity between healthy and discoloured tissue has been found. The changes in peroxidase activity may be associated with formation of lignin-like material from phenols and polyphenols and so contribute to the development of the pigment responsible for the visible symptoms of vascular discoloration.

Physiological changes also occur in stored roots which greatly reduce the extent of discoloration which occurs if they are subsequently inured (Marriott *et al.*, 1979). These changes have been observed in roots which were cured in moist sawdust at tropical ambient temperature after harvesting and injured by transverse cuts after storage periods of 5-10 days (Fig. 3). They cannot relate directly to curing and wound

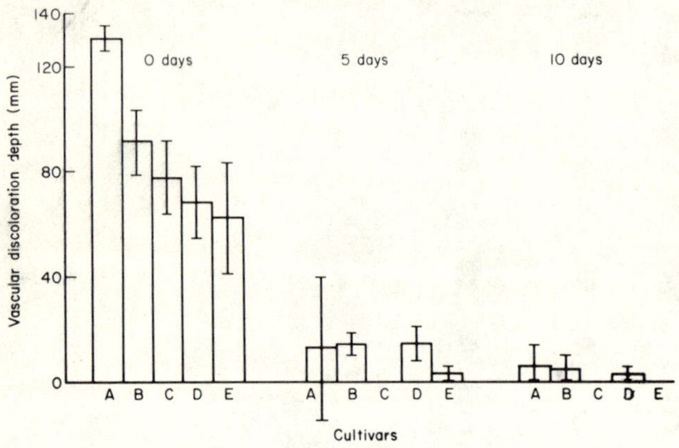

Fig. 3 Susceptibility to vascular discoloration at a transverse cut of cassava root pieces of different cultivars following storage of intact roots in moist sawdust at ambient temperature. Vertical bars indicate twice the standard errors of means for 10 pieces. Cultivars: A, M Col 22; B, Llanera: C, CMC 40; D, Mex 59; E, CMC 84. (From Marriott, Been and Perkins, 1979.)

periderm formation since if a severe fresh injury is inflicted, discoloration beneath such an injury is less than beneath similar injuries in freshly harvested roots. Similar changes in susceptibility occur when the aerial parts of plants are totally pruned 1, 2 or 3 weeks prior to harvest, and these changes markedly retard the deterioration of both intact roots (Lozano, Cock and Castano, 1978) and cut root pieces (Marriott *et al.*, 1979). Methods of combining this pruning treatment with fungicide application and storage in plastic-lined sacks are being tested and may provide a simple low-cost method to store cassava roots, even if they are severely damaged. Lozano *et al.* (1978) suggest that cassava leaves produce a compound which is translocated to the roots and which influences susceptibility to vascular discoloration, and this hypothesis is currently being tested further as part of a TPI/CIAT programme to develop practicable storage methods for fresh cassava.

There is little information on compositional or acceptability changes in stored cassava roots. Effects in cassava roots stored in moist sawdust and in clamps have been studied by Booth *et al.* (1976), who found that after storage at tropical ambient temperature for 1 week, content of total sugars increased from 5·2 per cent to 13·9 per cent with a corresponding decrease in starch content, although there was relatively little further change in storage up to 8 weeks. When compared to fresh roots, stored roots required longer cooking times and received lower, though acceptable, sensory ratings. Similar changes in starch-sugar conversion have been reported in the roots after the aerial parts of plants are pruned (Lorenzi, Gutierrez, Normanha and Cione, 1978).

All known cassava cultivars contain cyanide in the form of the cyanogenic glycosides linamarin and lotaustralin throughout their tissues. Tuber levels of HCN vary greatly between a few and several hundred mg kg^{-1} (Coursey, 1973). The tissues also contain the enzyme linamarase, which has been purified and utilised in an enzymatic assay for total cyanide content of cassava (Cooke, 1978; Cooke, Blake and Battershill, 1978). When the roots are cut or damaged in any way, HCN is rapidly released as a result of enzymatic action. Root cyanide content decreases parallel to changes in carbohydrates when plants are pruned (Lorenzi *et al.*, 1978). However total cyanide content and susceptibility to vascular discoloration have been compared between cultivars and no correlation was to be found, and it has not been possible to assign any role to cyanide in the deterioration of cassava. Freshly sliced tissue discs and mitochondria isolated from cassava tissue exhibit an unusually high level of cyanide-insensitive respiration (Passam, 1976), but no physiological role can yet be established for this alternative respiratory pathway in cassava or in other bulky storage tissues (Solomos, 1977).

Yams

The storage life of the organs of dormancy that form the edible yams is finally determined by the onset of sprouting. It is accompanied by a large increase in respiratory rate (Fig. 4). Attempts to prolong dormancy using sprout suppressants effective for potatoes have given negative or only weakly positive results (Passam, 1977). A probable explanation is that in yams, bud primordia are formed only late in

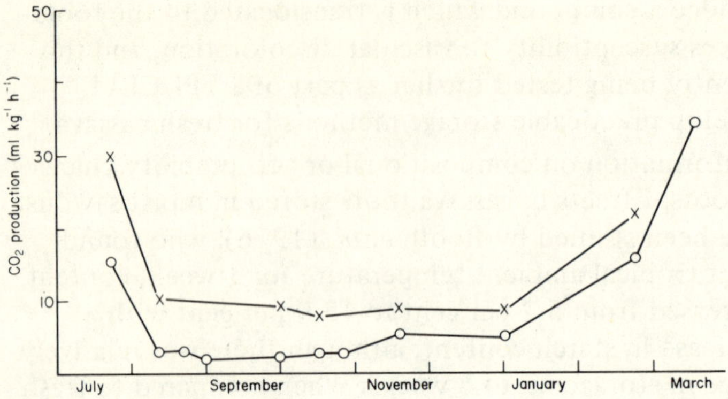

Fig. 4 Respiration of yam tubers during storage. O, at 25°C; X at 35°C. (From Passam, Read and Rickard, 1978.)

dormancy and originate beneath the periderm, thus being protected from treatment with sprout suppressants applied at harvest time (Passam, 1979). It is suggested that a more promising approach to dormancy is the application of growth regulators, in particular gibberellic acid, with which Martin (1977) and Mantell, Mohamed, Hayne, and Phelps (1977) have also recently obtained promising results.

During storage under tropical ambient conditions, respiratory rates are high and carbon dioxide and transpiratory water loss are important factors in loss of marketable weight. In *D. rotundata*, it was found that only 7 per cent of weight loss during the 5 month dormant period at 25° C was attributable to dry-matter loss by respiration, but in the periods immediately after harvesting and again before sprouting more than 25 per cent of the much more rapid weight loss was attributable to respiration (Passam, Read and Rickard, 1978).

Perhaps the major cause of loss within the natural storage life is damage caused by wound pathogens, mainly fungi, which occur in these and other crops (Coursey and Booth, 1972). Though the problem is essentially pathological, it is closely related to the physiology of the organ in that defence mechanisms, particularly suberization and wound periderm formation, can be crucial to limiting pathogenic invasion. These effects were described long ago for potatoes (Artschwager, 1927) and sweet potatoes (Artschwager and Starrett, 1931) and curing processes to induce them prior to storage are an established part of operational procedures in temperate latitudes. Recent work at TPI has shown that yams, although classified as monocotyledons and therefore not expected to form a true periderm, do exhibit a similar respiratory response and extensive cork formation on wounding. The rate and extent of the process and the effect of metabolic inhibitors on this response have been described (Passam, Hughes and Read, 1977; Passam, Read and Rickard, 1976 a,b).

Like many of the tropical products whose storage at low temperature has been investigated, chilling injury occurs in yams (Coursey, 1968). It has been found to be a major problem in shipments to temperate countries and to limit possible application of refrigeration storage. The effect is well known in sweet potato (Daines, Hammond, Haard and Ceponis, 1976; Lieberman, Craft, Audia and Wilcox, 1958) which is often grown as a temperate crop and stored through the winter season. Being the topic of a subsequent paper (Rhodes, 1980) it will not be discussed further here.

Other crops

Virtually nothing is known of the storage physiology of the aroid root crops. They appear to have only a short post-harvest life, rapid deterioration being caused mainly by pathogenic factors.

Potato and sweet potato, though extensively studied under temperate country conditions, are poorly known as to their behaviour under tropical conditions, and do not warrant futher discussion here.

Acknowledgement. We thank Mr D. G. Coursey for advice during preparation of this manuscript.

References

Anon. (1972). La yuca parafinda. Nueva tecnologia desarrollada por el Instituto de Investigaciones Tecnologicas. *Revista Instituto Investigaciones Tecnologicas*, Bogota **14**, 47-51.

Artschwager, E. (1927). Wound periderm formation in the potato as affected by temperature and humidity. *Journal of Agricultural Research* **35**, 995-1000.

Artschwager, E. & Starrett, R. C. (1931). Suberisation and wound periderm formation in sweet potato and gladiolus as affected by temperature and relative humidity. *Journal of Agricultural Research* **43**, 353-364.

Averre, C. W. (1967). Vascular streaking of cassava roots. *Proceedings of the 1st International Symposium on Tropical Root Crops, Trinidad* **2**, 31-35.

Booth, R. H. (1976). Storage of fresh cassava (*(Manihot esculenta*). I Post-harvest deterioration and its control. *Experimental Agriculture* **12**, 103-111.

Booth, R. H. (1977). Storage of fresh cassava (*Manihot esculenta*). II Simple storage techniques. *Experimental Agriculture* **13**, 119-128.

Booth, R. H., de Buckle, T. S., Cardenas, O. S., Gomez, G. & Hervas, E. (1976). Changes in quality in cassava roots during storage. *Journal of Food Technology* **11**, 245-264.

Cooke, R. D. (1978). An enzymatic assay for the total cyanide content of cassava (*Manihot esculenta* Crantz). *Journal of the Science of Food and Agriculture* **29**, 345-352.

Cooke, R. D., Blake, G. G. & Battershill, J. M. (1978). Purification of cassava linamarase. *Phytochemistry* **17**, 381-383.

Coursey, D. G. (1968). Low temperature injury in yams. *Journal of Food Technology* **3**, 143-150.

Coursey, D. G. (1973). Cassava as food: toxicity and technology. In *Chronic Cassava Toxicity: Proceedings of an Interdisciplinary Workshop, London, England.* International Development Centre Monograph IDRC-010e, 27-36.

Coursey, D. G. & Booth, R. H. (1972). The post-harvest phytopathology of perishable tropical produce. *Review of Plant Pathology* **51**, 751-765.

Coursey, D. G. & Haynes, P. H. (1970). Root crops and their potential as food in the tropics. *World Crops* **22**, 261-265.

Daines, R. H., Hammond, D. F., Haard, N. F. & Ceponis, M. J. (1976). Hardcore development in sweet potatoes. A response to chilling and its remission as influenced by cultivar, curing temperatures and time and duration of chilling. *Phytopathology* **66**, 582-587.

Drummond, O. A. (1953). Da etiologia das rajas pretas das raizes da mandioca. *Anals de Congresso Nacional de Sociedade Botanica do Brasil, Recife*, pp. 57-60.

Goering, T. J. (1979). *Tropical Root Crops and Rural Development.* World Bank Staff Working Paper No. 324, pp. 85. World Bank, Washington.

Gregory, R. P. F. (1966). A rapid assay for peroxidase activity. *Biochemical Journal* **101**, 582-583.

Lieberman, M., Craft, C. C., Audia, W. V. & Wilcox, M. S. (1958). Biochemical studies of chilling injury in sweet potatoes. *Plant Physiology* **33**, 307-311.

Lorenzi, J. O., Gutierrez, L. E., Normanha, E. S. & Cione, J. (1978). Variacao de carboidratos e acido cianidrico em raizes de mandioca, apos a poda da parte aerea. *Bragantia* **37**, 139-144.

Lozano, J. C., Cock, J. H. & Castano, J. (1978). New developments in cassava storage. In *Proceedings of a Cassava Protection Workshop 1977,* pp. 135-141. Eds. J. C. Lozano, A. Belloti & T. Martinez. Centro Internacional de Agricultura Tropical, Cali, Colombia.

Mantell, S. H., Mohamed, N., Hayne, S. Q. & Phelps, R. H. (1977). Virus diseases of yams in the Commonwealth Caribbean. *ODM Yam Virus Project Technical Report No. 3 Caribbean Agricultural Research and Development Institute,* 46 pp.

Marriott, J., Been, B. O. & Perkins, C. (1978). The aetiology of vascular discoloration in cassava roots after harvesting and association with water loss from wounds. *Physiologia Plantarum* **44**, 38-42.

Marriott, J., Been, B. O. & Perkins, C. (1979). The aetiology of vascular discoloration in cassava roots after harvesting: development of endogenous resistance in stored roots. *Physiologia Plantarum* **45**, 51-56.

Martin, F. W. (1977). Prolonging the storage life of yams by removal of shoots and by treatment with gibberellic acid and waxes. *Tropical Root and Tuber Crops Newsletter* **10**, 39-44.

Montaldo, A. (1973). Vascular streaking of cassava root tubers. *Tropical Science* **15**, 39-46.

Noon, R. A. & Booth, R. H. (1977). Nature of post-harvest deterioration of cassava roots. *Transactions of the British Mycological Society* **69**, 287-290.

Passam, H. C. (1976). Cyanide insensitive respiration in root tubers of cassava (*Manihot esculenta* Crantz). *Plant Science Letters* **7**, 211-218.

Passam, H. C. (1977). Sprouting and apical dominance of yam tubers. *Tropical Science* **19**, 29-39.

Passam, H. C. (1979). Dormancy of yams in relation to storage. In *International Foundation for Science Provisional Report No. 3 – Yams, Igname Buea, Cameroun, 1978,* pp. 357-370.

Passam, H. C., Hughes, P. A. & Read, S. J. (1977). Wound repair in yam tubers: the effects of inhibitors of respiration and protein synthesis on repair physiology. *New Phytologist* **79**, 593-603.

Passam, H. C., Read, S. J. & Rickard, J. E. (1976a). Wound repair in yam tubers: physiological processes during repair. *New Phytologist* **77**, 325-331.

Passam, H. C., Read, S. J. & Rickard, J. E. (1976b). Wound repair in yam tubers: the dependence of storage procedures on the nature of the wound and its repair. *Tropical Science* **18**, 1-11.

Passam, H. C., Read, S. J. & Richard, J. E. (1978). The respiration of yam tubers and its contribution to storage losses. *Tropical Agriculture, Trinidad* **55**, 207-214.

Rhodes, M. J. C. (1980). Chilling injury – some underlying biochemical mechanisms mechanisms. This volume, pp. 377-393.

Rickard, J. E. (1979). The microscopic study of primary deterioration on cassava roots (abstract only). *Proceedings of the Royal Microscopical Society* **14**, 234.

Rickard, J. E., Burden, O. J. & Coursey, D. G. (1978). Effects of insolation on tropical horticultural produce. *Acta Horticulturae* **84**, 115-122.

Rickard, J. E., Marriott, J. & Gahan, P. B. (1979). Occlusions in cassava xylem vessels associated with vascular discoloration. *Annals of Botany* **43**, 523-526.

Solomos, T. (1977). Cyanide resistant respiration in higher plants. *Annual Reviews Plant Physiology* **28**, 279-297.

Wang, S. C. & Pinckard, J. A. (1973). Peroxidase activity in the developing cotton boll and its relation to decay by *Diplodia gossypina. Phytopathology* **63**, 1095-1099.

Chilling injury — some underlying biochemical mechanisms

M. J. C. Rhodes
A.R.C. Food Research Institute, Colney Lane, Norwich

Introduction

Chilling injury is a physiological disorder which results from the exposure of sensitive plant species to temperatures in the range of 0-12°C (Lyons, 1973). The present paper is concerned with the problem of chilling injury in the commercially important fruits and vegetables since the onset of the disorder is a major factor in restricting the storage life of these commodities. This restriction in storage life is illustrated in the work of Kidd and West (1936) in Fig. 1 which shows the relationship between storage life, measured as the time taken to give 50 per cent wastage of the commodity, and temperature for three varieties of apples showing different degrees of sensitivity to chilling injury. In a variety which is insensitive to chilling injury such as Newton Wonder there is a major increase in storage life in the temperature range of +3 to −1°C while in the other two varieties shown storage life is at a maximum at a temperature of about 4°C and then falls as the temperature is lowered. This decrease in storage life is largely due to the increased incidence of chilling injury.

Characteristics of the disorder

Symtoms

The symptoms of chilling injury vary from commodity to commodity and are characteristic for that species. Chilling injury is generally associated with the necrosis of groups of cells situated either externally, leading to the formation of depressed areas, pitting and external discoloration, or internally, leading to internal browning. The breakdown of the surface of the commodity is frequently associated with increased susceptibility to microbial infection and hence the onset of chilling injury is associated with major wastage of the stored commodity. Table 1 summarizes the characteristic symptoms found in a number of fruits and vegetables susceptible to chilling injury. Pitting and internal or external browning are common symptoms, but waterlogging of tissues is observed in citrus fruits following exposure to chilling temperatures while in peaches, a dry mealy texture (woolliness) develops after exposure of unripe fruit to chilling temperatures. Failure to ripen normally is a characteristic of chilling injury in unripe fruit stored at low temperature, and this inhibition of ripening persists even when the temperature is subsequently raised to one at which ripening would normally occur.

Factors affecting the severity of the disorder

The severity of chilling injury is influenced by the temperature to which a commodity is exposed, by the time for which it is exposed to that temperature and by the sensitivity of the commodity to the low temperature. Most fruits and vegetables of tropical or sub-tropical origin are susceptible to chilling injury, while most crops of temperate origin are resistant to this disorder. There are some exceptions to this general rule. For example in the temperate crop, the apple, some varieties are susceptible to chilling injury (Fig. 1), but here injury only occurs

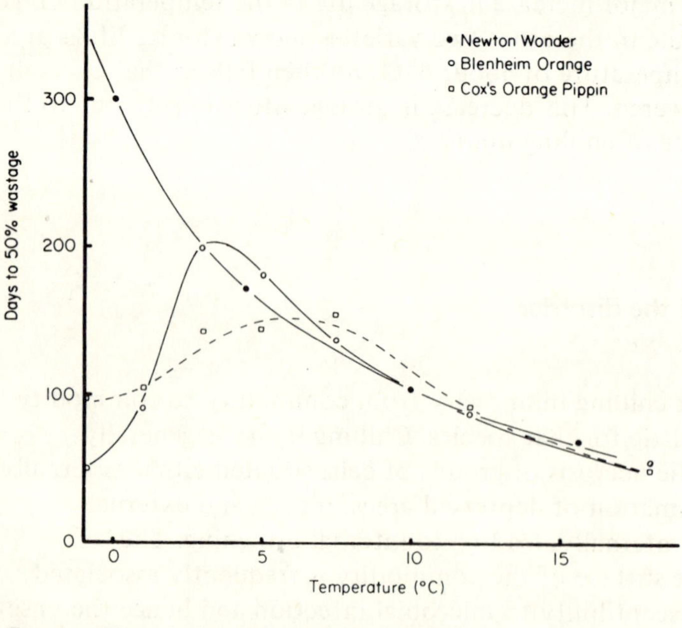

Fig. 1 The relationship between storage life and temperature for three varieties of apple (data after Kidd and West, 1936).

Table 1. Symptoms of chilling injury in fruits and vegetables

Commodity	Approximate lowest safe temperature of storage (°C)	Symptoms of injury
Apple – certain varieties only	3	Internal browning, scald
Avocado	7	Internal browning
Banana	13	Blackening of tissue
Grapefruit	7	Surface pitting, internal browning and water logging
Lemon	7	Pitting
Orange	1-2	Surface pitting, internal water logging
Pepper	7	Pitting
Sweet potato	12	Pitting of surface, internal browning
Tomato	12	Surface pitting, failure to ripen

following exposure to temperatures below 3° C. It is possible to define the susceptibility of sensitive species in terms of a threshold temperature or minimum storage temperature above which chilling injury does not occur. Table 1 shows that the minimum storage temperature for tropical products such as the banana, sweet potato and tomato is about 12° C, while for sub-tropical species such as the citrus fruit, capsicums and avocado it is about 7° C. It is interesting that even among the citrus fruits there is a marked difference in the temperature requirements for injury in oranges compared with lemons and grapefruit. These threshold temperatures are rough guides rather than absolute definitions of the temperature requirements to give injury, since there are differences in capacity to withstand small temperature differences below the threshold between varieties and even between different batches of a given variety.

Chilling sensitive commodities can be exposed for short periods to temperatures below their threshold temperatures without injury, but beyond a certain minimum period of exposure symptoms of injury became apparent and become increasingly severe as the period at the chilling temperature is extended. Frequently the symptoms of chilling injury do not become apparent during the period of cold storage and only develop if the commodity is returned to a warmer temperature. Figure 2 shows the time course of development of chilling injury in tomatoes picked at the breaker stage and stored for various periods at 2° C before transfer to 20° C. More than 11 days at 2° C were necessary for a significant incidence of injury, which was measured both by the number of fruit showing symptoms and by the retardation in ripening

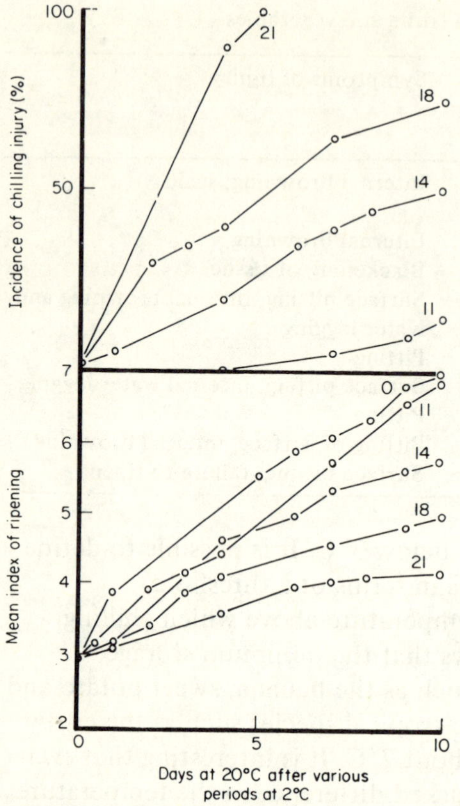

Fig. 2 Effect of storage at 2°C on ripening and the incidence of chilling injury in tomatoes (cultivar Winterbrid). Tomatoes were picked at the breaker stage and stored for various periods (time in days shown at the right-hand end of each line) at 2°C before transfer to 20°C. Chilling injury was recorded as the percentage of the population (30 fruits) showing symptoms of injury (upper curves) while changes in ripening occurring after transfer to 20°C are shown in the lower curves. Ripeness was measured on a scale of 1 to 7: 1, mature green; 2, breaker; 3, one-third coloured; 4, two-thirds coloured; 5, orange; 6, red; 7, red, soft.

(measured by colour change). The degree of injury became progressively more acute as the period at 2°C was extended up to 21 days, when all the sample showed symptoms of injury and when the fruit failed to ripen beyond an orange colour when returned to 20°C. Figure 3 shows the time course of injury in tomatoes stored at 0, 2 and 5°C, and indicates that the time of exposure to give 50 per cent incidence of injury ranged from 10 days at 0°C, 16 days at 2°C to about 22 days at 5°C. The period of exposure to low temperature required to induce symptoms can vary even between commodities which have the same threshold temperature. For example, both the

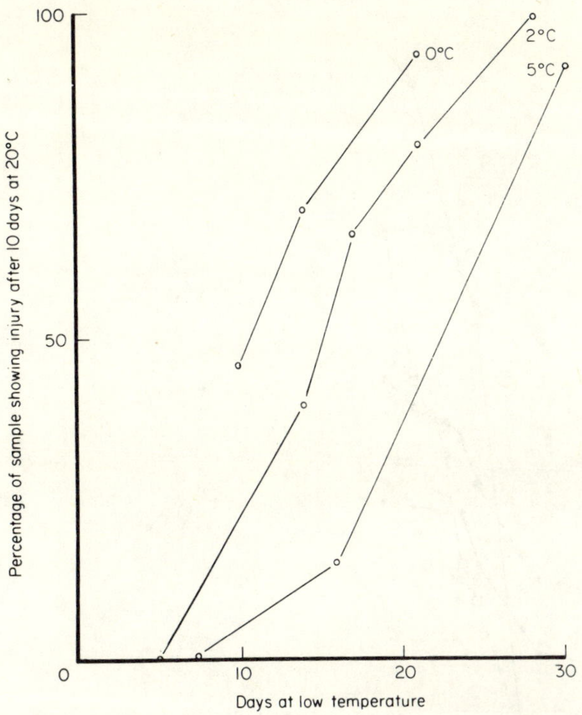

Fig. 3 Effect of a range of storage temperatures on the incidence of chilling injury in tomatoes (cultivar Eurocross BB). Tomatoes were picked at the breaker stage and stored at 0, 2 and 5°C for various periods of time before transfer to 20°C. Chilling injury was recorded after 10 days at 20°C.

banana and the tomato have a threshold temperature of about 12°C, but in bananas at 5°C a few hours exposure can cause considerable damage, while tomatoes would have to be exposed to this temperature for more than 2 weeks to achieve a similar degree of injury (Tomkins, 1963). With other species (e.g. peaches and plums) which develop symptoms of chilling injury during the period of exposure to the low temperature, the relationship between time and temperature on the speed of onset and severity of chilling injury is often complex (van der Plank and Davies, 1937). For instance, plums at 3°C developed symptoms of injury more rapidly than plums stored at 0°C, even though ultimately the injury was more severe at the lower temperature. Van der Plank and Davies (1937) considered that this was due to the interaction of two opposing factors; an equilibrium factor by which lowering the temperature increased the disposition towards injury and a kinetic factor by which higher temperature promoted the manifestation of the symptoms of injury. For maximum effect the

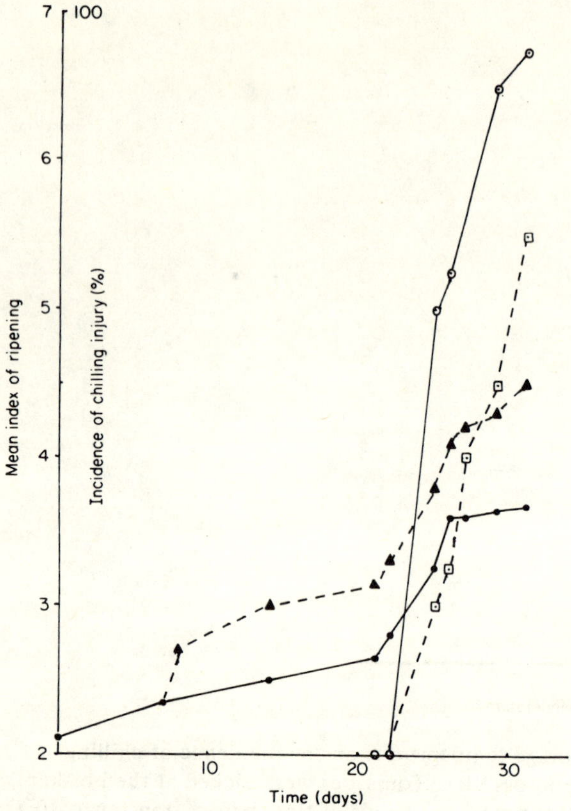

Fig. 4 Effect of interrupting storage at 2°C on the incidence of chilling injury in tomatoes (cultivar Eurocross BB). Tomatoes were picked at the breaker stage and stored either continuously at 2° for 21 days (solid lines) or for 20 days at 2° with 1 day at 20° on the 7th day of storage (dashed lines). Both groups of tomatoes were then transferred to 20°C for 10 days. Ripening (solid symbols) and chilling injury (open symbols) were recorded (as in Fig. 2) throughout the storage period.

exposure of the chilling-sensitive species to the injury-inducing temperature must be continuous; interruption of the cold period has in a number of cases been shown to reduce the severity of the disorder (Lyons, 1973). This is illustrated in Figure 4 for the tomato. Interrupting a 21-day period of exposure of tomatoes to 2°C by a 24 hour period at 20°C after 7 days was effective in both promoting more uniform ripening and in reducing the number of fruits showing symptoms of injury.

The degree of sensitivity of a commodity to chilling injury is essentially a characteristic of that species, but there are varietal effects, which within limits, lead to differences in sensitivity. The differences between apple varieties in the range of 0-3°C has already been illustrated. There is work in Australia to devise biochemical tests to assist breeders in the selection of new varieties (particularly of tomatoes) with improved resistance to chilling temperatures (Smillie and Nott, 1979). Factors such as maturity affect the sensitivity, particularly of fruit, to chilling injury. It is our experience with tomatoes that rotting occurs with shorter periods of exposure to the chilling temperature as the exposed fruit becomes more mature. However, other workers with other commodities find that the symptoms of chilling injury are generally more severe when immature fruits are exposed to low temperatures compared with more mature ones (see review of Lyons, 1973). The susceptibility of sensitive commodities can be modified by the environmental conditions under which the chilling temperature is applied. There are some experiments which show that the sensitivity of some species can be reduced if the material is 'conditioned' at a temperature slightly above the threshold for injury (Lyons, 1973). Similarly bananas transferred directly from 21°C to 5°C showed more injury than bananas whose temperature was similarly lowered but in steps of 3°C at 12-hourly intervals (Pantastico, Grierson and Soule, 1967) but generally such treatments are only effective against low degrees of chilling. Storage under hypobaric conditions (Burg, 1975) has been claimed to reduce the level of chilling injury while storage under high CO_2 may in some circumstances reduce chilling injury (Vakis, Grierson and Soule, 1970) although in others it may promote chilling injury (Tomkins, 1963). Our experience shows that high levels of CO_2 increase rotting in tomatoes stored at 2°C while hypobaric storage does not have a marked effect in reducing the sensitivity of tomato tissue to low temperature.

Underlying biochemical changes occurring at chilling temperatures

Sachs (1865) observed that protoplasmic streaming ceased in root hairs of cucumber and tomato plants at about 10-12°C while in chilling resistant species it continued even at about 0°C. Lewis (1956) and Wheaton (1963) have recently confirmed Sachs' observation. These studies were essentially of a qualitative nature and more recently Paterson and Graham (1977) have made quantitative measurements of the rates of protoplasmic streaming in four sensitive and two resistent species. They determined temperature coefficients for protoplasmic streaming for the six species over three temperature ranges, 15-20°C, 10-15°C, and 10-5°C. Differences in temperature coefficient between the six species were small in the upper temperature range but became marked in the lowest range. The largest temperature coefficients were found in the chilling-sensitive species and the smallest in the resistant species. There was no critical temperature below which streaming stopped abruptly but there was evidence that as well as a major reduction in the rate of streaming the number of organelles streaming also decreased in sensitive species at low temperatures. Protoplasmic streaming is an endergonic process dependent on the provision of cellular energy in the form of ATP and is affected by the physical state of the protoplasm and of the subcellular organelles. These findings suggest that exposure of sensitive species to chilling temperatures induces fundamental changes in cell physiology and that processes involved in energy production and in the state of the cellular membranes may be central to these changes.

In leaves, wilting is a major symptom of chilling injury (Simon, 1974). For instance cucumber leaves exposed to 5°C at 85 per cent RH began to wilt within a few hours. Wilson (1976) and Simon (1974) have shown that the severity of chilling injury can be reduced or eliminated by exposing the leaf to the chilling temperature in a water-saturated atmosphere. There have been many observations (see review of Simon, 1974) that the rate of leakage of solutes is also enhanced if leaves of sensitive species are exposed to chilling temperatures. These results suggest that there are changes in the properties of the cellular membranes at such temperatures. Simon (1974) has proposed that during extended periods in the cold, cell membranes become leaky allowing a mass outflow of solute-laden water from the cells to the cell walls. This flow of water can result in the leaves having a water-soaked appearance, a characteristic symptom of injury in the leaves. The water in the inter-cellular spaces can be lost by evaporation, leaving solutes in the cell walls from which they can readily diffuse if the leaves are placed in water, as in the standard procedures for measuring solute leakage. Fruits and vegetables have a low surface-to-volume ratio compared with leaves and they are generally coated with a thick and relatively impervious cuticle so that they have less tendency to dry out, and hence the loss of water is not an important symptom of chilling injury in fruits and vegetables. However, a water-soaked appearance is a symptom of chilling injury in oranges (Simon, 1974) and the severity of the disorder can be reduced in some commodities such as grapefruits (Morris and Platenius, 1938) by applying thin layers of surface waxes. It is pertinent also to point out that in apples the reverse situation holds in that high rates of water loss are correlated with a low incidence of chilling injury (Wills, Scott, and McGlasson, 1970). It has, however, been found for a number of fruits and vegetables (Lieberman, Croft, Audia and Wilcox, 1958; Fukushima and Tsugiyama, 1977) that exposure to chilling temperatures is associated with enhanced rates of solute leakage. These results suggest that there is an increase in the permeability of membranes in fruits and vegetables as well as in leaves when they are exposed to chilling temperatures.

The pattern of respiratory behaviour of fruits and vegetables is modified as a result of exposure to conditions which ultimately cause chilling injury. Biale (1941) showed that avocado fruits stored at chilling temperatures neither ripened normally nor developed the typical climacteric rise in respiration associated with the ripening process. Typically, sensitive commodities stored below the threshold temperature show an exaggerated rise in respiration on transfer to a normal temperature for ripening. This peak of respiration can be used as an index of the severity of chilling injury (Lyons, 1973). Such a respiratory stimulation on transfer to a higher temperature may be due to the rapid utilization of respiratory intermediates which accumulate under chilling conditions. Acetaldehyde, ethanol and α-keto acids such as oxaloacetic and pyruvic acid have been shown to accumulate in fruits during chilling storage (Hulme, Smith and Wooltorton, 1964; Murata, 1969; Pantastico, Soule and Grierson, 1968). There is evidence in cotton seedlings for an early decline in ATP levels on exposure to chilling temperatures, 50 per cent of the ATP being lost in 1 day at 5°C (Steward and Guinn, 1969). The work of Lewis and Workman (1964) with mature green tomatoes in which they demonstrated a marked decline in ^{32}P incorporation following storage of the fruit at 0°C indicates a reduced capacity for oxidative phosphorylation at chilling temperatures. Lyons and Raison (1970) studied the effect of temperature on the activity of mitochondria prepared from a range of chilling-sensitive and resistant species. They found that when the data were plotted in the form of an Arrhenius plot, the activity of mitochondria from sensitive species in oxidizing succinic acid showed a discontinuity or break point at a temperature which roughly corresponded with the threshold temperature for chilling injury in that commodity. Neither the capacity of mitochondria from resistant species to oxidize succinate nor the activity of soluble enzymes such as the enzymes of glycolysis from either sensitive or resistant species showed such a break point in the Arrhenius plot. In the chilling-sensitive species the change in the slope of the Arrhenius plot above and below the discontinuity was such as to suggest a roughly threefold increase in activation energy for succinate oxidation in the lower compared with the upper temperature range. Since mitochondria from sensitive species responded to ADP in the presence of substrate below as well as above the discontinuity, it seems likely that the capacity for ATP synthesis is maintained at all temperatures although its efficiency may be impaired at the lower temperatures. All these findings suggest that at chilling temperatures there are major changes in the respiratory metabolism of sensitive species which may lead to a relative reduction in the capacity for ATP synthesis.

Many other observations suggest that many divergent areas of intermediary metabolism are disturbed during storage at chilling temperatures (see review of Lyons, 1973) but there is no consensus as to whether these result from primary or secondary responses to the low temperature. One area in which we have worked is the modification of phenylpropanoid metabolism which occurs when some commodities are exposed to chilling temperatures. There is considerable evidence that there are striking responses in phenylpropanoid metabolism in plant tissues which are under various types of stress including mechanical, chemical or microbiological stress (Rhodes and Wooltorton, 1979). Low temperature storage of sensitive species can be regarded as a form of physiological stress. The oxidation of phenylpropanoid compounds, such as chlorogenic acid is involved in the browning reactions which are among the principal symptoms of chilling injury. However, we have been more interested in the promotion of phenylpropanoid biosynthesis which can occur in some commodities under chilling conditions. Lieberman, Croft and Wilcox (1959) showed an increased accumulation of chlorogenic acid in sweet potatoes stored at 7·5°C but not at 15°C. Mondy, Gedde-Dahl and Mobley (1966) showed a similar rise in chlorogenic acid in potatoes which were stored at 4·4°C but not at 10°C. We have recently confirmed these findings (Rhodes and Wooltorton, 1977, 1978) in a study of changes in activity of enzymes of chlorogenic acid biosynthesis. In both tomatoes and potatoes a number of these enzymes increase in activity when the commodity is stored at temperatures below the threshold for chilling injury but not above the threshold. A number of enzymes show a similar pattern of behaviour, but since it shows the greatest degree of change in activity we have chosen to study the enzyme hydroxycinnamoyl CoA; quinate hydroxycinnamoyl transferase (CQT) which catalyses the reversible thiolytic cleavage of chlorogenic acid (Rhodes and Wooltorton, 1976). Figures 5 and 6 show that the activity of this enzyme increased when the material was stored below 5°C in the case of the potato and below 12-15°C in the case of the tomato but not when storage was carried out

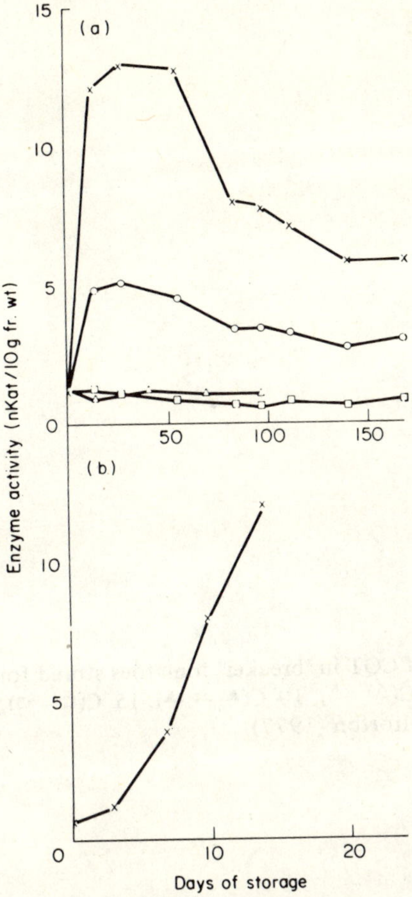

Fig. 5 Changes in the activity of CQT in potatoes (var. majestic) stored at 0°C(x—x), 2°C(○—○), 5°C(△—△) and 10°C(□—□) for (a) up to 24 weeks (b) for up to 14 days. Data from Rhodes & Wooltorton (1978).

at higher temperatures. If chilled tomato fruit was transferred to a higher temperature, the activity of CQT fell rapidly to its initial value over a two-day period. The rise in CQT activity occurred during storage at 2°C of fruit at all stages of maturity from mature green to total ripe (Fig. 7). Labelling studies have shown that the enhanced activity of these enzymes measured *in vitro* is associated with an enhanced rate of labelling of chlorogenic acid when tissue is fed with ^{14}C cinnanic acid and with an increase in its concentration (Rhodes and Wooltorton, unpublished data). The mechanism by which the activity of CQT is enhanced is at present unknown, but it is of interest that exposure to low temperatures can cause stimulation of this enzyme activity.

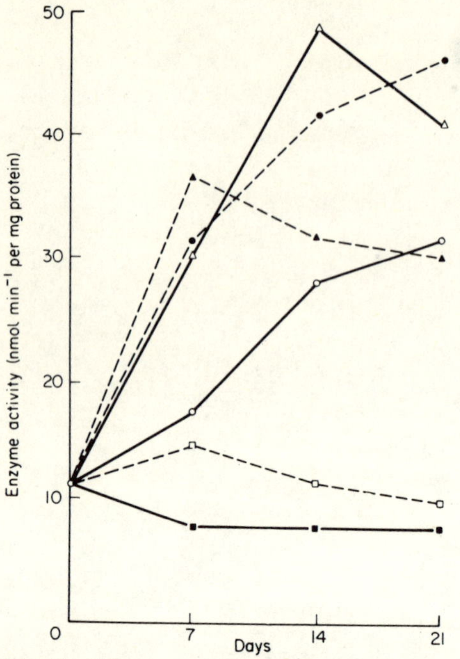

Fig. 6 Time course of changes in activity of CQT in 'breaker' tomatoes stored for up to 21 days at 0°C(○—○), 2°C(●------●), 5°C(△—△), 10°C(▲- - -▲), 15°C(□—□) and 20°C(■ - - - ■). Data from Rhodes & Wooltorton (1977).

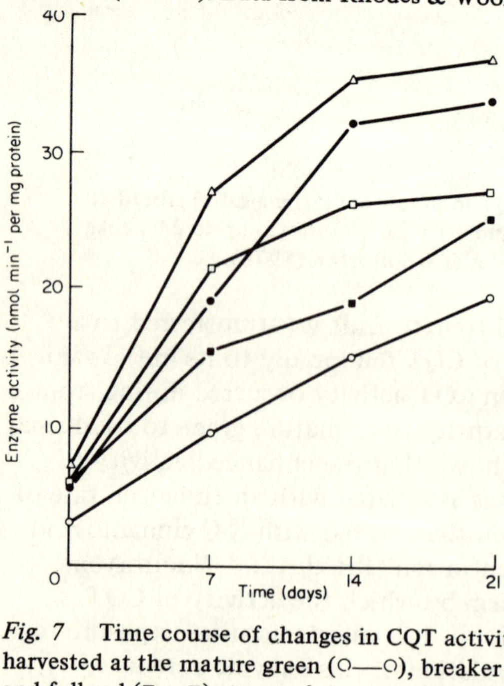

Fig. 7 Time course of changes in CQT activity during storage at 2°C of tomatoes harvested at the mature green (○—○), breaker (●—●), orange (△—△), pink (□—□) and full red (■—■) stages of ripeness. Data from Rhodes & Wooltorton (1977).

The mechanism of chilling injury

Any theory to explain chilling injury must accommodate the fact that many diverse areas of metabolism are ultimately affected following exposure to low temperatures and that the symptoms of injury vary to a considerable extent from commodity to commodity. Indeed doubt has been expressed as to whether it is possible to explain such a diverse effect in a single unifying theory with a single primary response to temperature upon which all the metabolic and physiological changes which ultimately lead to injury are based. A two-part theory of a direct response to temperature inducing secondary reactions which ultimately lead to injury appears essential to explain the partial reversibility and complex time-temperature relationships of the disorder. The primary direct response(s) to temperature must be reversible and not cause injury directly, while the secondary consequences of this response may be slow acting and ultimately irreversible. These secondary consequences of the responses of the tissue to low temperature are generally discussed in terms of metabolic imbalance (Fidler, 1968). However the nature of this imbalance and its physiological consequences are uncertain but the idea of the accumulation of cellular toxins which lead finally to cellular damage and injury has often been invoked.

Lyons and Raison (1970) proposed that changes in the state of the cellular membranes might be the principal primary response of tissues to chilling temperatures. They envisaged a phase transition occurring at a critical temperature at which the membrane changes from a normal flexible liquid crystalline state to a solid gel structure. This change in state would lead to contraction of the membrane so that cracks or channels form across it. This gelling of the membrane may not be continuous but may occur in localized areas leading to phase separations between regions of the membrane. Such changes in the membrane state can explain the increased permeability observed in sensitive species exposed to temperatures below the threshold for chilling and the changes in the energy of activation of membrane-bound enzymes observed by the appearance of discontinuities in the Arrhenius plots of their activities. The markedly reduced activity of the mitochondrial system at chilling temperatures could lead to imbalance between glycolysis and the tricarboxylic acid cycle and explain the accumulation of intermediates such as acetaldelyde and α-keto acids under conditions in which the soluble glycolytic system retained a relatively higher activity than the mitochondrial system at chilling temperatures. Changes in the rates of water loss in leaves can similarly be explained in terms of changed membrane properties. There are discontinuities in the Arrhenius plots of the activity of enzymes bound to chloroplast (Shneyour, Raison and Smillie, 1973) microsomal and glyoxysomal membranes (Lyons, 1972; Raison, 1973), as well as to mitochondrial ones. Thus it appears that changes occur in the whole range of cellular membranes and the occurrence of such changes is revealed more directly in electron-spin resonance studies coupled to the use of lipid-soluble spin labels such as 12-nitroxide stearate (Raison, Lyons, Mehlhorn and Keith, 1971) or spin labels such as TEMPO (2,2,5,5-tetramethyl piperidine-1-oxyl) which partitions between the aqueous and lipid phases of the cell (Fey, Workman, Marcellos and Burke, 1979). These spin-label studies confirm the general hypothesis that chilling-sensitive species undergo a change in the structure and functional integrity of their membranes at a temperature which correlates with the threshold temperature for chilling injury. Lyons, Wheaton and Pratt (1964) correlated sensitivity to chilling temperatures with the observation that chilling-sensitive species have a higher ratio of saturated to unsaturated fatty acids in their membranes than do chilling-resistant species. There are, however, a number of exceptions to the rule (Lyons and Asmundsen, 1965) and it is unlikely that the change in state of the membranes could be due entirely to the properties of the lipid components and it may well be that other membrane components such as proteins and sterols play a role in determining the responses of the intact membrane to temperature.

The exact nature of the changes in membrane structure that occur below the threshold for chilling injury are unknown and probably gelling of the membrane may represent an over simplification of the changes that occur. That changes do occur seems to be well established, and so is their ability to explain many of the biochemical and physiological changes that occur in chilled commodities, but it is still an open question as to whether they can explain all of the diverse metabolic changes occurring in the cold and the diversity of symptoms of the disorder. There are many other changes in the cold which could lead to imbalance between metabolic systems and hence to injury. Simple differences in the Q_{10} values for enzymes which limit particular related pathways could lead to imbalance between them as the temperature is lowered. Our work with the enzyme CQT shows that exposure to low temperatures can promote increases in activity of a soluble enzyme. There is evidence at least *in vitro* that some enzymes exhibit cold lability in that they are more stable at ambient temperatures than in the cold where their activity is rapidly lost. There are several examples of cold lability of enzymes described in the literature (Pullman, Penafsky, Datta and Racker, 1960; Norris and Fowden, 1974) but whether such cold lability is exhibited *in vivo* is an open question. However, the possibility does exist that some enzyme activities are promoted in the cold while others are diminished and this could lead to disturbances in metabolism. Some of these changes in activity may ultimately be controlled by changed permeability properties and the release of an activator or inactivator, but there is at present no evidence for this. The present state of knowledge of the mechanism of chilling injury suggests that the major direct response to temperature is a change in membrane structure and properties which lead to metabolic imbalance and ultimately to injury. There may well be additional direct response to temperature by which metabolic imbalance and injury may occur which are distinct and separate from changes in membrane structure. We know very little about such changes and the contribution they make to the diversity of symptoms revealed by sensitive species following exposure to chilling temperatures.

References

Biale, J. B. (1941). The climacteric rise in respiration rate of the Fuerte avocado. *Proceedings of the American Society for Horticultural Science* **39**, 137-142.

Burg, S. P. (1975). Hypobaric storage and transportation of fresh fruit and vegetables. In *Post Harvest Biology and Handling of Fruit and Vegetables*, pp. 172-188. Eds. N. F. Haard & D. K. Salunkhe, Connecticut: AVI Publishing Co.

Fey, R. L., Workman, M., Marcellos, H. & Burke, M. J. (1979). Electron spin resonance of 2,2,6,6-tetramethyl piperidine-1-oxyl(TEMPO)-labelled plant leaves. *Plant Physiology* **63**, 1220-1222.

Fidler, J. C. (1968). Low temperature injury to fruits and vegetables. In *Low Temperature Biology and Food Stuffs. Recent Advances in Food Science,* Vol. 4, pp. 271-281. Eds. J. Hawthorne & E. J. Rolfe. Oxford: Pergamon Press.

Fukushima, T. & Tsugiyama, T. (1977). Chilling injury in cucumber fruit II Chemical analysis of leakage substances and anatomical observations of symptoms. *Scientia Horticulturae* **6**, 199-206.

Hulme, A. C., Smith, W. H. & Wooltorton, L. S. C. (1964). Biochemical changes associated with the development of low-temperature breakdown in apples. *Journal of the Science of Food and Agriculture* **15**, 303-307.

Kidd, F. & West, C. A. (1936). Temperature and duration of life in apples. *Report of Food Investigation Board for 1935,* pp. 97-101.

Lewis, D. A. (1956). Protoplasmic streaming in plants sensitive and insensitive to chilling temperatures. *Science* **124**, 75-76.

Lewis, T. L. & Workman, M. (1964). The effect of low temperature on phosphate esterification and cell membrane permeability in tomato fruit and cabbage leaf tissue. *Australian Journal of Biological Science* **17**, 147-152.

Lieberman, M., Croft, C. C. & Wilcox, M. S. (1959). Effect of chilling on the chlorogenic acid and ascorbic acid content of Puerto Rico sweet potatoes. *Proceedings of the American Society for Horticultural Science* **74**, 642-648.

Lieberman, M., Croft, C. C., Audia, W. V. & Wilcox, M. S. (1958). Biochemical studies of chilling injury in sweet potato. *Plant Physiology* **33**, 307-311..

Lyons, J. M. (1972). Phase Transitions and control of cellular metabolism at low temperatures. *Cryobiology* **9**, 341-350.

Lyons, J. M. (1973). Chilling injury in plants. *Annual Review of Plant Physiology* **24**, 445-466.

Lyons, J. M. & Asmundsen, G. M. (1965). Solidification of unsaturated/saturated fatty acid mixtures and its relationship to chilling sensitivity in plants. *Journal of the American Oil Chemist's Society* **42**, 1056-1058.

Lyons, J. M. & Raison, J. K. (1970). Oxidative activity of mitochondria isolated from plant tissues sensitive and resistant to chilling injury. *Plant Physiology* **45**, 386-389.

Lyons, J. M., Wheaton, T. A. & Pratt, H. K. (1964). Relationships between the physical nature of mitochondrial membrane and chilling sensitivity in plants. *Plant Physiology* **39**, 262-268.

Mondy, N. I., Gedde-Dahl, S. B. & Mobley, E. O. (1966). Effect of storage temperature on the cytochrome oxidase and polyphenol oxidase activities and phenolic content of potatoes. *Journal of Food Science* **31**, 32-37.

Morris, L. L. & Platenius, H. (1938). Low temperature injury to certain vegetables after harvest. *Proceedings of the American Society for Horticultural Science* **36**, 609-613.

Murata, T. (1969). Physiological and biochemical studies of chilling injury in bananas. *Physiologia Plantarum* **22**, 401-411.

Norris, R. D. & Fowden, L. (1974). Cold lability of prolyl t-RNA synthetase from higher plants. *Phytochemistry* **13**, 1677-1687.

Pantastico, E. B., Grierson, W. B. & Soule, J. (1967). Chilling injury in tropical fruits 1. Bananas. *Proceedings of the American Society for Horticultural Science (Tropical Region)* **11**, 82-91.

Pantastico, E. B., Soule, J. & Grierson, W. (1968). Chilling injury in tropical and sub tropical fruits 2. Limes and grapefruits. *Proceedings of the American Society for Horticultural Science (Tropical Region)* **12**, 171-183.

Paterson, B. D. & Graham, D. (1977). Effect of chilling temperatures on the protoplasmic streaming of plants from different climates. *Journal of Experimental Botany* **28**, 736-743.

Pullman, M. E., Penafsky, H. S., Datta, A. & Racker, E. (1960). Purification and properties of soluble dinitrophenol-stimulated adenosine triphosphatase. *Journal of Biological Chemistry* **235**, 3322-3329.

Raison, J. K. (1973). Temperature-induced phase changes in membrane lipids and their influence on metabolic regulation. *Symposium of the Society of*

Experimental Biology **27**, 485-512.

Raison, J. K., Lyons, J. M., Mehlhorn, R. J. & Keith, A. D. (1971). Temperature-induced phase changes in mitochondrial membranes detected by spin labelling. *Journal of Biological Chemistry* **246**, 4036-4040.

Rhodes, M. J. C. & Wooltorton, L. S. C. (1976). The enzymic conversion of hydroxycinnamic acids to p-coumaroyl quinic and chlorogenic acids in tomato fruits. *Phytochemistry* **15**, 947-951.

Rhodes, M. J. C. & Wooltorton, L. S. C. (1977). Changes in the activity of enzymes of phenylpropanoid metabolism in tomato fruit stored at low temperatures. *Phytochemistry* **16**, 655-659.

Rhodes, M. J. C. & Wooltorton, L. S. C. (1978). Changes in the activity of hydroxycinnamoyl CoA: quinate hydroxycinnamoyl transferase and in the levels of chlorogenic acid in potatoes and sweet potatoes stored at various temperatures. *Phytochemistry* **17**, 1225-1229.

Rhodes, M. J. C. & Wooltorton, L. S. C. (1979). The biosynthesis of phenolic compounds in wounded plant storage tissues. In *The Biochemistry of Wounded Plant Tissues,* pp. 243-286. Ed. G. Kahl. Berlin: Walter de Gruyter.

Sachs, J. Von (1865). *Handbuch der Experimental - Physiologie der Pflanzen.* Leipzig: Engelman.

Shneyour, A., Raison, J. K. & Smillie, R. M. (1973). The effect of temperature on the rate of photosynthetic electron transport in chloroplasts of chilling sensitive and chilling resistant plants. *Biochimica et Biophysica Acta* **292**, 152-161.

Simon, E. W. (1974). Phospholipids and plant membranes permeability. *New Phytologist* **73**, 377-420.

Smillie, R. M. & Nott, R. (1979). Assay of chilling injury in wild and domestic tomatoes based on photosystem activity of the chilled leaves. *Plant Physiology* **63**, 796-801.

Steward, J. McD. & Guinn, G. (1969). Chilling injury and changes in adenosine triphosphate of cotton seedlings. *Plant Physiology* **44**, 605-608.

Tomkins, R. G. (1963). The effects of temperature, extent of evaporation and restriction of ventilation on the storage life of tomatoes. *Journal of Horticultural Science* **38**, 335-347.

Vakis, N., Grierson, W. & Soule, J. (1970). Chilling injury in tropical and sub tropical fruits 3. The role of CO_2 in suppressing chilling injury of grapefruits and avocadoes. *Proceedings of the American Society for Horticultural Science (Tropical Region)* **14**, 89-100.

Van der Plank, J. E. & Davies, R. (1937). Temperature in cold injury curves of fruit. *Journal of Pomology* **15**, 226-247.

Wheaton, T. A. (1963). Physiological comparison of plants sensitive and insensitive to chilling temperatures. *Ph. D. Thesis,* University of California, Davis, 93 pp.

Wills, R. B. H., Scott, R. J. & McGlasson, W. G. (1970). A role for acetate in the development of low temperature breakdown in apples. *Journal of the Science of Food and Agriculture* **21**, 42-44.

Wilson, J. M. (1976). The mechanism of chill and drought hardening of *Phaseolus vulgaris* leaves. *New Phytologist* **76**, 257-270.

Phenolic metabolism and oxidation in relation to the post-harvest physiology of fruit

V. S. Butt
Botany School, Oxford University, South Parks Road, Oxford, OX1 3RA

Introduction

With the possible exception of size in some fruits, marketability is largely determined by flavour and colour. Phenolic substances contribute in many fruit to both of these properties and hence the factors which determine the content of these substances, whether as a group or individually are, from the economic standpoint, of prime importance. There is surprisingly little information available about the molecular events determining their levels and which may be associated with the changes in fruit; at best, they are described in terms of changes in net phenolic content measured under a variety of different controlled conditions. In recent years, the biosynthesis and biodegradation of phenolic compounds in plant organs or cell suspensions, growing under or subjected to various types of environmental stress, have received very considerable attention (see, for example, Hahlbrock and Grisebach, 1979), and this expansion of our knowledge can be usefully applied to the changes taking place in developing and post-harvest fruit. A picture of these changes can now be presented from which some of the regulatory mechanisms operating in fruit can be inferred and the possibilities for their environmental control better understood.

Fruit phenolics

The major groups of phenolic compounds found in fruit are cinnamic acid derivatives and their conjugates, such as chlorogenic acid, flavonols, anthocyanidins and leuco-anthocyanidins, catechins and condensed tannins. The variety of molecular structure is wide, but the members most frequently found in each group have the common structural feature of *ortho*-dihydroxylation (I), or sometimes *vic*-trihydroxylation (II).

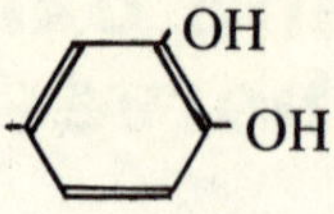

(I)

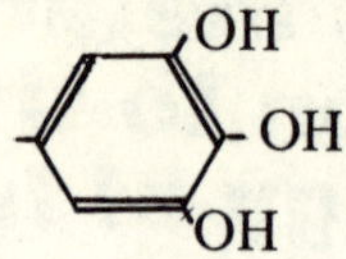

(II)

These structural features have particular relevance to the susceptibility of phenolics to atmospheric oxidation, whether enzymically or non-enzymically catalyzed.

The distribution of these compounds in a range of common fruits is shown in Table 1. The anthocyanins and flavonoids are important in fruit colour, although carotenoids or chlorophyll have greater significance in some fruits such as tomatoes, apples and citrus fruits. The catechins and chlorogenic acid provide the taste, although the flavonoid glycoside, naringin (III), is responsible for the bitterness of grapefruit, while a simple structural change by ring opening and reduction gives naringin dihydrochalcone (IV), which is 500 times sweeter than sucrose (Horowitz, 1964).

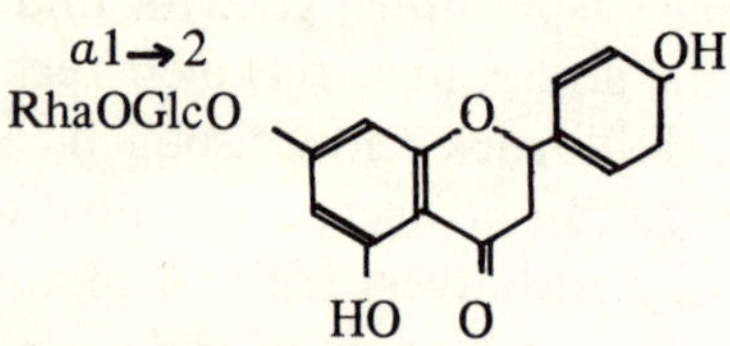

(III)

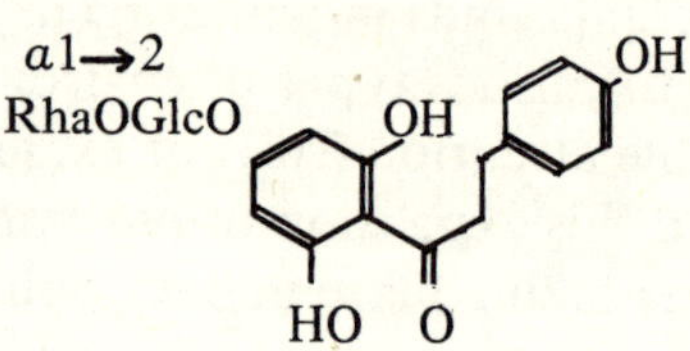

(IV)

The rather tasteless banana fruit has a low phenolic content; the dopamine (V) present does not contribute to the taste but is responsible for the discoloration of the fruit (Griffiths, 1959).

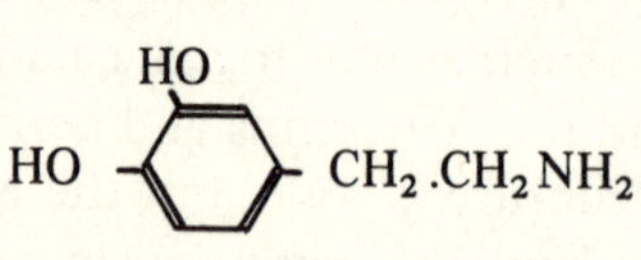

(V)

Table 1. Phenolic compounds of fruit

	Cinnamic acids	Chlorogenic acid	Flavonols	Catechins	Leucoanthocyanins	Anthocyanins	Others
Apple	+	+	+	+	+	(+)	Phloretin
Pear	+	+	–	+	+	(+)	
Peach	–	+	–	+	+	+	
Grape	+	–	+	+*	–	(+)	Benzoic acids
Banana	–	–	–	–	–	–	Dopamine
Apricot	+	+	+	+	–	–	
Tomato	+	+	–	–	–	–	
Blackcurrant	+	+	+	–	–	+	

* Flavan-3,4-diol

Induction of phenolic production in plants

The accumulation of phenolic compounds is usually a characteristic of a particular stage in the development of a plant tissue or organ. In leaves, anthocyanins accumulate when the plastid pigments are undergoing rapid decomposition (Sanger, 1971) and in many fruit a similar sequence is observed. A detailed study of the changes taking place as strawberry fruit matures shows chlorophyll and carotenoid contents to have declined each to 25 per cent of their maximum values at about 22 days after petal fall before a subsequent rapid rise in anthocyanin content, mainly pelargonidin 3-glucoside, over the next 28 days (Woodward, 1972). For the full development of colour in apple and strawberry fruit, light is essential. The flavonoid content of fruit is a feature of the developmental stage of the fruit and the environmental conditions to which it has been subjected.

Phenolic accumulation characteristically follows a sigmoid curve of the type shown in Fig. 1, in which there is a lag period, followed by a rapid rise before reaching a relatively constant level or even going into decline (see Woodward, 1972; Stickland, 1972). This is shown in the changes in the flavone glycoside content of the cotyledons and leaves of parsley (Hahlbrock, Sutter *et al.*, 1971), and can be mimicked in cell suspensions of parsley, subjected to illumination, the response depending upon the age of the culture (Hahlbrock, 1976). The accumulation of chlorogenic acid in potato tuber discs, triggered by excision and especially after further illumination (Lamb and Rubery, 1976), and of lignin in the stems of 6-day pea seedlings after illumination (Butt and Wilkinson, 1979) follow the same pattern. In pea

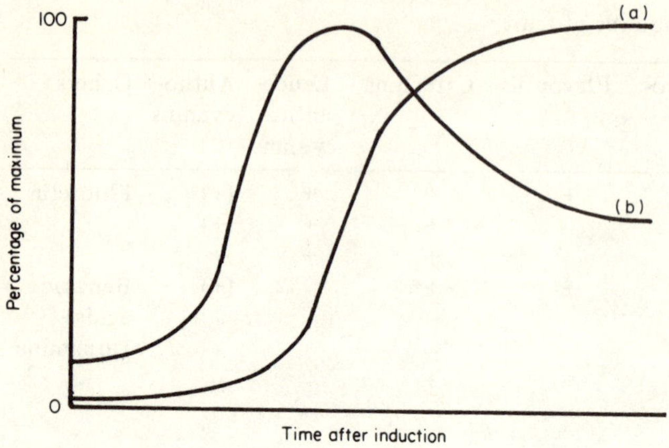

Fig. 1. Generalized curves for time course for induction of (a) phenolic accumulation and (b) phenylalanine ammonia-lyase activity.

seedlings, there is a qualitative difference between stems and apical buds, in which illumination induces the accumulation of mainly lignin and flavonoids respectively (Butt and Wilkinson, 1979). Thus, the internal controls of the plant prescribe the changes which may take place; an environmental change may be necessary to initiate them or raise them to their full potential.

Enzyme induction in phenolic metabolism

Increases in the content of phenolic substances in plant tissues have been found to be accompanied by increases in the activity of the biosynthetic enzymes. The first enzyme initiating synthesis from phenylalanine is phenylalanine ammonia-lyase (PAL), which catalyses the deamination of phenylalanine to cinnamic acid. In parsley seedlings (Hahlbrock, Sutter *et al.*, 1971) and cell suspensions (Hahlbrock, Ebel *et al.*, 1971), potato tuber discs (Lamb and Rubery, 1976) and pea stems and buds (Butt and Wilkinson, 1979), PAL activity increases rapidly to a maximum corresponding with the period during which the rate of phenolic accumulation is maximum (Fig. 1). This coincides with similar proportionate increases in cinnamate 4-hydroxylase activity in potato tuber discs, parsley cell suspensions and pea seedlings, and in the latter two plant systems, hydroxycinnamate-CoA ligase also increases, exactly parallel with PAL. The developmental and environmental triggers appear to lead to increases in the activities of a block of enzymes which initiate the flux of carbon into the phenolic products.

Later enzymes may increase, though not co-ordinately with the earlier enzymes, as in parsley cell suspension (Hahlbrock, Ebel *et al.*, 1971), or these enzymes may already be present in activities sufficient to account for the rates of phenolic accumulation (Ebel *et al.*, 1974; Butt and Wilkinson, 1979).

The increases in enzyme activity may in some cases arise from the activation of enzymes already present in some latent form (Smith, Billett and Giles, 1977), but the internal triggers probably lead more usually to *de novo* enzyme synthesis. This is indicated, for example, by the induction of PAL in potato tuber discs, in which (1) the induction is inhibited by cycloheximide, cordycepin and actinomycin (Lamb, 1977), and (2) when carried out in D_20, the newly formed PAL can be recognized on a density gradient by its heavy deuteration (Lamb, Merritt and Butt, 1979).

The enzyme thereby produced is not, however, stable, for following the increase in PAL or cinnamate 4-hydroxylase activities, a decline sets in down to a level about half of the maximum induced activity (Fig. 1). The fall is invariably associated with a decline in the rate of phenolic accumulation, which ceases when the enzyme activities reach the lower steady-state level. The fall appears to be due to some form of irreversible destruction. This deactivation may be due to an inherent, proteolytic activity of the cells, the changes in net activity being due to alterations in the rate of synthesis (Betz, Schäfer and Hahlbrock, 1978) or the rate of destruction increases with time, induced either by the initiating trigger or by the enzymes themselves, to establish a new steady-state enzyme activity. The latter has been demonstrated in excised potato tubers by pulse-chase experiments with D_20, in which measurements of enzyme labelling and activity at successive stages after induction have shown, by computer analysis, that after the increase in rate of enzyme synthesis, the rate of enzyme destruction increases and both then decline (Lamb *et al.*, 1979).

The onset of phenolic accumulation during fruit development can therefore be interpreted as initiated by the *de novo* synthesis of enzymes, triggered by specific internal conditions, probably hormonal, characteristic of the state of maturation. This would be consistent with observations correlating naringenin accumulation in developing grapefruit with increasing PAL activity with a time-course indentical with that shown in Fig. 1 (Maier and Hasegawa, 1979); in grapefruit callus, an increase in PAL activity was shown to be sensitive to growth regulators and inhibited by cycloheximide (Thorpe, Maier and Hasegawa, 1971).

Turnover of phenolic compounds

When phenolic compounds have been synthesized, they are usually held at constant concentration over a relatively long period of time, though they may slowly disappear, as reported, for instance, for leaf anthocyanins (Sanger, 1971). The concentration of flavonoids especially may remain constant for periods of days or weeks. This fact is, at first sight, difficult to reconcile with the observation that a substantial proportion of the biosynthetic enzymes induced by the conditions leading to phenolic accumulation remain present in the tissue.

In tissues and cell suspensions, however, there is good reason to believe that phenolic compounds undergo turnover, in which synthesis is balanced by destruction (Barz and Hösel, 1979). For instance, the kaempferol content falls at once on transferring chickpea (*Cicer arietinum*) cell suspensions from light to darkness; it is rapidly restored on transfer to light (Barz, 1975). By pulse-chase experiments, it is possible to calculate a half-life for many flavonoids: in roots of chickpea, the isoflavones formononetin and biochanin A have half-lives of 50 h and more than 320 h respectively; in leaves and stems, 78 h and 220 h; in cell suspension cultures, 67 h and 110 h (Barz and Hösel, 1979). Organs and cells thus differ in their capacity for biosynthesis and degradation, and further, the relative stabilities of different flavonoids differ with the organs, cells or species examined.

In consequence, phenolic compounds attain a maximum steady-state concentration, at which synthesis and breakdown are balanced; both the enzymes and the substrates are undergoing dynamic turnover. Three major agents have been listed (Barz and Hösel, 1979) as responsible for phenolic degradation or disappearance: (1) conjugation, e.g. peptidation, glycosylation, (2) degradation by dioxygenase or peroxidase action, (3) oxidation, with polymer formation. The fruit physiologist is particularly concerned with the last of these, because of the brown colour of most of the polymers produced (Mathew and Parpia, 1971).

Phenolic oxidation and polymerization

The oxidative polymerization of phenols is primarily catalyzed by polyphenol oxidase or *o*-diphenol oxidase. Substituted *o*-diphenols are first oxidized by molecular oxygen to the corresponding *o*-quinones:

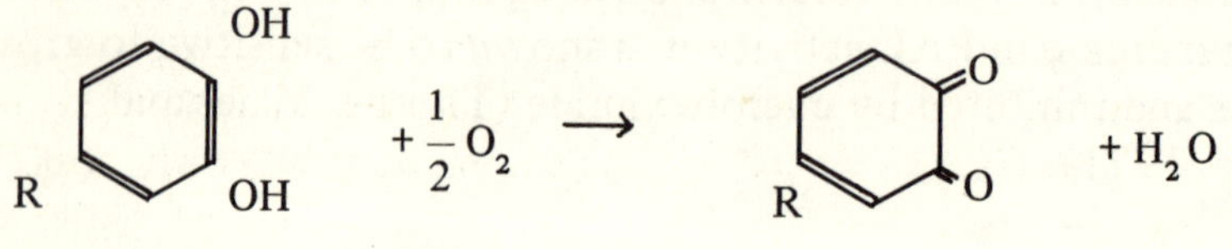

The primacy of *o*-diphenols among fruit phenolics has already been mentioned, and *o*-diphenol oxidase has been demonstrated in almost all fruits, often localized in particles, e.g. in apple chloroplasts (Harel, Mayer and Shain, 1965), and has been purified from apple (Harel and Mayer, 1968), grape (Kidron, Harel and Mayer, 1977), olive (Ben-Shalom, Kahn, Harel and Mayer, 1977) and peach (Wong, Luh and Whitaker, 1971).

The browning polymerization reaction characteristic of fruits, especially when subject to mechanical injury after harvesting or during processing, is a consequence of the reactivity of the *o*-quinones at the nucleophilic centre engendered in the molecule:

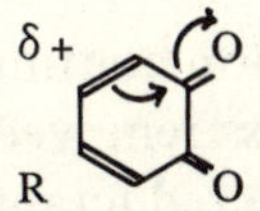

so that carbon-carbon or carbon-oxygen bridges can be formed. In consequence, the enzyme catalyzes the oxidative polymerization of catechins (see Wong, 1976), the browning reaction in bananas due to the oxidation of dopamine (Palmer, 1963; Singleton, 1972), and the decolouration and browning of anthocyanins, in which however, catalytic traces of other soluble catechols are necessary to sustain, by non-enzymic interaction, the oxidation at a measurable rate (Peng and Markakis, 1963).

Observations with fruits, wines and model systems (see Simpson, Lee, Rodriguez and Chichester, 1976) show anthocyanidins to decompose in the order:

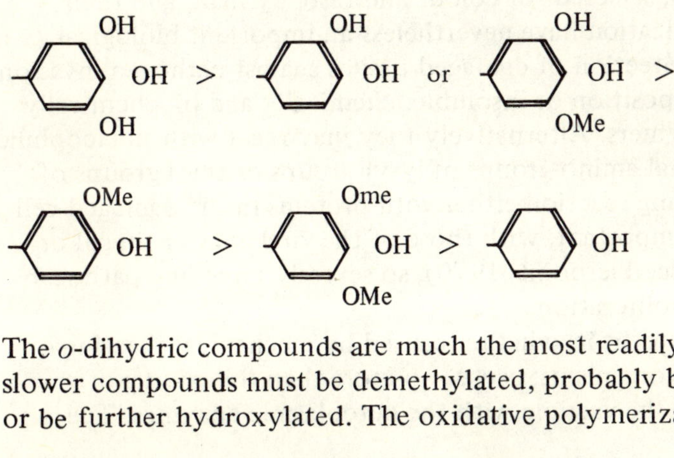

The *o*-dihydric compounds are much the most readily oxidized; the slower compounds must be demethylated, probably by oxidative action, or be further hydroxylated. The oxidative polymerization of phenolic substrates fed to mung bean cells shows a similar relationship (Barz, 1975).

The biosynthesis of *o*-diphenols such as cyanidin, or the further oxidation of monophenols such as pelargonidin, requires the introduction of a second hydroxyl group. This reaction (sometimes known as their cresolase activity) can be catalyzed by many phenolases according to the equation:

in which H_2A is a reducing agent. The reaction continues until H_2A is exhausted, when the phenolic product, unless removed from the site of enzyme action, will be further oxidized. The reductant most commonly employed *in vitro* is ascorbate, but this can be substituted by NADPH, NADH or tetrahydrofolate.

The phenolases are thus potent destructive agents for natural phenolic substances, yet are almost ubiquitous in plant tissues containing phenolics. This may be due to their biosynthetic function. Thus, hydroxylase activity has been shown to be triggered by light concomitantly with PAL in broad bean and spinach-beet seedlings (Butt, 1979) and during the induced production of polyphenols in excised sweet potato root (Hyodo and Uritani, 1966). The phenolase appears to be essential for the synthesis of phenolic compounds, which the enzyme is then likely to destroy unless they are protected from its action, especially when ascorbate and other reducing agents within the cell have been exhausted.

These phenolics, selected for colour and taste by man, and their oxidative polymerization have nevertheless an important biological function in the protection of damaged tissues against pathogen invasion by the physical deposition of insoluble, chemically and biochemically indestructible polymers. Alternatively they may react with nucleophilic groups, e.g. terminal amino-groups of lysyl groups or thiol groups of proteins, in a tanning reaction either with proteins in the damaged cell or, perhaps more important, with those of the viral coat or fungal or bacterial surface (see Pierpoint, 1970), so severely impeding pathogen penetration and proliferation.

Surface damage to fruit, whether caused by mechanical shearing or attack by invading organisms, causes no more than the mixing of substrate stored in the vacuole with the cytoplasmic enzymes. The oxidation and browning take place in the broken cells at the surface or in those near to it. The breakdown of cells under senescence has the same consequences. Some of the slow polymerization processes taking place already in the cell are accelerated by the release and mingling of the vacuolar contents. By contrast with this passive response to damage, storage organs, such as the tubers of potato or Jerusalem artichoke or the roots of sweet potato, react by producing phenolases and phenolic compounds in the surface cells *de novo* – a slower but more effective response adjusted quantitatively to the degree of physical damage. The post-climacteric fruit has before it only decay and death, and despite the processes of selection and domestication, possesses no mechanism to ensure its long-term survival; the fruit chemist can do little more than delay senescence and death by controlling the environment. The storage tissue has the prospect and necessity for future development, and hence is programmed to protect itself actively from damage.

Other mechanisms of phenolic destruction

At least two other mechanisms for the destruction of phenolics in fruit are known to exist, presented here for completeness. Loss of colour and anthocyanin disappearance are affected by alterations in cell pH from pH 3 towards neutrality. The neutralization of the flavylium charge leads to 2-hydroxylation, ketonization and ring fission with the formation of $\alpha\alpha$-diketones (Harper, 1968), which break down to substituted benzoic acids and a second aromatic fragment (Jurd, 1972; see Simpson *et al.*, 1976). The final stage ensures the irreversibility of these reactions.

The presence of peroxidases in some fruit suggests the existence of oxidative processes analogous to those in cell suspension cultures, which either by the action of hydrogen peroxide or by molecular oxygen, degrade exogenous flavonoids to their aromatic components (Barz and Hösel, 1979).

Conclusions

From this discussion emerges a picture of dynamic physiological change during the development, climacteric and senescence of fruit. The production of phenolics is a consequence of the induction of enzymes by hormonal and environmental triggers. These enzymes are in a state of active turnover, probably because of the co-induction of proteolytic enzymes which destroy them. The level of phenolic compounds, and the types which are present, are a function of the activities of these enzymes and also of those enzymes which destroy phenolics. Active among these are phenolases, essential for their synthesis but which may be removing them continuously by oxidative polymerization, a process greatly accelerated by damage to the tissue and, ultimately, by senescence and cell autolysis. Both processes of synthesis and degradation are chemically selective so that taste and colour can be differentially affected by storage, the conditions for which have to be optimized to ensure that the biological controls are effective in the balance sought for marketing. The fruit, because of the particular character of phenolases as both biosynthetic and degradative enzymes, tends towards breakdown and senescence during the post-harvest stage. A study of the nutritional and environmental conditions arresting the degradative action of phenolase suggests itself as a first approach to a rational study of food storage.

References

Barz, W. (1975). Abbau von Flavonoiden und Isaflavonoiden – ein Überblick. *Berichte der Deutschen Botanischen Gesellschaft* **88**, 71-81.

Barz, W. & Hösel, W. (1979). Metabolism and degradation of phenolic compounds in plants. In *Biochemistry of Plant Phenolics,* pp. 339-369. Eds. T. Swain, J. B. Harborne & C. F. van Sumere. New York: Plenum Press.

Ben-Shalom, N., Kahn, V., Harel, E. & Mayer, A. M. (1977). Catechol oxidase from green olives: properties and partial purification. *Phytochemistry* **16**, 1153-1158.

Betz, B., Schäfer, E. & Hahlbrock, K. (1978). Light-induced phenylalanine ammonia-lyase in cell-suspension cultures of *Petroselinum hortense. Archives of Biochemistry and Biophysics* **190**, 129-135.

Butt, V. S. (1979). Oxidases in aromatic metabolism. In *Biochemistry of Plant Phenolics,* pp. 433-456. Eds. T. Swain, J. B. Harborne & C. F. van Sumere, New York: Plenum Press.

Butt, V. S. & Wilkinson, E. M. (1979). Enzyme changes accompanying lignification and flavonoid synthesis in illuminated pea shoots. In *Regulation of Secondary Product and Plant Hormone Metabolism.* Eds. M. Luckner & K. Schreiber. *Proceedings of 12th FEBS Meeting,* Vol. 55, pp. 147-154. Oxford: Pergamon Press.

Ebel, J., Schaller-Hekeler, B., Knobloch, K. H., Wellman, E., Grisebach, H. & Hahlbrock, K. (1974). Coordinated changes in enzyme activities of phenylpropanoid metabolism during the growth of soybean cell suspension cultures. *Biochimica et Biophysica Acta* **362**, 417-424.

Griffiths, L. A. (1959). Detection and identification of the polyphenoloxidase substrate of the banana. *Nature, London* **184**, 58-59.

Hahlbrock, K. (1976). Regulation of phenylalanine ammonia-lyase activity in cell-suspension cultures of *Petroselinum hortense.* Apparent rates of enzyme synthesis and degradation. *European Journal of Biochemistry* **63**, 137-145.

Hahlbrock, K. & Grisebach, H. (1979). Enzymic controls in the biosynthesis of lignin and flavonoids. *Annual Review of Plant Physiology* **30**, 105-130.

Hahlbrock, K., Sutter, A., Wellmann, E., Ortmann, R. & Grisebach, H. (1971). Relationship between organ development and activity of enzymes involved in flavone glycoside biosynthesis in young parsley plants. *Phytochemistry* **10**, 109-116.

Hahlbrock, K., Ebel, J., Ortmann, R., Sutter, A., Wellmann, E. & Grisebach, H. (1971). Regulation of enzyme activities related to the biosynthesis of flavone glycosides in cell suspension cultures of parsley (*Petroselinum hortense*). *Biochemica et Biophysica Acta* **244**, 7-15.

Harel, E. & Mayer, A. M. (1968). Interconversion of sub-units of catechol oxidase from apple chloroplasts. *Phytochemistry* **7**, 199-204.

Harel, E., Mayer, A. M.& Shain, Y. (1965). Purification and multiplicity of catechol oxidase from apple chloroplasts. *Phytochemistry* **4**, 783-790.

Harper, K. A. (1968). Structural changes of flavylium salts. IV. Polarographic and spectrometric examination of pelargonidin chloride. *Australian Journal of Chemistry* **21**, 221-227.

Horowitz, R. M. (1964). Relations between the taste and structure of some phenolic glycosides. In *Biochemistry of Phenolic Compounds*, pp. 545-571. Ed. J. B. Harborne. London: Academic Press.

Hyodo, H. & Uritani, I. (1966). A study on increase in *o*-diphenol oxidase activity during incubation of sliced sweet potato tissue. *Plant and Cell Physiology* **7**, 137-144.

Jurd, L. (1972). Some advances in the chemistry of anthocyanin-type plant pigments. In *The Chemistry of Plant Pigments.* Ed. C. O. Chichester. *Advances in Food Research,* Suppl. 3, pp. 123-142.

Kidron, M., Harel, E. & Mayer, A. M. (1977). Copper content and amino acid composition of catechol oxidase from Clairette grapes. *Phytochemistry* **16**, 1050-1057.

Lamb, C. J. (1977). Phenylalanine ammonia-lyase and cinnamic acid 4-hydroxylase: characterisation of the concomitant changes in enzyme activities in illuminated potato tuber discs. *Planta* **135**, 169-175.

Lamb, C. J. & Rubery, P. H. (1976). Photocontrol of chlorogenic acid biosynthesis in potato tuber discs. *Phytochemistry* **15**, 665-668.

Lamb, C. J., Merritt, T. K. & Butt, V. S. (1979). Synthesis and removal of phenylalanine ammonia-lyase activity in illuminated discs of potato tuber parenchyme. *Biochimica et Biophysica Acta* **582**, 196-212.

Maier, V. P. & Hasegawa, S. (1970). L-Phenylalanine ammonia-lyase activity and naringenin glycoside accumulation in developing grapefruit. *Phytochemistry* **9**, 139-144.

Mathew, A. G. & Parpia, H. A. B. (1971). Food browning as a polyphenol reaction. *Advances in Food Research* **19**, 75-145.

Palmer, J. K. (1963). Banana polyphenol oxidase. Preparation and properties. *Plant Physiology, Lancaster* **38**, 508-513.

Peng, C. Y. & Markakis, P. (1963). Effect of phenolase on anthocyanins. *Nature, London* **199**, 597-598.

Pierpoint, W. S. (1970). Formation and behaviour of *o*-quinones in some processes of agricultural importance. *Rothamsted Experimental Station. Report for 1970, Part II*, pp. 199-218.

Sanger, J. E. (1971). Quantitative investigations of leaf pigments from their inception in buds through autumn coloration to decomposition in falling leaves. *Ecology* **52**, 1075-1092.

Simpson, K. L., Lee, T. -C., Rodriguez, D. B. & Chichester, C. O. (1976). Metabolism in senescent and stored tissues. In *Chemistry and Biochemistry of Plant Pigments*, Vol. I., pp. 779-842. Ed. T. W. Goodwin. London: Academic Press.

Singleton, V. L. (1972). Common plant phenols other than anthocyanins, contributions to coloration and discoloration. In *The Chemistry of Plant Pigments.* Ed. C. O. Chichester. *Advances in Food Research*, Suppl. 3., pp. 143-191.

Smith, H., Billett, E. E. & Giles, A. B. (1977). The photocontrol of gene expression in higher plants. In *Regulation of Enzyme Synthesis and Activity in Higher Plants*, pp. 93-127. Ed. H. Smith. London: Academic Press.

Stickland, R. G. (1972). Changes in anthocyanin, carotenoid, chlorophyll, and protein in developing florets of the Chrysanthemum. *Annals of Botany* **36**, 459-469.

Thorpe, T. A., Maier, V. P. & Hasegawa, S. (1971). Phenylalanine ammonia-lyase activity in citrus fruit tissue cultured *in vitro*. *Phytochemistry* **10**, 711-718.

Wong, E. (1976). Biosynthesis of flavonoids. In *Chemistry and Biochemistry of Plant Pigments*, Vol. I, pp. 464-526. Ed. T. W. Goodwin. London: Academic Press.

Wong, T. C., Luh, B. S. & Whitaker, J. R. (1971). Isolation and characterization of polyphenol oxidase isoenzymes of clingstone peach. *Plant Physiology, Lancaster* **48**, 19-23.

Woodward, J. R. (1972). Physical and chemical changes in developing strawberry fruits. *Journal of the Science of Food and Agriculture* **23**, 465-473.

Subject index